Joel L. Schiff

Topics in Complex Analysis

De Gruyter Studies in Mathematics

Volume 88

Joel L. Schiff

Topics in Complex Analysis

DE GRUYTER

Mathematics Subject Classification 2020
Primary: 30-01; Secondary: 31-01

Author
Dr. Joel L. Schiff
62 Penzance Rd
Auckland 0630
New Zealand
j.schiff@auckland.ac.nz

ISBN 978-3-11-075769-9
e-ISBN (PDF) 978-3-11-075782-8
e-ISBN (EPUB) 978-3-11-075787-3
ISSN 0179-0986

Library of Congress Control Number: 2022941383

Bibliographic information published by the Deutsche Nationalbibliothek
The Deutsche Nationalbibliothek lists this publication in the Deutsche Nationalbibliografie; detailed bibliographic data are available on the Internet at http://dnb.dnb.de.

Typesetting: VTeX UAB, Lithuania
Printing and binding: CPI books GmbH, Leck

www.degruyter.com

God made the universe out of complex numbers…[1]
Richard Hamming

1 From the article: The Unreasonable Effectiveness of Mathematics, Indeed, recent research (2021–2022) has indicated that complex numbers are actually indispensible to the standard theory of quantum mechanics.

Preface

Complex analysis is a form of surrealism and is one of the greatest creations of the human mind (see Figure 1 for another form of surrealism). It is both of this world and also of a netherworld of infinite beauty.

Some of the behavior to be found in complex analysis is truly extraordinary, such as that of an analytic function in the neighborhood of an essential singularity taking every complex value infinitely many times with one possible exception. Or a family of meromorphic functions is normal if it omits three distinct values. All this originates from the primitive first steps of letting $z = x + iy$ and a few rules to deal with its arithmetic. But therein are contained the seeds of the infinite complexity of Julia sets, where such astonishing sets are even found lurking in the innocuous looking Newton's method once a change to a complex variable is made. As well we have such profound results as the Riemann mapping theorem, the Euler product formula, the second fundamental theorem of Nevanlinna, or such deep open questions as the Riemann hypothesis. If the latter turns out to be true, then something equally profound can be said about the distribution of prime numbers.

In spite of its existence in a seemingly netherworld of unreality, complex analysis has found such real-world applications in fluid flow, electrical engineering, Fourier analysis/signal processing, and the surreal worlds of quantum mechanics and string

Figure 1: *Les Promenades d'Euclide*, 1955, from the surrealist mind of René Magritte. © Rene Magritte. ADAGP/Copyright Agency, 2022.

https://doi.org/10.1515/9783110757828-201

theory, among a myriad of other scientific arenas. A small sampling of real-world applications is discussed at various points in the text to bring the reader back to Earth. This text is a presentation of the wonderous nature of the subject beyond the basic fundamentals. In some instances, the fundamentals will be mentioned, so it is clearly understood what has been taken as already known. As with real life, complex analysis embraces known knowns, known unknowns, and unknown unknowns.

One of the main themes of the book is the Dirichlet problem in its many guises and formulations. This has been a significant theoretical aspect since Riemann first used the Dirichlet principle to establish his celebrated mapping theorem in his thesis of 1851. The fact that the original Dirichlet principle was inherently flawed has proved over the years to be a source of considerable research in not only complex analysis, but also in the calculus of variations. Another theme of the book is the notion of a normal family and how it is an instrumental feature in many of the results discussed in this text. Defined only a few years prior, this notion lies at the very core of the work of Julia and Fatou, which has become so prominent on the mathematical landscape. With the advent of modern computers, a heretofore unseen world of stunning complexity has been revealed arising from the iteration of the simplest of functions. All the while, these sets have been there waiting for the science to be developed to reveal their presence.

There is more, much more, but of course, it is not possible to present all of the advanced topics in a single text. Some topics are presented, which the author finds particularly interesting and full of beauty, elegance, and magic. We hope that the reader will feel the same.

The reader will also notice that many of the well-known theorems are referenced in footnotes. This is in part to note that not everything was done by Euler, Gauss, and Riemann (although much was) and that many significant theorems, as well as their authors, are not really that ancient. The creator of the theory of normal families, Paul Montel, passed away in 1975 when this author was actually teaching about normal families. The same goes for Rolf Nevanlinna, who created the profound modern theory of meromorphic functions and passed away in 1980, and the author had the opportunity to meet with him. Of course, this also indicates the rather mature nature of the author, but the point is that many significant developments in the theory of complex analysis are of relatively recent vintage.

Remarks

- Some undergraduate prior knowledge of complex analysis is assumed.
- Constants that are of no significance in an equation and whose values become affected by another constant will occasionally remain as in the original for simplicity.

- A few selected theorems are given without proof as their proofs are either too technical for this text or would take the discourse too far afield. One such example is Nevanlinna's second fundamental theorem. However, we will use such theorems for further discussion.
- Paths and curves will always be such that any requisite integration can be taken over them. The more specifically defined term contour is also used in this context.

Auckland University, 2022 JLS

Acknowledgments

The author gratefully acknowledges the assistance of the following individuals who helped to make this book possible:

José Ibrahim Villanueva Gutiérrez, Anita Kean, Chris King, and Nick Dudley Ward for their fine computer images, Katy Metcalf for her excellent graphics, Aaron Schiff for his technical assistance, and Shaun Cooper, Mark Harmer, Aimo Hinkkanen, Chris King, Norm Levenberg, Gaven Martin, Nick Dudley Ward, and Julian Wilson for reading portions of the manuscript and making many valuable suggestions.

https://doi.org/10.1515/9783110757828-202

Contents

1 Analytic functions

Analytic continuation

In this chapter, we discuss some of the fundamental notions that will be utilized in the text or are simply interesting in their own right.

It is sometimes possible to extend a function that is analytic in some neighborhood of a point to an analytic function in a much larger domain. Suppose that $f_1(z)$ and $f_2(z)$ are analytic in two open disks D_1 and D_2, respectively, and $D_1 \cap D_2 \neq \emptyset$. If $f_1(z) = f_2(z)$ for all $z \in D_1 \cap D_2$, then we say that the functions $f_1(z)$ and $f_2(z)$ are *(direct) analytic continuations* of each other. Moreover, each analytic continuation of the other is unique by the Identity theorem. This results in an analytic function defined on $D_1 \cup D_2$ with the pairs $(f_i(z), D_i)$, $i = 1, 2$, called *function elements.*

Of course, this notion can be extended to a collection of open disks D_i, $i = 1, 2, \ldots, n$, covering a straight-line segment. That is, we have functions $f_i(z)$ that are analytic in open disks D_i for $i = 1, 2, \ldots, n$ with $D_i \cap D_{i+1} \neq \emptyset$ and such that $f_{i+1}(z)$ is a direct analytic continuation of $f_i(z)$. The aggregate of the $f_i(z)$ is an analytic function $f(z)$ defined on $\bigcup_{i=1}^{n} D_i$.[1]

For example, if $f_1(z)$ is analytic in an open disk $D_1(z_1, r_1)$ whose power series representation has a radius of convergence r_1, and if z_1 is a point on the circle of convergence at which $f_1(z)$ is analytic, then there is a power series expansion in an open disk $D_2(z_1, r_2)$ centered at the point z_1 representing an analytic function $f_2(z)$ that is the direct analytic continuation of $f_1(z)$.

To guarantee the uniqueness of our extension, we have the following:

1.1 Monodromy theorem. *If $f(z)$ is analytic in a disk contained in a simply connected domain Ω and can be continued analytically along every polygonal path contained in Ω, then $f(z)$ can be defined as a single-valued analytic function on the whole of Ω.*

Proof. Since Ω is simply connected, it is also polygonally connected so that any two points in Ω can be connected by a polygonal path that lies in Ω. Let us assume that $f(z)$ is analytic in a neighborhood of a point A and consider the triangle $ABCA$ that lies wholly in the interior of Ω. By the hypothesis, $f(z)$ can be continued analytically along the line segment AP which yields a single-valued analytic function in a neighborhood of the segment Ap, which we again denote by $f(z)$. Then we can find points p_1 and p_2 along the line segment BC such that $f(z)$ can be analytically continued along the triangular path Ap_1pp_2A as in Figure 1.1. By construction this analytic continuation is the same as $f(z)$ in a neighborhood of the point A.

1 For a more complete discussion of analytic continuation, see Ahlfors (1979, Chapter 8).

https://doi.org/10.1515/9783110757828-001

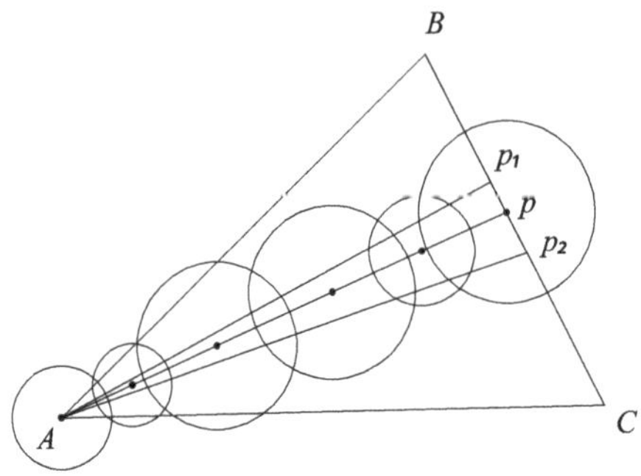

Figure 1.1: The polygonal figure *ABCA* described in the text starting with the function $f(z)$ that is analytic in a neighborhood of the point *A*. Courtesy Katy Metcalf.

Since the line segment BC is compact, it can be covered by a finite number of segments of the form p_1p_2. It follows that $f(z)$ can be analytically continued along the triangular path $ABCA$ such that the continued function and $f(z)$ coincide at A.

We can apply induction on the number of sides of our figure to extend the preceding argument. In fact, suppose that for all closed n-gons in Ω, we have the requisite analytic continuation. For an arbitrary $(n+1)$-gon, we can take a suitably chosen diagonal to obtain two polygons such that neither one has more than n sides, allowing the use of the induction hypothesis.

We conclude that for any two points z_1 and z_2 in Ω, the analytic continuation along any two polygonal paths in Ω joining z_1 to z_2 determines the same function at z_2, thus establishing the theorem.[2] □

Another useful result regarding analytic functions in a disk we will make use of is the following.

1.2 Cauchy inequality. *If $f(z)$ is analytic in $|z - z_0| < R$ with Taylor series*

$$f(z) = \sum_{n=0}^{\infty} a_n (z - z_0)^n,$$

then

$$\frac{|f^{(n)}(z_0)|}{n!} = |a_n| \leq \frac{M(r)}{r^n}, \quad n = 1, 2, 3, \ldots,$$

where $M(r) = \max_{|z-z_0|=r} |f(z)|$ and $r < R$.

2 Instead of analytic continuation along polygonal paths, the result can be extended to the case of two Jordan arcs connecting any two points z_1 and z_2 in Ω by considering sufficiently close polygonal approximations to the two arcs.

The proof is an immediate consequence of the formula for the Taylor coefficients

$$a_n = \frac{1}{2\pi i} \int_{|z-z_0|=r} \frac{f(z)}{(z-z_0)^{n+1}} dz.$$

Differentiation under the Integral sign

There will be some occasions when our analytic function involves two variables as in the discussion of the gamma function in Chapter 10. Throughout the text, we have occasion to consider *contours*, that is, a finite sum of piecewise smooth (continuously differentiable) arcs and the next result is easily extendable to a contour replacing the interval $[a, b]$.

1.3 Theorem. *Given a domain Ω let $f(z, t)$ be a complex-valued function where $z \in \Omega$ and $t \in [a, b]$ Suppose that $f(z, t)$ is continuous on $\Omega \times [a, b]$ and likewise for $\frac{\partial f}{\partial z}$. Then*

$$F(z) = \int_a^b f(z, t) dt$$

defines an analytic function in Ω, and

$$F'(z) = \int_a^b \frac{\partial f(z,t)}{\partial z} dt.$$

Proof. Fix a point $z_0 \in \Omega$. By the continuity of $f(z, t)$, given $\varepsilon > 0$, there is $\delta > 0$ such that for $|z_0 - z_1| < \delta$, we have $|f(z_0, t) - f(z_1, t)| \le \frac{\varepsilon}{b-a}$, for all $t \in [a, b]$. With the given $\delta > 0$,

$$\begin{aligned} |F(z_0) - F(z_1)| &\le \int_a^b |f(z_0, t) - f(z_1, t)| dt \\ &\le (b-a) \cdot \max_{t \in [a,b]} |f(z_0, t) - f(z_1, t)| \le \varepsilon, \end{aligned}$$

establishing the continuity of $F(z)$.

Similarly, by the continuity of $\frac{\partial f}{\partial z}$ and $t \in [a, b]$

$$\left| \frac{f(z+h, t) - f(z, t)}{h} - \frac{\partial f(z,t)}{\partial z} \right| \to 0$$

as $h \to 0$. Therefore

$$\left| \frac{F(z+h, t) - F(z, t)}{h} - \int_a^b \frac{\partial f(z,t)}{\partial z} \right|$$

$$= \left| \int_a^b \left(\frac{f(z+h,t) - f(z,t)}{h} - \frac{\partial f(z,t)}{\partial z} \right) dt \right|$$

$$\leq (b-a) \cdot \max_{t \in [a,b]} \left| \frac{f(z+h,t) - f(z,t)}{h} - \frac{\partial f(z,t)}{\partial z} \right| \to 0$$

as $h \to 0$, which proves the theorem. □

1.4 Corollary. *Let $f(z,t)$ be continuous for $z \in \Omega \times [a,\infty)$ with continuous partial derivative $\frac{\partial f}{\partial z}$ there. Assuming that the integral*

$$F(z) = \int_a^\infty f(z,t)dt$$

converges uniformly on compact subsets of Ω, then $F(z)$ is an analytic function in Ω, and

$$F'(z) = \int_a^\infty \frac{\partial f(z,t)}{\partial z} dt.$$

Indeed, for the proof, we take our interval to be $[a,n]$ and set

$$f_n(z) = \int_a^n f(z,t)dt.$$

By the theorem, $f_n(z)$ is an analytic function with

$$f_n'(z) = \int_a^n \frac{\partial f(z,t)}{\partial z} dt.$$

In view of our assumption, it follows that $f_n \to f = F(z)$ converges uniformly on compact subsets of Ω, and hence by the Weierstrass Theorem 1.27 in the sequel, $F(z)$ is analytic in Ω whose derivative is given by the preceding theorem as desired.

In general, the maximum modulus principle is not applicable to unbounded domains Ω.[3] Although, if we include the point at infinity by setting $\partial_\infty \Omega = \partial\Omega \cup \{\infty\}$, which is the *extended boundary* of Ω in the extended complex plane $\hat{\mathbb{C}} = \mathbb{C} \cup \{\infty\}$ discussed in Chapter 3, we do have the following version of the MMP.

1.5 Extended maximum principle. *If $f(z)$ is analytic in a domain Ω such that $\overline{\lim}_{z\to\zeta} |f(z)| \leq M < \infty$ for all $\zeta \in \partial_\infty \Omega$, then $|f(z)| \leq M$ for all $z \in \Omega$.*

3 The function $f(z) = e^z$ is unbounded in the right half-plane, but $|f(z)| = 1$ on the boundary line $x = 0$.

Proof. Let us assume on the contrary that, for some point $z_0 \in \Omega$, we have $|f(z_0)| > M$ and take an exhaustion[4] of Ω by subdomains $\{\Omega_n\}$, each containing the point z_0. Then $|f(z)|$ attains its maximum on $\overline{\Omega_n}$ at some point z_n on $\partial\Omega_n$ for all n. It follows that

$$M < |f(z_0)| < |f(z_1)| < \cdots < |f(z_n)| < \cdots.$$

Any accumulation point of the sequence $\{z_n\}$ must lie outside all the subdomains Ω_n and so lies on $\partial_\infty\Omega$. Taking an accumulation point $\zeta \in \partial_\infty\Omega$, let $\{z_{n_k}\}$ be a subsequence such that $z_{n_k} \to \zeta$. Therefore

$$\overline{\lim_{z_{n_k}\to\zeta}}|f(z_{n_k})| > M,$$

which contradicts our hypothesis, proving the result. □

The argument employed above will be typical in the context of several sequences of functions that form normal families (Chapter 4).

The next theorem is another useful extension of the MMP that subject to a restriction on the growth of an analytic function can be applied to unbounded domains. The theorem actually illustrates a general principle that is applicable in various scenarios involving unbounded domains.

1.6 Phragmén–Lindelöf theorem.[5] *Let Ω be a simply connected domain with $\partial_\infty\Omega = A \cup B$, and let $f(z)$ be an analytic function in Ω. Suppose that there exists a nonzero analytic function $\omega(z)$ on Ω satisfying $|\omega(z)| \le 1$ for all $z \in \Omega$ and*

(i) $\overline{\lim}_{z\to a}|f(z)| \le M$ *for all* $a \in A$;

(ii) $\overline{\lim}_{z\to b}|\omega(z)|^{\varepsilon}|f(z)| \le M$ *for all* $b \in B$ *and* $\varepsilon > 0$.

Then $|f(z)| \le M$ for all $z \in \Omega$.

Proof. Since Ω is simply connected and $\omega \ne 0$, we take a single-valued analytic branch of the function $(\omega(z))^{\varepsilon}$ and define $F(z) = (\omega(z))^{\varepsilon} f(z)$. Then $F(z)$ is analytic in Ω, and

$$\overline{\lim_{z\to\zeta}}|F(z)| \le M$$

for all $\zeta \in \partial_\infty\Omega$ by (i) and (ii). Hence, by the preceding extended maximum principle, $|F(z)| \le M$ in Ω, that is,

$$|f(z)| \le M|\omega(z)|^{-\varepsilon},$$

and the conclusion follows by letting $\varepsilon \to 0$. □

---

4 For any domain Ω in the complex plane, there always exists a sequence of subdomains $\{\Omega_n\}$, called an *exhaustion* of Ω, such that $\overline{\Omega_n} \subset \Omega_{n+1}$ and $\bigcup_{n=1}^{\infty}\Omega_n = \Omega$. This is a topological fact. The exhaustion can also be chosen such that respective boundaries are piecewise smooth.

5 Sur une extension d'un principe classique de l'analyse et sur propriétés des fonctions monogènes dans le voisinage d'un point singulier, *Acta Math.* 31 (1908), 381–406.

1.7 Exercise. Let $f(z)$ be analytic in the region $R = \{z : -\frac{\pi}{4} \le \operatorname{Arg}(z) \le \frac{\pi}{4}\}$ and satisfy the condition $\overline{\lim}_{z\to\zeta}|f(z)| \le M$ for all $\zeta \in \partial R$. Suppose further that

$$f(z) = O(e^{r^\beta})^6$$

as $|z| = r \to \infty$, where $0 < \beta < 1$.

(i) Taking the function $\omega(z) = e^{-z^\alpha}$, $0 < \alpha < 1$, show that $|\omega(z)| \le 1$. As a consequence, for $F(z) = (\omega(z))^\varepsilon f(z)$ and $\zeta \in \partial R$,

$$\overline{\lim_{z\to\zeta}}|F(z)| \le M.$$

(ii) Show that for α satisfying $\beta < \alpha < 1$, $|F(z)| \to 0$ as $|z| = r \to \infty$. Conclude that $|f(z)| \le M$ for $z \in R$.

Fundamental to a great many theorems in complex function theory is the following:

1.8 Schwarz lemma. *If $f(z)$ is analytic in the unit disk U and $|f(z)| \le 1$, $f(0) = 0$, then $|f(z)| \le |z|$ and $|f'(0)| \le 1$. Equality holds if and only if $f(z) = cz$ with $|c| = 1$ (a rotation).*

Indeed, $f(0) = 0$ means that $f(z) = zg(z)$, where $g(z)$ is analytic in U. Moreover, $f'(z) = zg'(z) + g(z)$, so that $f'(0) = g(0)$. Thus

$$g(z) = \begin{cases} \frac{f(z)}{z} & z \ne 0 \\ f'(0) & z = 0, \end{cases}$$

and on $|z| = r$, we have $|g(z)| = |\frac{f(z)}{z}| \le \frac{1}{r}$. By the maximum modulus principle, $|g(z)| \le 1/r$ in $|z| \le r$. Letting $r \to 1$ implies $|g(z)| \le 1$, i. e., $|f(z)| \le |z|$.

Finally, $|f'(0)| = |g(0)| \le 1$. If $|f(z_0)| = |z_0|$ for any $z_0 \in U$, then $|g(z)|$ attains its maximum in U and so must be constant. Therefore $f(z) = cz$, $|c| = 1$.

1.9 Corollary. *If $|f(z)| \le M$ in the disk $|z - z_0| < R$, and $f(z_0) = 0$, then*

$$|f(z)| \le \frac{M}{R}|z - z_0|.$$

There is an interesting variation of the Schwarz lemma that involves the bound on the real harmonic part of a function $f = u + iv$ that is analytic in a disk. Denote

$$A(r) = \max_{|z|=r} u(z) = \max_{|z|\le r} u(z).$$

---

6 We will frequently use the notation $f(z) = O(g(x))$ as $|z| = r \to \infty$ to mean that there are positive constants C and r_0 such that $|f(z)| \le Cg(x)$ for all $r \ge r_0$. Here $g(x)$ is real-valued.

Although, in general, $u(z) \le |f(z)|$, we have the following:

1.10 Hadamard–Borel–Carathéodory theorem. *If $f(z)$ is analytic in $|z| \le R$, then for $0 < r < R$,*

$$M(r) \le \frac{2r}{R-r}A(R) + \frac{R+r}{R-r}|f(0)|, \tag{1.1}$$

where $M(r) = \max_{|z|=r} |f(z)|$.

Proof. If $f(z) = c_1 + ic_2 \equiv$ constant, then inequality (1.1) follows via a straightforward calculation. So we may assume that $f(z)$ is nonconstant, and initially we assume that $f(0) = 0$. Note that $0 = A(0) < A(R)$ since $f(z) = u + iv$ is analytic.

Next, consider the function

$$F(z) = \frac{f(z)}{2A(R) - f(z)}. \tag{1.2}$$

Since the denominator cannot vanish, $F(z)$ is analytic in $|z| < R$, and $F(0) = 0$. Moreover,

$$|F(z)|^2 = \frac{u^2 + v^2}{(2A(R) - u)^2 + v^2}.$$

Since $u^2 \le (2A(R) - u)^2$, it follows that $|F(z)| \le 1$. Thus by the Schwarz lemma,

$$|F(z)| \le \frac{r}{R},$$

and unwinding expression (1.2), we have

$$|f(z)| \le \left|\frac{2A(R)F(z)}{1 + F(z)}\right| \le \frac{(2A(R)(\frac{r}{R}))}{1 - (\frac{r}{R})} = \frac{2rA(R)}{R-r},$$

as desired.

If $f(0) \ne 0$, then the function $g(z) = f(z) - f(0)$ will satisfy the preceding inequality, that is,

$$|g(z)| \le \frac{2r \max_{|z|=r} \mathrm{Re}(g(z))}{R-r} \le \frac{2r(A(R) + |f(0)|)}{R-r},$$

and again the result follows. Furthermore, if $A(R) \ge 0$, then it is evident that

$$M(r) \le \frac{R+r}{R-r}(A(R) + |f(0)|).$$

□

The following consequence will be used in Chapter 2.

1.11 Corollary. *As above, with $A(R) \geq 0$,*

$$\max_{|z|=r} |f^{(n)}(z)| \leq \frac{2^{n+2} n! R}{(R-r)^{n+1}} (A(R) + |f(0)|).$$

In fact, we just need to apply the Cauchy formula

$$|f^{(n)}(z)| = \frac{n!}{2\pi i} \int_C \frac{f(\zeta)}{(\zeta - z)^{n+1}} d\zeta$$

to the circle C with center z ($|z| = r$) and radius $\rho = (R-r)/2$. For ζ on C,

$$|\zeta| \leq |z| + \frac{R-r}{2} = \frac{R+r}{2}.$$

By the preceding theorem, for $\zeta \in C$,

$$\max |f(\zeta)| \leq \frac{R + (R+r)/2}{R - (R+r)/2} (A(R) + |f(0)|).$$

Putting this into the Cauchy formula gives

$$\begin{aligned} |f^{(n)}(z)| &\leq \frac{n!}{\rho^n} \frac{4R}{(R-r)} (A(R) + |f(0)|) \\ &\leq \frac{2^{n+2} n! R}{(R-r)^{n+1}} (A(R) + |f(0)|), \end{aligned}$$

establishing the result.

The following result for an analytic function defined in an annular region was stated without proof by Jacques Hadamard (1865–1963).[7]

1.12 Hadamard three-circles theorem. *Let $f(z)$ be analytic in the closed annulus $\mathcal{A}$: $r_1 \leq |z| \leq r_2$ with $r_1 < |z| = r < r_2$. If $M(r) = \max_{|z|=r} |f(z)|$, then*

$$M(r)^{\log\left(\frac{r_2}{r_1}\right)} \leq M(r_1)^{\log\left(\frac{r_2}{r}\right)} M(r_2)^{\log\left(\frac{r}{r_1}\right)}.$$

Proof. Consider the function $F(z) = z^\alpha f(z)$ for some yet to be determined real constant α, where $z^\alpha = e^{\alpha \log z}$. Note that $|z^\alpha| = r^\alpha$, so that this modulus is independent of any branch of $\log z$. Furthermore, if the maximum modulus of any branch of $F(z)$ were attained at a point z_0 interior to the annulus, then in some a small disk about z_0, we would obtain a contradiction to the maximum modulus principle. As a consequence, the maximum modulus of $F(z)$ is attained on the boundary of $\mathcal{A}$, that is,

7 Sur les fonctions entières, *Bull. Soc. Math. France* 24 (1896), 186–187. The name of the theorem is due to Edmund Landau.

$$|F(z)| \le \max\{r_1^{\alpha}M(r_1), r_2^{\alpha}M(r_2)\}.$$

Then on a third circle $|z| = r$ in $\mathcal{A}$ with $r_1 < r < r_2$,

$$|f(z)| \le \max\{r^{-\alpha}r_1^{\alpha}M(r_1), r^{-\alpha}r_2^{\alpha}M(r_2)\}. \tag{1.3}$$

At this stage the optimal choice of the value α is such that the two components in the preceding bracketed expression are equal, that is to say,

$$r_1^{\alpha}M(r_1) = r_2^{\alpha}M(r_2).$$

Solving for α gives

$$\alpha = -\left(\log \frac{M(r_2)}{M(r_1)}\right) \Big/ \left(\log \frac{r_2}{r_1}\right),$$

and substituting this value into inequality (1.3) yields

$$M(r) \le \left(\frac{r}{r_1}\right)^{-\alpha} M(r_1).$$

From this inequality the preceding expression for α then gives

$$\begin{aligned} M(r)^{\log\left(\frac{r_2}{r_1}\right)} &\le \left(\frac{r}{r_1}\right)^{\log \frac{M(r_2)}{M(r_1)}} M(r_1)^{\log\left(\frac{r_2}{r_1}\right)} \\ &= M(r_1)^{\log\left(\frac{r_2}{r}\right)} M(r_2)^{\log\left(\frac{r}{r_1}\right)} \end{aligned}$$

via the properties of logarithms,[8] as required. Note that we have equality only when $F(z) = c$, that is, $f(z) = cz^{\beta}$ for some real β. □

Observe that the result can be written in the form

$$\log M(r) \le \frac{\log r_2 - \log r}{\log r_2 - \log r_1} \log M(r_1) + \frac{\log r - \log r_1}{\log r_2 - \log r_1} \log M(r_2). \tag{1.4}$$

Recall that a function $\phi(x)$ on an interval I is *convex* if for any two points $x_1 < x_2$ in the domain I, $\phi(x)$ lies beneath the straight-line chord connecting the two points $\phi(x_1)$ and $\phi(x_2)$. This is expressed as

$$\phi(tx_1 + (1-t)x_2) \le t\phi(x_1) + (1-t)\phi(x_2)$$

for $0 \le t \le 1$. Setting $t = (x_2 - x)/(x_2 - x_1)$ gives another expression of convexity:

---

8 We have used the fact that $a^{\log b} = b^{\log a}$.

$$\phi(x) \le \frac{x_2 - x}{x_2 - x_1}\phi(x_1) + \frac{x - x_1}{x_2 - x_1}\phi(x_2),$$

from which and inequality (1.4) it is clear that $\log M(r)$ is a convex function of $\log r$.

Our next result for analytic functions will be extended to meromorphic functions in Chapter 3. It is interesting that it relates the integral mean of an analytic function to the zeros of the function (in its present form) and has numerous applications.

1.13 Jensen Formula. Suppose that an analytic function $f(z)$ in $|z| < R$ has finitely many zeros listed according to multiplicity[9] at the points $a_1, a_2, \ldots, a_n \neq 0$ in the disk $|z| < r < R$ with $|a_1| \le |a_2| \le \cdots \le |a_n|$. Then we have a version of the *Jensen formula*

$$\log|f(0)| = \frac{1}{2\pi}\int_0^{2\pi} \log|f(re^{i\phi})|d\phi - \sum_{i=1}^{n} \log\frac{r}{|a_i|} \tag{1.5}$$

for $|z| < r < R$. There are various proofs of this result, and one will be given in Chapter 3.

Next, let $n(t, 0)$ be the *counting function* of the number of zeros listed according to multiplicity in $|z| \le t < R$. Then, assuming that $f(z)$ is analytic in $|z| < R$ and considering all zeros $a_1, a_2, \ldots, a_n$,

$$\begin{aligned}\sum_{k=1}^{n} \log\frac{r}{|a_k|} &= n\log r - \sum_{k=1}^{n} \log|a_k| \\ &= n(\log r - \log|a_n|) + \sum_{k=1}^{n-1} k(\log|a_{k+1}| - \log|a_k|) \\ &= \sum_{k=1}^{n-1} k \int_{|a_k|}^{|a_{k+1}|} \frac{dt}{t} + n\int_{|a_n|}^{r} \frac{dt}{t}.\end{aligned} \tag{1.6}$$

Considering the last two terms, for $|a_k| \le t < |a_{k+1}|$, $k = 1, 2, \ldots, (n-1)$, we have $k = n(t, 0)$, and with $|a_n| \le t < r$, we have $n = n(t, 0)$. Therefore we can write equation (1.6) as[10]

$$\sum_{k=1}^{n} \log\frac{r}{|a_k|} = \int_0^{r} \frac{n(t,0)}{t}dt. \tag{1.7}$$

9 When a zero/pole of order m is listed according to multiplicity, this means that it is repeated m times in the list.

10 In writing the integral as in (1.7), note that $n(t, 0) = 0$ for $0 \le t < |a_1|$ for $f(0) \neq 0$. This will be our understanding henceforth when encountering an integral such as in (1.7).

A more direct approach is via a Stieltjes integral.[11] In either case, Jensen's formula can be rewritten as

$$\frac{1}{2\pi}\int_0^{2\pi}\log|f(re^{i\phi})|d\phi = \int_0^r \frac{n(t,0)}{t}dt + \log|f(0)|. \tag{1.8}$$

1.14 Corollary. *If $f(z)$ is as above and $|f(z)| \le M$ in $|z| < r$, then the number of zeros N in $|z| < \alpha r$ for $0 < \alpha < 1$ is bounded by*

$$N \le \frac{1}{\log(1/\alpha)}\log\frac{M}{|f(0)|}.$$

Proof. Let $a_1, a_2, \ldots, a_N (\ne 0)$ be the zeros in $|z| < \alpha r$, so that by formula (1.5) we have

$$\log\frac{r^N}{|a_1||a_2|\cdots|a_N|} = \frac{1}{2\pi}\int_0^{2\pi}\log|f(re^{i\phi})|d\phi - \log|f(0)| \le \log\frac{M}{|f(0)|}.$$

From the inequality

$$\log\frac{r^N}{|a_1||a_2|\ldots|a_N|} \ge \log\frac{1}{\alpha^N} = N\log\left(\frac{1}{\alpha}\right),$$

the result then follows. □

Note that the number of zeros N is controlled by three key constraints: the size of the disk, the upper bound M, and the value of $|f(0)|$. Larger bounds M or smaller values of $|f(0)|$ allow for the existence of more zeros in the disk.

Infinite products

Like their counterparts infinite series, various forms of infinite products have been widely used in complex analysis, and they will appear from time to time through the text. One of the most beautiful infinite products going back to the English clergyman and mathematician John Wallis in 1656 is the expression

$$\frac{\pi}{2} = \frac{2\cdot 2\cdot 4\cdot 4\cdot 6\cdot 6\cdot 8\cdots}{1\cdot 3\cdot 3\cdot 5\cdot 5\cdot 7\cdot 7\cdots},$$

---

11 Using integration by parts applied to the Stieltjes integral,

$$\sum_{k=1}^n \log\frac{r}{|a_k|} = \sum_{k=1}^n \log\frac{r}{t_k} = \int_0^r \log\frac{r}{t}dn(t,0) = \log\frac{r}{t}\cdot n(t,0)\Big|_0^r - \int_0^r n(t,0)d\log\frac{r}{t} = \int_0^r \frac{n(t,0)}{t}dt.$$

which is almost too good to be true. We will come across other such marvels in Chapter 10 as infinite products form a major component of analytic number theory.

Recall that the zeros of a nonconstant analytic function in a domain Ω have no accumulation point in Ω. However, taking a sequence of points $\{a_n\}$ in Ω, can these be the zero set of some analytic function in Ω? In the case where $|a_n| \to \infty$, Karl Weierstrass (1815–1897) found an entire function having the zeros precisely at the points a_n, which is discussed in Chapter 2. In the case where $|a_n| \to 1$, Wilhelm Blaschke (1885–1962) found the answer discussed in the sequel. As both solutions involve infinite products, let us first recall some preliminary results.

Given a sequence of complex numbers $\{a_n\}$, we form the *partial products* $p_n = (1 + a_1)(1 + a_2)\cdots(1 + a_n)$, and initially suppose that $a_n \neq -1$ for all $n = 1, 2, 3, \ldots$. If $\lim_{n\to\infty} p_n = p \neq 0$ exists, then we say that the infinite product

$$\prod_{n=1}^{\infty}(1 + a_n)$$

converges to p and write $p = \prod_{n=1}^{\infty}(1 + a_n)$. Otherwise, we say that the infinite product *diverges*. Relaxing the conditions slightly can sometimes be useful: if a finite number of terms $a_k = -1$, then we say that $\prod_{n=1}^{\infty}(1 + a_n)$ converges to zero if the partial products of the nonzero terms converge to some $p \neq 0$.

As a consequence (omitting any terms that are zero), if $\prod_{n=1}^{\infty}(1 + a_n)$ converges to p, then

$$1 + a_n = \frac{p_n}{p_{n-1}} \to \frac{p}{p} = 1$$

as $n \to \infty$, that is, $a_n \to 0$ as $n \to \infty$, analogously to a convergent infinite series.

As an example, the infinite product $\prod_{n=1}^{\infty}(1 + \frac{1}{n})$ has partial products $p_n = \frac{2}{1}\frac{3}{2}\cdots\frac{n+1}{n} = n + 1$, so that the infinite product diverges although $a_n = \frac{1}{n} \to 0$ as $n \to \infty$. Therefore $a_n \to 0$ as $n \to \infty$ is a necessary but not sufficient condition for the convergence of $\prod_{n=1}^{\infty}(1 + a_n)$, again like in the case of infinite series.

1.15 Theorem. *If $a_n \geq 0$, $n = 1, 2, 3, \ldots$, then $\prod_{n=1}^{\infty}(1+a_n)$ and $\sum_{n=1}^{\infty} a_n$ converge or diverge together.*

Proof. For the partial product $p_n = \prod_{k=1}^{n}(1 + a_k)$, since $1 + x \leq e^x$ for $x \geq 0$, we obtain

$$\begin{aligned} a_1 + a_2 + \cdots + a_n &\leq (1 + a_1)(1 + a_2)\cdots(1 + a_n) \\ &\leq e^{a_1+a_2+\cdots+a_n}, \end{aligned}$$

and the result follows. □

1.16 Corollary. *The infinite product $\prod_{n=1}^{\infty}(1 + a_n)$ converges absolutely, that is, $\prod_{n=1}^{\infty}(1 + |a_n|)$ converges, if and only if $\sum_{n=1}^{\infty} a_n$ converges absolutely, i. e., $\sum_{n=1}^{\infty} |a_n|$ converges.*

For example, the infinite product

$$\prod_{n=1}^{\infty}\left(1-\frac{z^2}{n^2}\right)$$

converges absolutely for all $z \in \mathbb{C}$ since $\sum_{n=1}^{\infty}|z^2/n^2| = |z|^2\sum_{n=1}^{\infty}\frac{1}{n^2} < \infty$.

1.17 Theorem. *If* $\prod_{n=1}^{\infty}(1+|a_n|)$ *converges* ($a_n \neq -1$), *then* $\prod_{n=1}^{\infty}(1+a_n)$ *converges to some* $p \neq 0$.

Proof. We include the proof of this elementary fact as the details are applicable to the proof of the next theorem.

Let $p_n(z) = \prod_{k=1}^{n}(1+a_k)$ and $q_n = \prod_{k=1}^{n}(1+|a_k|)$, so that

$$\begin{aligned} p_n - p_{n-1} &= (1+a_1)\cdots(1+a_{n-1})a_n, \\ q_n - q_{n-1} &= (1+|a_1|)\cdots(1+|a_{n-1}|)|a_n|. \end{aligned}$$

Therefore

$$|p_n - p_{n-1}| \leq q_n - q_{n-1}.$$

By assumption the partial products q_n converge to some nonzero limit, which implies that $\sum_{n=2}^{\infty}(q_n - q_{n-1})$ converges since its partial sums are $s_n = q_n - q_1$. Hence $\sum_{n=2}^{\infty}(p_n - p_{n-1})$ converges by the comparison test, and therefore $\lim_{n\to\infty} p_n = p$ exists.

It remains to show that $p \neq 0$. By Corollary 1.16, $\sum_{n=1}^{\infty}|a_n|$ converges, so that terms $1 + a_n \to 1$, implying that $\sum_{n=1}^{\infty}\left|\frac{a_n}{1+a_n}\right|$ converges. Again by Corollary 1.16 and the above discussion, the partial products

$$P_n = \prod_{k=1}^{n}\left(1-\frac{a_k}{1+a_k}\right) = \frac{1}{p_n}$$

converge to some limit P. We conclude that $\lim_{n\to\infty} p_n = p \neq 0$, proving the theorem. □

It will be further convenient to take the logarithm of an infinite product, but it is necessary to show that this can be done legitimately to obtain the desired infinite sum of the logarithms.

1.18 Theorem. *For the infinite product* $\prod_{n=1}^{\infty}(1+a_n)$ ($a_n \neq -1$) *to converge, it is necessary and sufficient that* $\sum_{n=1}^{\infty}\log(1+a_n)$ *converges, where we take the principal value of the logarithm.*

Proof. For $p_n = \prod_{k=1}^{n}(1+a_k)$ and $s_n = \sum_{k=1}^{n}\log(1+a_k)$, we have $p_n = e^{s_n}$ for $n = 1, 2, 3, \ldots$. In the first instance, if $s_n \to s$, then $p_n \to e^s = p \neq 0$, proving the sufficiency.

On the other hand, suppose that $p_n \to p = \rho e^{i\varphi}$. Let $p_n = \rho_n e^{i\varphi_n}$ with the argument satisfying $\varphi - \pi < \varphi_n \le \pi + \varphi$ and define $\operatorname{Log} p_n = \log|p_n| + i\varphi_n$. Since $e^{s_n} = p_n$, we have

$$s_n = \operatorname{Log} p_n + 2\pi i k_n, \tag{1.9}$$

where k_n is some integer. Considering only the imaginary parts of (1.9) with $\theta_n = \arg(1+a_n)$, we have

$$\theta_1 + \theta_2 + \cdots + \theta_n = \varphi_n + 2\pi k_n.$$

Consequently,

$$2\pi(k_n - k_{n-1}) = \theta_n - (\varphi_n - \varphi_{n-1}).$$

Since the infinite product converges, $a_n \to 0$, implying $\theta_n \to 0$ as $n \to \infty$; as well $\varphi_n \to \varphi$ as $n \to \infty$. Therefore $k_n - k_{n-1} \to 0$, and thus there is some n_0 such that for $n \ge n_0$,

$$|k_{n+1} - k_n| < 1.$$

As each k_n is an integer, it follows that $k_n = k$ is a constant and thus $s_n = \operatorname{Log} p_n + 2\pi i k$ for $n \ge n_0$. We conclude that $s_n \to s = \operatorname{Log} p + 2\pi i k$, proving the necessity. □

In the real case, we remark that for $a_n \ge 0$, it is clear that if $p = \prod_{n=1}^{\infty}(1 + a_n)$ and $s = \sum_{n=1}^{\infty} \log(1 + a_n)$, then $\log p = s$. Moreover, there is an analytic function version of this theorem. Let us do one such for Theorem 1.17, and we leave the analytic version of Theorem 1.18 for the reader. Both are applied in the text.

1.19 Theorem. *If $\{f_n\}$ is a sequence of analytic functions defined in a domain Ω, and $\sum_{n=1}^{\infty} |f_n(z)|$ converges uniformly on compact subsets, then the product $\prod_{n=1}^{\infty}(1 + f_n(z))$ converges uniformly on compact subsets to an analytic function $f(z)$ on Ω.*

(By the conclusion we mean that the partial products $p_n(z) \to p(z) \ne 0$ uniformly on compact subsets, assuming without loss of generality that $f_n(z) \ne -1$ for all n).[12]

Proof. On any compact subset $K \subset \Omega$, the hypothesis implies that the partial sums

$$s_n(z) = \sum_{k=1}^{n} |f_k(z)| \to s(z)$$

converge uniformly on K to a continuous function $s(z)$ and $s(z) \le M < \infty$ on K. Regarding the corresponding partial products,

12 If $f_n(z_0) = -1$ for some n, then we take the product of the remaining functions f_n at z_0. Since $\sum_{n=1}^{\infty} |f_n(z)|$ converges at each $z \in \Omega$, $f_n(z_0) = -1$ for only a finite number of values n.

$$Q_n(z) = (1 + |f_1(z)|) \cdots (1 + |f_n(z)|) \le e^{|f_1(x)| + \cdots + |f_n(x)|} \le e^M.$$

Therefore

$$Q_n(z) - Q_{n-1}(z) = (1 + |f_1(z)|) \cdots (1 + |f_{n-1}(z)|)|f_n(z)| \le e^M |f_n(z)|.$$

Since $\sum_{n=1}^{\infty} |f_n(z)|$ converges uniformly on K, it follows that

$$\sum_{n=2}^{\infty} (Q_n(z) - Q_{n-1}(z))$$

converges uniformly on K. Setting

$$P_n(z) = (1 + f_1(z)) \cdots (1 + f_n(z)),$$

the remainder of the proof follows *mutatis mutandis* as in Theorem 1.17, and the analyticity follows by the uniform convergence of $P_n(z)$ on K in view of Theorem 1.27. □

1.20 Corollary. *Let $\{f_n\}$ be a sequence of analytic functions in a domain Ω such that no f_n is $\equiv 0$ (but may have isolated zeros) and*

$$\sum_{n=1}^{\infty} |1 - f_n(z)|$$

converges uniformly on compact subsets of Ω. Then the product

$$\text{(i)} \quad \prod_{n=1}^{\infty} f_n(z) = f(z)$$

converges uniformly on compact subsets of Ω to an analytic function $f(z)$ in Ω. Moreover,

$$\text{(ii)} \quad m(f; z) = \sum_{n=1}^{\infty} m(f_n; z) \quad \textit{for } z \in \Omega,$$

where $m(f; z)$ is the multiplicity of the zero of f at z.

Indeed, writing $F_n = -1 + f_n$ implies that $\sum_{n=1}^{\infty} |F_n|$ converges uniformly on compact subsets of Ω to a *bounded* sum (on each subset). Since $1 + F_n(z) = f_n(z)$, it follows from the theorem that $\prod_{n=1}^{\infty} f_n(z)$ converges uniformly on compact subsets to some $f(z)$. Since each partial product is analytic, so is $f(z)$, since the convergence is uniform on compact subsets, establishing statement (i).

To verify (ii), note that $f(z_0) = 0$ if and only if $f_n(z_0) = 0$ for some n. Now the fact that for any $z \in \Omega$, the sum $\sum_{n=1}^{\infty} |1 - f_n(z)|$ converges implies that only finitely many of the f_n can be zero at $z \in \Omega$. The product of the remaining f_n has a nonzero limit from the preceding theorem. Therefore $m(f; z)$ is a finite sum of multiplicities of the f_n that vanish at the point z.

Blaschke products

An interesting and useful infinite product is constructed from conformal mappings of the unit disk to itself such that each term has a zero in the disk with zeros tending to the boundary.

1.21 Definition. A function of the form

$$B(z) = e^{iy} z^m \prod_{n=1}^{\infty} \frac{|a_n|}{a_n} \left(\frac{a_n - z}{1 - \overline{a_n} z} \right)$$

is called a *Blaschke product*, where $z \in U, y$ is real, $m \geq 0$ is an integer, and $\sum_{n=1}^{\infty}(1 - |a_n|) < \infty$. The sequence $\{a_n\}$ in U may be finite or infinite. These products arise again in Chapter 3.

1.22 Blaschke product theorem.[13] *Let $a_1, a_2, \ldots$ be a sequence of numbers such that $0 < |a_1| \leq |a_2| \leq \cdots < 1$ subject to the constraint $\sum_{n=1}^{\infty}(1 - |a_n|) < \infty$, and let $m \geq 0$ be an integer. Then*

$$B(z) = z^m \prod_{n=1}^{\infty} \frac{|a_n|}{a_n} \left(\frac{a_n - z}{1 - \overline{a_n} z} \right)$$

converges uniformly on each disk $|z| \leq \rho < 1$, $B(z)$ has zeros only at the points $z = a_n$, $n = 1, 2, \ldots$, and possibly at the origin, and $|B(z)| < 1$ for all $z \in U$, that is, $B(z)$ is a bounded analytic function in U.

Proof. The integer m merely accounts for any zeros at the origin, so we can assume that $m = 0$. For $|z| \leq \rho < 1$,

$$\begin{aligned} |1 - A_n(z)| &= \left| 1 - \left(\frac{|a_n|}{a_n} \cdot \frac{a_n - z}{1 - \overline{a_n} z} \right) \right| = \left| \frac{(a_n + |a_n| z)(1 - |a_n|)}{a_n (1 - \overline{a_n} z)} \right| \\ &\leq \frac{(1 + |z|)(1 - |a_n|)}{|1 - \overline{a_n} z|} \leq \frac{2(1 - |a_n|)}{1 - \rho}. \end{aligned}$$

By the Weierstrass M-test the fact that $\sum_{n=1}^{\infty}(1 - |a_n|) < \infty$ implies that $\sum_{n=1}^{\infty} |1 - A_n(z)|$ converges uniformly on $|z| \leq \rho < 1$ and hence uniformly on compact subsets of U. Since each term $A_n(z)$ is analytic in U, the preceding theorem implies that $\prod_{n=1}^{\infty} A_n(z)$ defines an analytic function $B(z)$ in U with zeros only at the points $z = a_n$, $n = 1, 2, 3 \ldots$.

Furthermore, note that for $|z| < 1$ and each N, the partial products $P_N(z)$ satisfy

$$|P_N(z)| = \prod_{n=1}^{N} \left| \frac{a_n - z}{1 - \overline{a_n} z} \right| < 1,$$

13 Eine Erweiterung des Satzes von Vitali über Folgen analytischer Funktionen, *Ber. Verhandl. Kön. Sächs. Gesell. Wiss. Leipzig* 67 (1915), 194–200.

so that $|B(z)| = \lim_{N\to\infty} |P_N(z)| < 1$ in U in view of the maximum modulus principle. Thus the theorem is proved. □

The requirement on the zeros in U that $\sum_{n=1}^{\infty}(1 - |a_n|) < \infty$ turns out to be a necessary condition for the zeros of a bounded analytic function in U, as we will see in a moment. First, we need the following:

1.23 Lemma. *Suppose that* $0 < b_n < 1$, $n = 1, 2, 3, \ldots$. *Then*

$$\prod_{n=1}^{\infty}(1 - b_n) > 0$$

if and only if

$$\sum_{n=1}^{\infty} b_n < \infty.$$

Proof. First, observe that

$$p_n = (1 - b_1)(1 - b_2)\cdots(1 - b_n),$$

so that $p_1 \geq p_2 \geq p_3 \geq \ldots$, implying that $\lim_{n\to\infty} p_n = p$ exists.

If we first assume that $\sum_{n=1}^{\infty} b_n < \infty$, then $\sum_{n=1}^{\infty} |-b_n| < \infty$, implying by Corollary 1.16 that $\prod_{n=1}^{\infty}(1 + |-b_n|)$ converges, so that $\prod_{n=1}^{\infty}(1 - b_n)$ converges to some $p \neq 0$ by Theorem 1.17.

On the other hand, if we assume that $\sum_{n=1}^{\infty} b_n = \infty$, then using the inequality $1 - x \leq e^{-x}$ for $0 < x < 1$, we have

$$p \leq p_n = \prod_{k=1}^{n}(1 - b_k) \leq e^{-b_1 - b_2 - \cdots - b_n} \to 0$$

as $n \to \infty$, that is, $p = 0$, establishing the result. □

The significance of the Blaschke product is that it characterizes all bounded analytic functions $f(z)$ in the unit disk U in terms of their zeros. First, we prove a condition on the zeros that must prevail.

1.24 Theorem. *If* $f(z) \not\equiv 0$ *is a bounded analytic function in* U, *and* $a_1, a_2, a_3, \ldots$ *are the zeros of* $f(z)$ *listed according to multiplicity, then*

$$\sum_{n=1}^{\infty}(1 - |a_n|) < \infty.$$

Proof. We may assume without loss of generality that $f(z)$ has infinitely many zeros and $|a_n| \leq |a_{n+1}|$. Furthermore, we will assume that $f(0) \neq 0$, for otherwise consider $f(z)/z^m$ for $m > 0$.

Let $n(r)$ = number of zeros of $f(z)$ in $|z| < r < 1$ and define for $f(z) \neq 0$ on $|z| = r$,

$$F_r(z) = f(z) \prod_{k=1}^{n(r)} \frac{r^2 - \overline{a_k} z}{r(z - a_k)}, \tag{1.10}$$

which is analytic in $|z| \leq r$, and $|F_r(z)| = |f(z)|$ on $|z| = r$. Since $|f(z)| < M$ in U, $|F_r(z)| < M$ in $|z| \leq r$ by the maximum modulus principle. In particular, for $z = 0$,

$$|f(0)| \prod_{k=1}^{n(r)} \frac{r}{|a_k|} \leq M.$$

Replacing $n(r)$ by N for $n(r) \geq N$, the preceding inequality remains valid since $|a_k| < r$, so that

$$p_N = \prod_{k=1}^{N} |a_k| \geq \frac{|f(0)| r^N}{M},$$

which holds for each N as $r \to 1$. Letting $r \to 1$ and then $N \to \infty$ implies that

$$\prod_{k=1}^{\infty} |a_k| \geq \frac{|f(0)|}{M} > 0.$$

In view of Lemma 1.23, with $|a_k| = 1 - b_k$, we obtain

$$\sum_{k=1}^{\infty} (1 - |a_k|) < \infty.$$

□

1.25 Corollary. *If $f(z)$ is a bounded analytic function in U with zeros at $a_1, a_2, a_3, \ldots$ in U such that $\sum_{n=1}^{\infty}(1 - |a_n|) = \infty$, then $f(z) \equiv 0$.*

So, for example, if $a_n = 1 - \frac{1}{n^p}$ for $p \geq 1$, then $\sum_{n=1}^{\infty}(1 - |a_n|) = \sum_{n=1}^{\infty} \frac{1}{n^p}$. For $p = 1$, we conclude that there is no bounded analytic function in U with zeros at the points a_n except the identically zero function. For $p > 1$, we can construct a bounded analytic function in U with zeros at the points a_n in accordance with the Blaschke product theorem.

In the context of bounded analytic functions, without loss of generality, we consider $|f(z)| < 1$ for if not, then we can take $|f(z)|/M$ where $M = \sup_{z \in U} |f(z)|$. Here we see that all bounded analytic functions in the unit disk are Blaschke products modulo an entire function.

1.26 Theorem. *Every bounded analytic function $f(z)$ in U with $|f(z)| < 1$ can be expressed in the form*

$$f(z) = e^{g(z)} z^m \prod_{n=1}^{\infty} \frac{|a_n|}{a_n} \left(\frac{a_n - z}{1 - \overline{a_n} z} \right),$$

where $g(z)$ is analytic in U with $\mathrm{Re}(g) \leq 0$.

In fact, let $f(z)$ have zeros at the points $a_1, a_2, a_3, \dots$ as in Theorem 1.22 and assume for a moment that $f(0) \neq 0$. Now let $B_0(z) = \prod_{n=1}^{\infty} \frac{|a_n|}{a_n}\left(\frac{a_n - z}{1-\overline{a}_n z}\right)$, and let $P_N(z)$ be the corresponding partial products.

Then the function $f(z)/B_0(z)$ is analytic and nonzero in U. As in the preceding theorem, we deduce from (1.10) that

$$|F_r(z)| = |f(z)| \prod_{k=1}^{n(r)} \left| \frac{r^2 - \overline{a}_k z}{r(z - a_k)} \right| < 1$$

for $|z| \leq r$. Again, replacing $n(r)$ by N for $n(r) \geq N$, the preceding inequality remains valid, and letting $r \to 1$ means that $|f(z)/P_N(z)| \leq 1$ and

$$\left| \frac{f(z)}{B_0(z)} \right| = \lim_{n\to\infty} \left| \frac{f(z)}{P_N(z)} \right| \leq 1.$$

Taking a branch of $g(z) = \log\left(\frac{f(z)}{B_0(z)}\right)$, we can express $f(z)$ in the form

$$f(z) = e^{g(z)} z^m \prod_{n=1}^{\infty} \frac{|a_n|}{a_n} \left(\frac{a_n - z}{1 - \overline{a}_n z} \right),$$

where $g(z)$ is analytic in U with $\mathrm{Re}(g) \leq 0$, and m is a nonnegative integer.

This theorem has a meromorphic counterpart in Proposition 3.30.

Sequences of analytic functions

We will often have occasion to consider sequences of analytic functions on a domain in the complex plane, especially, in the study of normal families. When the convergence is uniform on compact subsets, we have an important consequence regarding the limit function.

1.27 Weierstrass theorem. *If $f_n \to f$ uniformly on compact subsets of a domain Ω and f_n is analytic in Ω, then f is analytic in Ω, and the derivatives $f_n^{(k)} \to f^{(k)}$ uniformly on compact subsets of Ω.*

Proof. For any $z_0 \in \Omega$, we consider a small closed disk $\overline{D}(r, z_0) = \{|z - z_0| \leq r\} \subset \Omega$.[14] By the uniform convergence, for any $\varepsilon > 0$, there is n_0 such that

$$|f(\zeta) - f_n(\zeta)| < \varepsilon$$

for all $n \geq n_0$ and all $\zeta \in \overline{D}(r, z_0)$. Now for $k = 0, 1, 2, \dots$, consider the function

14 Throughout the text, we will use the notation $D(r, z_0) = \{|z - z_0| < r\}$.

$$g_k(z) = \frac{k!}{2\pi i} \int_{C_r} \frac{f(\zeta)d\zeta}{(\zeta - z)^{k+1}},$$

where the integration is taken along the circle $c_r = \{|z - z_0| = r\}$, and we restrict z to the disk $D(\frac{r}{2}, z_0)$. Then by the Cauchy integral formula for derivatives

$$|g_k(z) - f_n^{(k)}(z)| \leq \frac{k!}{2\pi} \int_{C_r} \frac{|f(\zeta) - f_n(\zeta)||d\zeta|}{|\zeta - z|^{k+1}} \leq \frac{k!\varepsilon 2^{k+1}}{r^k}$$

for $n \geq n_0$. This implies that $f_n^{(k)} \to g_k$ uniformly on $D(\frac{r}{2}, z_0)$ for each $k = 0, 1, 2, \ldots$. Taking $k = 0$, we have $f_n \to f \equiv g_0$, which is analytic in $D(\frac{r}{2}, z_0)$ and hence in Ω since z_0 was arbitrary. Finally, $f_n^{(k)} \to f^{(k)} \equiv g_k(z)$ uniformly in $D(\frac{r}{2}, z_0)$, and the theorem is complete as a consequence of a standard compactness argument. □

The next result has many important consequences, which will come to light in the next chapter.

1.28 Hurwitz theorem. *Let $\{f_n\}$ be a sequence of analytic functions on a domain Ω such that $f_n \to f$ uniformly on compact subsets and $f(z_0) = 0$ for some $z_0 \in \Omega$, but $f \not\equiv$ constant. Then for each sufficiently small disk $|z - z_0| < r$ in Ω, there is a number $n_0(r)$ such that for all $n > n_0(r)$, the functions f_n and f have the same number of zeros in $|z - z_0| < r$ counted according to multiplicity.*

Proof. We can choose r sufficiently small so that the point z_0 is the only zero of $f(z)$ in the disk $|z - z_0| \leq r$. Then $|f(z)| > m > 0$ for some constant m on the compact boundary $|z - z_0| = r$ by the continuity of $f(z)$. Moreover, by the uniform convergence of $f_n \to f$ on $|z - z_0| = r$

$$|f_n(z) - f(z)| < m < |f(z)|$$

for all n sufficiently large. The result then follows by an application of Rouché's theorem.[15] □

Throughout this text, a *univalent* function is one that is analytic and one-to-one. The term *analytic univalent* is sometimes used for emphasis as occasionally "univalent" can also mean that the function is meromorphic and one-to-one.

1.29 Corollary. *Let $\{f_n\}$ be a sequence of univalent functions on a domain Ω that converge uniformly on compact subsets to an analytic function f. Then f is univalent in Ω.*

Proof. We already know that the function f is analytic by the Weierstrass theorem 1.27. Now suppose that $f(z_1) = f(z_2)$ for $z_1 \neq z_2$ and for each $n = 1, 2, 3, \ldots$, define

15 Rouché theorem: If $f(z)$ and $g(z)$ are analytic within and on a simple closed contour C and if $|g(z)| < |f(z)|$ for all z on C, then the functions $f(z)$ and $f(z) + g(z)$ have the same number of zeros inside C.

$$g_n(z) = f_n(z) - f_n(z_1),$$

so that each g_n is zero at the point z_1. Moreover, the analytic limit function

$$g(z) = f(z) - f(z_1)$$

vanishes at the point z_2, so that by the Hurwitz theorem there is a sufficiently small neighborhood of z_2 (which excludes the point z_1) in which for n sufficiently large each $g_n(z)$ has a zero. The corresponding functions $f_n(z)$ would then take the value $f_n(z_1)$ for points $z \neq z_1$, which contradicts the univalence of f_n. Hence f is univalent. □

Local boundedness

In dealing with families of functions as, say, sequences, we require an extension of the notion of boundedness of a single function. In general, we do not require the entire family of functions to be bounded on their common domain by a single finite constant but only bounded in a neighborhood of each point.

1.30 Definition. Let $\mathcal{F}$ be a family of analytic functions defined on a common domain Ω. Then $\mathcal{F}$ is said to be *locally bounded* (*locally uniformly bounded*) on Ω if for each point $z_0 \in \Omega$ and a neighborhood $D(z_0, r) \subseteq \Omega$, there is a constant $M = M(z_0, r) > 0$ such that for all $f \in \mathcal{F}$,

$$|f(z)| \leq M$$

at all points $z \in D(z_0, r)$.

1.31 Remark. When the family $\mathcal{F}$ is locally bounded in a domain Ω, then for any compact set $K \subset \Omega$, the family is bounded in an open disk about each point $z_0 \in K$ forming an open covering of K. Taking a finite subcovering implies that $|f(z)| \leq M_K < \infty$ for all $f \in \mathcal{F}$, that is, the family $\mathcal{F}$ is *uniformly bounded on compact subsets of* Ω. The converse is evident.

It is a simple exercise to show that the family of analytic functions

$$\mathcal{F} = \left\{ f_y(z) = \frac{1}{z - e^{iy}} : y \in \mathbb{R} \right\}$$

is locally bounded but not uniformly bounded in the unit disk U.

The property of local boundedness is passed on to the family of derivatives.

1.32 Theorem. *If $\mathcal{F}$ is a locally bounded family of analytic functions on some domain Ω, then the same holds for the corresponding family of derivatives $\mathcal{F}'$.*

In fact, for each $f \in \mathcal{F}$, we can find a closed neighborhood $\overline{D}(z_0, r) \subset \Omega$ on which $|f(z)| \leq M$, so that by the Cauchy inequality 1.2

$$|f'(z)| \le \frac{1}{2\pi} \int\limits_{|z-z_0|=r} \frac{|f(\zeta)|}{|\zeta - z|^2} |d\zeta| \le \frac{4M}{r}$$

for all $z \in D(z_0, \frac{r}{2})$ and all $f' \in \mathcal{F}'$, establishing that $\mathcal{F}'$ is locally bounded in Ω.

If our family $\mathcal{F}$ is comprised of the functions $f_n(z) = n$, $n = 1, 2, 3, \ldots$, then it is clear that the converse of the preceding theorem does not hold, although we do have the following partial converse.

1.33 Theorem. *If $\mathcal{F}$ is a family of analytic functions on a domain Ω and $\mathcal{F}'$ is the corresponding family of derivatives that are locally bounded with the proviso that for some $z_0 \in \Omega$, $|f(z_0)| \le M < \infty$ for each $f \in \mathcal{F}$, then $\mathcal{F}$ is locally bounded in Ω.*

Proof. If we take a path γ from the point z_0 to any point $z \in \Omega$ and integrate over this path, then for all $f \in \mathcal{F}$, we obtain

$$|f(z)| \le |f(z_0)| + \int\limits_{\gamma} |f'(\zeta)||d\zeta| \le M + M' \cdot L(\gamma)$$

in view of Remark 1.31. The proof then follows from the corresponding inequality obtained by taking a neighborhood about the point z in which the family $\mathcal{F}'$ is uniformly bounded and integrating over a straight-line path from the point z to any point z' in the neighborhood. This then provides the local boundedness of the family $\mathcal{F}$. □

Equicontinuity

As for boundedness, there is also a natural extension of the notion of continuity to a family of functions that has very important consequences and is related to local boundedness. The concept goes back to the work of Guilio Ascoli.[16]

1.34 Definition. A family $\mathcal{F}$ of analytic functions defined on a domain Ω is said to be *equicontinuous at a point* $z_0 \in \Omega$ if for each $\varepsilon > 0$, there exists $\delta = \delta(\varepsilon, z_0) > 0$ such that for every $f \in \mathcal{F}$,

$$|f(z) - f(z_0)| < \varepsilon$$

whenever $|z - z_0| < \delta$. Furthermore, the family $\mathcal{F}$ is *equicontinuous on a subset* $S \subseteq \Omega$ if it is equicontinuous at each point of S.

1.35 Remark. Since a continuous function on a compact set is uniformly continuous, equivalent reasoning entails that if a family $\mathcal{F}$ is equicontinuous on a compact subset K, then for each $\varepsilon > 0$, there exists $\delta = \delta(\varepsilon, K) > 0$ such that for every $f \in \mathcal{F}$,

16 Le curve limite di una varietà data di curvi, *Mem. Accad. Lincei* (3) 18 (1883–84), 521–586.

$$|f(z) - f(z')| < \varepsilon$$

whenever $z, z' \in K$, $|z - z'| < \delta$. This is sometimes taken as the definition of equicontinuity on sets K that are not necessarily compact.

In general, we do not expect any relationship between the notions of a function being bounded and being continuous. However, for families of analytic functions, there is a very compelling relation between the two.

1.36 Theorem. *If $\mathcal{F}$ is a locally bounded family of analytic functions on a domain Ω, then $\mathcal{F}$ is equicontinuous on compact subsets of Ω.*

Proof. The local boundedness of $\mathcal{F}$ implies that the family of derivatives $\mathcal{F}'$ is uniformly bounded on compact subsets of Ω from the preceding considerations. Hence for any closed disk K in Ω, there is a constant $M = M(K)$ such that

$$|f'(z)| \leq M < \infty$$

for all $z \in K$ and $f' \in \mathcal{F}'$. Now taking points $z, z' \in K$ and integrating over a straight-line path from z to z', we have

$$|f(z) - f(z')| \leq \int_{z}^{z'} |f'(\zeta)||d\zeta| \leq M|z - z'|.$$

We conclude that given any $\varepsilon > 0$ and setting $0 < \delta = \delta(\epsilon, K) < \varepsilon/M$, then whenever $|z - z'| < \delta$ with $z, z' \in K$, we have

$$|f(z) - f(z')| < \varepsilon,$$

establishing the equicontinuity of the family $\mathcal{F}$. □

1.37 Example. If we consider the family of analytic functions on the unit disk U,

$$\mathcal{F} = \{f_n(z) = z + n : n = 1, 2, 3, \ldots\},$$

then $\mathcal{F}$ is clearly equicontinuous in U but not locally bounded. Thus the converse of the preceding theorem is not valid.

Möbius transformations

We consider transformations from $T(z) : \hat{\mathbb{C}} \to \hat{\mathbb{C}}$ of the form

$$w = T(z) = \frac{az + b}{cz + d} \tag{1.11}$$

with $a, b, c, d \in \mathbb{C}$, $ad - bc \neq 0$, and $\hat{\mathbb{C}} = \mathbb{C} \cup \{\infty\}$.[17] They are called either *Möbius transformations*, *linear fractional transformations*, or *bilinear transformations* and map $\hat{\mathbb{C}}$ one-to-one conformally onto itself with $T(\infty) = a/c\ (c \neq 0)$ or $T(\infty) = \infty (c = 0)$.

An example is the *doubly periodic group*

$$\{T(z) = z + n\omega + m\omega' : n, m \in \mathbb{Z}\},$$

which will feature in our discussion of the Weierstrass $\wp$-function in Chapter 3.

Observe that the matrices associated with the Möbius transformations $\frac{az+b}{cz+d}$ form a group under the composition of functions (exercise) called the *general linear group* $GL(2, \mathbb{C})$ and that the matrices

$$\begin{pmatrix} a & b \\ c & d \end{pmatrix} \quad \text{and} \quad \eta \begin{pmatrix} a & b \\ c & d \end{pmatrix}$$

represent the same transformation for any nonzero $\eta \in \mathbb{C}$. Since we can divide each of a, b, c, d by $\sqrt{ad - bc}$, it is usually more convenient to deal with Möbius transformations having the normalization $ad - bc = 1$, which determines exactly the same transformation. So we define

$$\text{Möb}(\hat{\mathbb{C}}) = \left\{ T(z) = \left(\frac{az + b}{cz + d} \right) : a, b, c, d \in \mathbb{C}, ab - cd = 1 \right\}$$

with the corresponding matrices forming the subgroup (*special linear group*)

$$SL(2, \mathbb{C}) = \left\{ \begin{pmatrix} a & b \\ c & d \end{pmatrix} : a, b, c, d \in \mathbb{C}, ad - bc = 1 \right\}.$$

1.38 Theorem. *The mapping $\phi : SL(2, \mathbb{C}) \to \text{Möb}(\hat{\mathbb{C}})$ defined as*

$$\phi : \begin{pmatrix} a & b \\ c & d \end{pmatrix} \to \frac{az + b}{cz + d}$$

is a group homomorphism, that is,

$$\phi\left(\begin{pmatrix} a & b \\ c & d \end{pmatrix} \begin{pmatrix} e & f \\ g & h \end{pmatrix} \right) = \phi \begin{pmatrix} a & b \\ c & d \end{pmatrix} \circ \phi \begin{pmatrix} e & f \\ g & h \end{pmatrix}.$$

Proof. Exercise. □

1.39 Corollary. *Möb($\hat{\mathbb{C}}$) is isomorphic to $SL(2, \mathbb{C})/\{I, -I\}$ where $I = \left(\begin{smallmatrix} 1 & 0 \\ 0 & 1 \end{smallmatrix}\right)$.*

---

17 We will take up the extended complex plane in more detail in Chapter 3.

Indeed, it is easy to see that $\ker\phi = \{I, -I\}$, and the result follows by the first isomorphism theorem. $SL(2,\mathbb{C})/\{I,-I\}$ is also denoted by $PSL(2,\mathbb{C})$, the projective special linear group, but we will not discuss the latter.

An important matrix subgroup is $SL(2,\mathbb{R})$ where the coefficients $a, b, c, d \in \mathbb{R}$. Note that the associated Möbius transformations with $a, b, c, d \in \mathbb{R}$, $ad - bc = 1$, satisfy

$$\operatorname{Im}\left(\frac{az+b}{cz+d}\right) = \frac{y}{|cz+d|^2}, \qquad z = x + iy,$$

and hence map the upper half-plane/lower half-plane/real axis onto themselves. The converse is also true.

1.40 Proposition. *Any Möbius transformation from the upper half-plane to itself can be written as* $T(z) = \frac{az+b}{cz+d}$ *with* $a, b, c, d \in \mathbb{R}$, $ad - bc = 1$.

Indeed, since

$$\operatorname{Im}\left(\frac{az+b}{cz+d}\right) = \frac{(ad-bc)\operatorname{Im}(z)}{|cz+d|^2},$$

the hypothesis implies that $ad - bc > 0$ and that the real axis is mapped to the real axis. For $z \in \mathbb{R}$, we have $z = \bar{z}$, and therefore $T(z) = \overline{T(\bar{z})}$. By identifying the four constants we obtain $a, b, c, d \in \mathbb{R}$.[18] Furthermore, via a normalization we can assume that $ad - bc = 1$.

1.41 Examples. The *modular group* $\boldsymbol{\Gamma}$ of Möbius transformations are those defined by

$$T(z) = \frac{az+b}{cz+d}$$

with $ad - bc = 1$ and $a, b, c, d \in \mathbb{Z}$. By the preceding discussion these transformations are mappings from the upper half-plane to itself.

The *Picard group* $\mathbf{P}$ of Möbius transformations is defined by

$$T(z) = \frac{az+b}{cz+d}$$

with $ad - bc = 1$ and a, b, c, d of the form $m + in$, $m, n \in \mathbb{Z}$.

1.42 Definition. A point z is a *fixed point* of a mapping $f(z)$ if $f(z) = z$.

The fixed points of any $T \in \text{Möb}(\hat{\mathbb{C}})$ that is not the identity transformation are given by the roots of the equation $cz^2 + (d-a)z - b = 0$:

---

18 We are justified in identifying the four constants of each transformation as it will be shown as a consequence of the next exercise that even two Möbius transformations agreeing on three points must be identical.

$$\xi_1, \xi_2 = \frac{(a-d) \pm \sqrt{(a+d)^2 - 4}}{2c}$$

whenever $c \neq 0$. Thus there are at most two fixed points, but we need to carefully consider all possible cases in determining them.

1.43 Exercise. Show that for any $T \in \text{Möb}(\hat{\mathbb{C}})$ that is not the identity, if $D = (a+d)^2 - 4 \neq 0$, then there are exactly two fixed points and one fixed point if $D = 0$.

As a consequence, we deduce that if $T_1(z_i) = T_2(z_i)$ for $i = 1, 2, 3$, then $T_2 \circ T_1^{-1}(z_i) = z_i$ has three fixed points and hence must be the identity transformation, that is, $T_1 \equiv T_2$, that is, *two Möbius transformations that agree on three points are identical.*

One of the very useful basic facts is that given distinct points $z_1, z_2, z_3 \in \hat{\mathbb{C}}$ and $w_1, w_2, w_3 \in \hat{\mathbb{C}}$, there is a unique Möbius transformation $T \in \text{Möb}(\hat{\mathbb{C}})$ such that $T(z_i) = w_i$, $i = 1, 2, 3$, a property called *transitivity*. In fact, the transformation

$$T(z) = \frac{(z - z_1)(z_2 - z_3)}{(z - z_3)(z_2 - z_1)} \tag{1.12}$$

maps the points z_1, z_2, z_3 to the points 0, 1, ∞ respectively, and likewise for the transformation $S(w) = \frac{(w-w_1)(w_2-w_3)}{(w-w_3)(w_2-w_1)}$. Then the transformation $S^{-1} \circ T$ is the required mapping; it is instructive to verify that T is indeed a Möbius transformation.

To this end, rewriting (1.12) in the standard form of (1.11)

$$T(z) = \frac{(z_2 - z_3)z - z_1(z_2 - z_3)}{(z_2 - z_1)z - z_3(z_2 - z_1)},$$

we find that $ad - bc$ satisfies

$$(z_2 - z_3)(z_1 - z_3)(z_2 - z_1) \neq 0.$$

Taking a normalization of T gives us a member of $\text{Möb}(\hat{\mathbb{C}})$ with the desired requirements.

The *cross-ratio* for any four distinct points is denoted by

$$(z_0, z_1, z_2, z_3) = \frac{(z_0 - z_1)(z_2 - z_3)}{(z_0 - z_3)(z_2 - z_1)}.$$

Then the preceding result demonstrates the following:

1.44 Proposition. *Given three distinct points $z_1, z_2, z_3 \in \hat{\mathbb{C}}$, if $T \in \text{Möb}(\hat{\mathbb{C}})$, then the cross-ratio is invariant under the transformation T, that is, $(z_0, z_1, z_2, z_3) = (T(z_0), T(z_1), T(z_2), T(z_3))$.*

Proof. Exercise. □

Möbius transformations belonging to Möb($\hat{\mathbb{C}}$) are classified by the quantity $\tau^2 = (a+d)^2$, which represents the square of the *trace* τ[19] of the matrix $\left(\begin{smallmatrix} a & b \\ c & d \end{smallmatrix}\right)$. There are four cases:

Parabolic: $\tau^2 = 4$ (this is the case of one fixed point);
Elliptic: τ real, $0 \le \tau^2 < 4$;
Hyperbolic: τ real, $\tau^2 > 4$;
Loxodromic: $\tau^2 \notin [0,4]$.

Note that if each a, b, c, d of $T(z) = \frac{az+b}{cz+d}$ is real, then there can only be parabolic, elliptic, and hyperbolic transformations. When iterated, these transformations have very interesting characteristics, which will be explored in some detail in Chapter 9.

19 For any square matrix, the trace is the sum of the terms on the principal diagonal. Whereas $\left(\begin{smallmatrix} a & b \\ c & d \end{smallmatrix}\right)$ and $\left(\begin{smallmatrix} -a & -b \\ -c & -d \end{smallmatrix}\right)$ both represent the same Möbius transformation $T(z) = \frac{az+b}{cz+d}$ but have different trace values, the square of the trace is the same for each matrix and so is uniquely associated with $T(z)$.

2 Entire functions

Entire functions are analytic in the entire complex plane, which makes them sound somewhat pedestrian, but a revolutionary theorem from 1879 regarding these functions makes them something very much out of the ordinary.

2.1 Picard little theorem.[1] *A nonconstant entire function takes every value in* $\mathbb{C}$ *with one possible exception.*

The proof of this celebrated result given by Picard was based on the elliptic modular function $\mu(z)$ that will be discussed in Chapter 3 and will be postponed until then. Another proof based on Nevanlinna's theory of meromorphic functions will also be given in that chapter, and a proof by Montel via a normal family argument will be given in Chapter 4. Other proofs over the years have been given, including one derived using Bloch's theorem (Chapter 6). This tends to illustrate the intertwined nature of complex function theory and what a fundamental result is the Picard theorem.

Elementary factors

As a consequence of the fundamental theorem of algebra, a polynomial $p(z)$ of degree n having zeros at $a_1, a_2, \ldots, a_n$ can be expressed in the form

$$p(z) = p(0)\left(1 - \frac{z}{a_1}\right)\left(1 - \frac{z}{a_2}\right)\cdots\left(1 - \frac{z}{a_n}\right).$$

However, for an entire function having infinitely many zeros $a_1, a_2, a_3, \ldots$, it is necessary to ensure that the right-hand product will converge. This requires some additional factors in the infinite product. To this end, we begin with *elementary factors* introduced by Weierstrass, namely the entire functions

$$E_p(z) = \begin{cases} 1 - z & p = 0 \\ (1 - z)e^{z + \frac{z^2}{2} + \frac{z^3}{3} + \cdots + \frac{z^p}{p}} & p = 1, 2, 3, \ldots. \end{cases}$$

The motivation for doing this is that the exponential term for $p \geq 1$ represents the partial sums of the power series

$$\log(1 - z) = -\sum_{n=1}^{\infty} \frac{z^n}{n} = -z - \frac{z^2}{2} - \frac{z^3}{3} - \cdots - \frac{z^p}{p} - \cdots \tag{2.1}$$

for $|z| < 1$ and with the imaginary part θ of the logarithm satisfying $-\pi \leq \theta < \pi$. Therefore, as $p \to \infty$, for $|z| < 1$,

1 Sur une proprieté des fonctions entières, *C. R. Acad. Sci. Paris* 88 (1879), 1024–1027.

https://doi.org/10.1515/9783110757828-002

$$e^{z+\frac{z^2}{2}+\frac{z^3}{3}+\cdots+\frac{z^p}{p}} \to \frac{1}{1-z},$$

and so $E_p(z) \to 1$ as $p \to \infty$, which is the desired effect as we will see.

We are now in a position to prove the following famous theorem.

2.2 Weierstrass product theorem. *Let $\{a_n\}$ be a sequence of nonzero complex numbers such that $|a_n| \to \infty$ as $n \to \infty$. Then there exists a sequence of nonnegative integers $\{p_n\}$ such that the function*

$$P(z) = \prod_{n=1}^{\infty}\left(1-\frac{z}{a_n}\right)e^{\frac{z}{a_n}+\frac{1}{2}\left(\frac{z}{a_n}\right)^2+\cdots+\frac{1}{p_n}\left(\frac{z}{a_n}\right)^{p_n}} \tag{2.2}$$

is an entire function with zeros at the points a_n ($\neq 0$) and with no other zeros in $\mathbb{C}$.

Proof. Formally, let

$$P_n(z) = \left(1-\frac{z}{a_n}\right)e^{\frac{z}{a_n}+\frac{1}{2}\left(\frac{z}{a_n}\right)^2+\cdots+\frac{1}{p_n}\left(\frac{z}{a_n}\right)^{p_n}}$$

with the sequence $\{p_n\}$ yet to be chosen so that

$$\log P_n(z) = \log\left(1-\frac{z}{a_n}\right) + \left[\frac{z}{a_n}+\frac{1}{2}\left(\frac{z}{a_n}\right)^2+\cdots+\frac{1}{p_n}\left(\frac{z}{a_n}\right)^{p_n}\right]$$

with the imaginary part θ of the logarithm satisfying $-\pi \le \theta < \pi$. From expansion (2.1) we find that

$$\log P_n(z) = -\frac{1}{p_n+1}\left(\frac{z}{a_n}\right)^{p_n+1} - \frac{1}{p_n+2}\left(\frac{z}{a_n}\right)^{p_n+2} - \cdots,$$

which gives us the bound for $|z| \le R < \infty$

$$\left|\log P_n(z)\right| \le \frac{\frac{1}{p_n+1}\left(\frac{R}{|a_n|}\right)^{p_n+1}}{\left(1-\frac{R}{|a_n|}\right)}. \tag{2.3}$$

Next, we observe that for any $R > 0$ and $|a_n| > 2R$,

$$\sum_{|a_n|>2R}\left(\frac{R}{|a_n|}\right)^{p_n+1} < \infty, \tag{2.4}$$

is certainly valid with $p_n = n-1$. Hence there is always a sequence of nonnegative integers $\{p_n\}$ for which (2.4) holds. Then for any such sequence with $|z| \le R$ and $|a_n| > 2R$, inequality (2.3) implies that

$$\sum_{|a_n|>2R}\left|\log P_n(z)\right| < 2\sum_{|a_n|>2R}\left(\frac{R}{|a_n|}\right)^{p_n+1} < \infty, \tag{2.5}$$

which gives the absolute and uniform convergence on compact sets in $\mathbb{C}$ of the series, and likewise, in view of the analytic version of Theorem 1.18, the infinite product of (2.2) converges uniformly on compacts subsets of $\mathbb{C}$ to the entire function $P(z)$ possessing the desired zeros and no others, thus proving the theorem. □

An important consequence of the theorem is the following characterization of meromorphic functions in the plane.

2.3 Theorem. *Every meromorphic function $f(z)$ in the entire complex plane is the quotient of two entire functions.*

In fact, considering the poles of $f(z)$, which necessarily tend to infinity, we can construct a single entire function $g(z)$ having all its zeros at exactly the same points with the same order.[2] Then $f(z)g(z) = h(z)$ has removable singularities at these points and hence is entire, so that $f(z) = h(z)/g(z)$.

As for ordinary polynomials that can be expressed as a product of linear factors in terms of its zeros, the same can be done with an entire function via the preceding theorem of Weierstrass.

2.4 Weierstrass factorization theorem.[3] *If $f(z)$ is an entire function having a zero of order $m \geq 0$ at the origin and other zeros at $a_1, a_2, a_3, \ldots$ listed according to multiplicity, then there exists a sequence $\{p_n\}$ of nonnegative integers such that*

$$f(z) = z^m e^{g(z)} \prod_{n=1}^{\infty} \left(1 - \frac{z}{a_n}\right) e^{\frac{z}{a_n} + \frac{1}{2}\left(\frac{z}{a_n}\right)^2 + \cdots + \frac{1}{p_n}\left(\frac{z}{a_n}\right)^{p_n}},$$

where $g(z)$ is an entire function.

Proof. If $f(z)$ has a zero of order m at the origin, then we consider the function $f(z)/z^m$. Otherwise, the function $f(z)/P(z) = h(z)$ is entire and nonzero, where $P(z)$ is as in (2.2). Then the function $h'(z)/h(z)$ is analytic in $\mathbb{C}$, and therefore $h'(z)/h(z) = g'(z)$ for some entire function $g(z)$. Since $(e^{g(z)}/h(z))' = 0$, $h(z) = ce^{g(z)}$ for some constant c, which we take inside the function and simply write $h(z) = e^{g(z)}$.[4] It follows that

$$f(z) = e^{g(z)} P(z),$$

as desired. □

In summary, the most general form of an entire function is given by

$$f(z) = z^m e^{g(z)} P(z)$$

2 Any pole at the origin is similarly dispensed with.

3 Zur Theorie der eindeutigen analytischen Funktionen, *Math. Abhand. Akad. Wissen. Berlin*, (1876), 11–60.

4 The entire function $g(z)$ is a primitive of $h'(z)/h(z)$.

for $m \geq 0$ and some entire function $g(z)$, and $P(z)$ is the infinite product $\prod_{n=1}^{\infty} E_{p_n}(z/a_n)$. If $f(z)$ has no zeros, then simply $f(z) = e^{g(z)}$.

Genus/canonical product

Sometimes it is possible to find a single integer value p in place of the sequence $\{p_n\}$ in (2.2). In such cases the smallest such value of p is called the *genus* of the entire function $f(z)$, and the *canonical product* is given by

$$P(z) = \prod_{n=1}^{\infty} E_p\left(\frac{z}{a_n}\right),$$

which is convergent for all $z \in \mathbb{C}$, and (2.4) holds with $p = p_n$, or, equivalently, $\sum 1/|a_n|^{p+1} < \infty$.

Thus, for example, if $a_n = n^2$, $n = 1, 2, 3, \ldots$, then $\sum 1/n^2 < \infty$, and the genus of the canonical product

$$P(z) = \prod_{n=1}^{\infty}\left(1 - \frac{z}{n^2}\right)$$

is $p = 0$.

The following is an important example as it occurs in the Weierstrass definition of the gamma function of Chapter 10.

2.5 Example. If we specify zeros at the points $a_n = -n$, $n = 1, 2, 3, \ldots$, then the genus is $p = 1$ since $\sum 1/|a_n| = \infty$ but $\sum 1/|a_n|^2 < \infty$. Then the *canonical product* is

$$P(z) = \prod_{n=1}^{\infty}\left(1 + \frac{z}{n}\right)e^{-z/n}, \tag{2.6}$$

and for future reference,

$$\frac{1}{P(z)} = \prod_{n=1}^{\infty}\left(1 + \frac{z}{n}\right)^{-1} e^{z/n} \neq 0.$$

Furthermore, note that if the *Euler–Mascheroni constant* is defined by $\gamma = \lim_{n\to\infty}(1 + \frac{1}{2} + \cdots + \frac{1}{n} - \log n) = 0.577\ldots,$[5] then

$$P(1) = \lim_{n\to\infty} \prod_{k=1}^{n}\left(1 + \frac{1}{k}\right)e^{-1/k}$$

5 It is left as an exercise for the reader to establish that the limit for γ exists.

$$
\begin{aligned}
&= \lim_{n\to\infty}(n+1)e^{-1-\frac{1}{2}-\dots-\frac{1}{n}} \\
&= \lim_{n\to\infty}\frac{n+1}{n}e^{\log n-1-\frac{1}{2}-\dots-\frac{1}{n}} \\
&= e^{-\gamma}.
\end{aligned}
$$

Another useful relation regarding $P(z)$ in the form (2.6) is the following.

2.6 Proposition. $P(z-1) = ze^{\gamma}P(z)$.

Proof. Note that the function $P(z-1)$ has the same zeros as $P(z)$ with one more at the origin. So let us write

$$P(z-1) = ze^{g(z)}P(z)$$

for some entire function $g(z)$. It turns out that we can find $g(z)$ by taking the logarithmic derivatives of both sides in view of the uniform convergence on compact subsets not containing a zero, namely

$$\sum_{n=1}^{\infty}\left(\frac{1}{z-1+n}-\frac{1}{n}\right) = \frac{1}{z}+g'(z)+\sum_{n=1}^{\infty}\left(\frac{1}{z+n}-\frac{1}{n}\right). \tag{2.7}$$

To compare the terms of the series on each side of equation (2.7), we replace n by $n+1$ in the left-hand series, which yields

$$
\begin{aligned}
\sum_{n=1}^{\infty}\left(\frac{1}{z-1+n}-\frac{1}{n}\right) &= \frac{1}{z}-1+\sum_{n=1}^{\infty}\left(\frac{1}{z+n}-\frac{1}{n+1}\right) \\
&= \frac{1}{z}-1+\sum_{n=1}^{\infty}\left(\frac{1}{z+n}-\frac{1}{n}\right)+\sum_{n=1}^{\infty}\left(\frac{1}{n}-\frac{1}{n+1}\right).
\end{aligned}
$$

The rearrangement of the series to obtain the last expression is valid since each term in the parentheses is $O(\frac{1}{n^2})$ and thus converges absolutely provided that $z \neq -1,-2,-3,\dots$. Since the partial sums of the last infinite sum converge to 1, equation (2.7) reduces to $g'(z) = 0$, so that $g(z) = c$, a constant. Therefore, for $z = 1$, equation (2.6) implies that

$$1 = P(0) = e^{c}e^{-\gamma},$$

and the result follows. □

2.7 Example. The Euler product for the sine[6] is given by

$$\sin \pi z = \pi z\prod_{n=1}^{\infty}\left(1-\frac{z^2}{n^2}\right). \tag{2.8}$$

6 Euler (1707–1783) used this formula (for real variables) to solve the *Basel problem* to find the sum in closed form of $\sum_{n=1}^{\infty}\frac{1}{n^2}$, which Euler proved to be $\pi^2/6$.

In fact, the zeros of $\sin \pi z$ are at all the integers $z = 0, \pm 1, \pm 2 \ldots$. Therefore the canonical product has genus $p = 1$, and we can write

$$\sin \pi z = z e^{g(z)} \prod_{n \neq 0} \left(1 - \frac{z}{n}\right) e^{z/n}.$$

Due to the uniform convergence of the product on compact sets not containing the points $z = \pm n$, that is, for n sufficiently large, we can take the logarithmic derivative on both sides to obtain

$$\pi \cot \pi z = \frac{1}{z} + g'(z) + \sum_{n \neq 0} \left(\frac{1}{z - n} + \frac{1}{n} \right).$$ [7]

Next, we are going to borrow the Mittag–Leffler representation for the function $f(z) = \pi \cot \pi z$ given in equation (3.4) of the next chapter, namely $\pi \cot \pi z = \frac{1}{z} + \sum_{n \neq 0} (\frac{1}{z-n} + \frac{1}{n})$. This also demonstrates the intimate connection between entire and meromorphic functions. It follows that $g'(z) = 0$, and therefore $g(z)$ is constant. Since $\lim_{z \to 0} \sin(\pi z)/\pi z = 1$, we have $e^{g(z)} = \pi$, and combining the terms involving n and $-n$ establishes the Euler product for the sine.

In terms of the canonical product $P(z)$ of equation (2.6), we therefore have another useful relation

$$\frac{\sin \pi z}{\pi z} = P(-z) P(z). \tag{2.9}$$

Order

Entire functions that do not grow sufficiently rapidly are polynomials and are not particularly interesting. More specifically, we maintain

2.8 Theorem. *If $f(z)$ is an entire function satisfying $f(z) = O(r^k)$ as $|z| = r \to \infty$, then $f(z)$ is a polynomial of degree less than or equal to k.*

Proof. For $f(z) = \sum_{n=0}^{\infty} a_n z^n$, $z \in \mathbb{C}$, Cauchy's inequality gives

$$|a_n| \leq \frac{M(r)}{r^n}$$

for $n = 0, 1, 2, \ldots$, where $M(r) = \max_{|z|=r} |f(z)|$. Moreover, by the hypothesis, for some $C > 0$ and $r_0 > 0$, we have $|f(z)| \leq C r^k$ for $|z| = r \geq r_0$. Hence

7 Note the importance of the convergence factor in this formulation. The terms for $\pm n$ in the infinite sum on the right can be bracketed, so that it can also be expressed as $\sum_{n=1}^{\infty} (\frac{2z}{z^2 - n^2})$.

$$|a_n| \le Cr^{k-n},$$

and it follows that $a_n = 0$ for all $n > k$ by letting $r \to \infty$. This proves the theorem. □

Therefore, to get beyond simple polynomials, we wish to investigate the behavior of more rapidly growing functions. An entire function $f(z)$ is of *finite order* if there is a number $\beta > 0$ such that

$$|f(z)| = O(e^{r^\beta})$$

as $|z| = r \to \infty$. Otherwise, the function is of *infinite order*. The infimum of all such values β is the *order* of $f(z)$, which we denote by λ, and it follows that for any $\varepsilon > 0$,

$$f(z) = O(e^{r^{\lambda+\varepsilon}})$$

as $|z| = r \to \infty$. In other words, there is a positive constant C and a positive real number r_0 such that

$$|f(z)| \le Ce^{r^{\lambda+\varepsilon}}$$

for $r \ge r_0$.

Note that the function $\sin z$ has order 1 (why?) and the function e^{z^n} has order $\lambda = n$. On the other hand, the function e^{e^z} has infinite order.

An equivalent characterization of the order of an entire function can be given in terms of its maximum modulus $M(r)$ on the circle $|z| = r$:

$$\lambda = \overline{\lim_{r\to\infty}} \frac{\log\log M(r)}{\log r}.$$

The proof of the equivalence of the two definitions is left as an exercise.

We know that an entire function with no zeros has the form $f(z) = e^{g(z)}$ for some entire function $g(z)$. If we know the order of $f(z)$, then something more can be said.

2.9 Theorem. *If $f(z)$ is an entire function of order λ having no zeros, then $f(z) = e^{g(z)}$ where $g(z)$ is a polynomial of degree less than or equal to λ.*

Proof. Since $f(z)$ has finite order λ, then for $|z| = r$, we have

$$e^{\mathrm{Re}(g(z))} = |f(z)| \le Ce^{r^{\lambda+\varepsilon}},$$

and, consequently, $\mathrm{Re}(g(z)) \le Kr^{\lambda+\varepsilon}$ as $r \to \infty$. Applying the Hadamard–Borel–Carathéodory Theorem 1.10 to the function $g(z)$ with $R = 2r$ gives

$$\begin{aligned} M(r) &\le \frac{2r}{R-r}A(R) + \frac{R+r}{R-r}|g(0)| \\ &\le 2A(2r) + 3|g(0)| \\ &= O(r^{\lambda+\varepsilon}) \end{aligned}$$

as $r \to \infty$. The result now follows by the preceding theorem. □

It is not completely obvious what the order of the derivative of an entire function should be. In fact, it is the same order as that for the function itself.

2.10 Theorem. *The order of the entire function $f'(z)$ is the same as the order of $f(z)$.*

Proof. Let $M'(r) = \max_{|z|=r} |f'(z)|$. For a point $z \in \mathbb{C}$, integrating along the straight line joining z to the origin, we have

$$f(z) - f(0) = \int_0^z f'(\zeta)d\zeta,$$

so that

$$M(r) \le rM'(r) + |f(0)|.$$

Moreover, taking a point z_0 for which $|f'(z_0)| = M'(r)$, we get

$$f'(z_0) = \frac{1}{2\pi i}\int_C \frac{f(\zeta)}{(\zeta - z_0)^2}d\zeta, \tag{2.10}$$

where $C : |\zeta - z_0| = R - r$ for $r < R$. We deduce from (2.10) that

$$M'(r) \le \frac{M(R)}{R - r}.$$

Letting $R = 2r$ gives

$$\frac{M(r) - |f(0)|}{r} \le M'(r) \le \frac{M(2r)}{r},$$

and it follows that the orders of f and f' are the same. □

Relation to counting function

Although seemingly unrelated, the rate of growth of the *counting function* $n(t) = n(t, 0)$ and the order of an entire function are indeed connected.

2.11 Theorem. *If $f(z)$ is an entire function of order λ, then its counting function $n(t)$ satisfies $n(t) = O(r^{\lambda+\varepsilon})$ for any $\varepsilon > 0$.*

Proof. First, assume that $f(0) \neq 0$. By the hypothesis we have that for any $\varepsilon > 0$,

$$\log|f(re^{i\theta})| < Ar^{\lambda+\varepsilon}$$

as $|z| = r \to \infty$. An application of Jensen's formula in the form (1.8) yields

$$\int_0^r \frac{n(t)}{t}\,dt < Ar^{\lambda+\varepsilon}$$

for all r sufficiently large. Note that

$$\int_0^{2r} \frac{n(t)}{t}\,dt \geq n(r)\int_r^{2r} \frac{dt}{t} = n(r)\log 2$$

since $n(t)$ is nondecreasing. It follows that

$$n(r) \leq \frac{1}{\log 2}\int_0^{2r} \frac{n(t)}{t}\,dt < Ar^{\lambda+\varepsilon}.$$

If $f(z)$ has a zero of order k at the origin, then we have to use the more general version of Jensen's formula, equation (3.15) of Chapter 3. Details are left to the reader. □

This means that entire functions of higher order can possess more zeros in all sufficiently large regions, which is a rather interesting phenomenon. Even more can be said about the zeros.

2.12 Corollary. *If $a_1, a_2, \ldots \neq 0$ are the zeros of the entire function $f(z)$ of order λ, then the series*

$$\sum \frac{1}{|a_n|^\rho}$$

is convergent for any given $\rho > \lambda$.

Indeed, we need only be concerned with the case of infinitely many zeros a_n such that $|a_n| \to \infty$. Choose some α such that $\lambda < \alpha < \rho$, so that there is a constant A with

$$n(r) < Ar^\alpha$$

for all large r. For $r = |a_n|$, we obtain $n(|a_n|) < A|a_n|^\alpha$. Consequently,

$$\frac{1}{|a_n|^\rho} < \left(\frac{A}{n}\right)^{\rho/\alpha},$$

proving the convergence.

Exponent of convergence

Therefore, in view of the preceding corollary, if we take ρ_1 to be the infimum of all positive numbers $\{\rho\}$ such that $\sum \frac{1}{|a_n|^\rho}$ converges, then the above corollary establishes that $\rho_1 \leq \lambda$. Indeed, we have $\rho_1 < \lambda$ in the case of the function e^{z^n} $(n \geq 1)$, for which $\rho_1 = 0$, but $\lambda = n$. The number ρ_1 is the *exponent of convergence of the zeros* of $f(z)$.

To determine the relationship between the genus p and the exponent of convergence ρ_1, note that:
If ρ_1 is an integer and $\sum \frac{1}{|a_n|^{\rho_1}}$ diverges, then $p = \rho_1$.
If ρ_1 is an integer and $\sum \frac{1}{|a_n|^{\rho_1}}$ converges, then $p = \rho_1 - 1$.
If ρ_1 is not an integer, then $p = [\rho_1]$ (the integer part of ρ_1).
In all cases, we have the relation

$$p \leq \rho_1 \leq \lambda,$$

where λ is the order.

Hadamard factorization theorem

We now prove an extension of the Weierstrass factorization theorem based on the order of the entire function due to Hadamard.[8]

2.13 Hadamard factorization theorem. *Suppose that $f(z)$ is an entire function of order λ having zeros at $a_1, a_2, a_3, \ldots$ $(\neq 0)$. Then*

$$f(z) = e^{q(z)} \prod_{n=1}^{\infty} \left(1 - \frac{z}{a_n}\right) e^{\frac{z}{a_n} + \frac{1}{2}\left(\frac{z}{a_n}\right)^2 + \cdots + \frac{1}{p}\left(\frac{z}{a_n}\right)^p}, \tag{2.11}$$

where $q(z)$ is a polynomial of degree at most λ, and the infinite product is given by the canonical product $P(z)$ with $p \leq [\lambda]$. If $f(0) = 0$, then the term z^m is appended to the product.

Proof. In view of the preceding considerations, the infinite product in the Weierstrass factorization theorem can be taken to be the canonical product, and we need only show that the entire function $g(z)$ in the theorem is indeed a polynomial $q(z)$ of degree at most λ. This ingenious proof is due to Edmund Landau (1877–1938).[9] To this end, define $\eta = [\lambda]$, so that $p \leq \eta \leq \lambda$. Considering $f(z)$ in the form given by (2.11), that is, $f(z) = e^{q(z)} P(z)$, taking the logarithmic derivative of both sides $\eta + 1$ times has the desired effect of isolating $q(z)$ and resulting in one of the terms vanishing, namely,

$$\frac{d^{\eta+1}}{dz^{\eta+1}} \sum_{n=1}^{\infty} \left(\frac{z}{a_n} + \frac{1}{2}\left(\frac{z}{a_n}\right)^2 + \cdots + \frac{1}{p}\left(\frac{z}{a_n}\right)^p \right) = 0$$

8 Études sur les propriétés des fonctions entières et en particulier d'une function considérée par Riemann, *J. Math.* (4) 9 (1893), 171–215.

9 Über die Zetafunktion und die Hadamardsche Theorie der ganzen Funktionen, *Math. Zeit.* 26 (1927), 170–175.

since $p \leq \eta$. What remains is

$$\frac{d^\eta}{dz^\eta}\left(\frac{f'(z)}{f(z)}\right) = q^{(\eta+1)}(z) - \eta! \sum_{n=1}^{\infty} \frac{1}{(a_n - z)^{\eta+1}}, \tag{2.12}$$

and we proceed to show that $q^{(\eta+1)}(z) = 0$.

We now define

$$F_R(z) = \frac{f(z)}{f(0)} \prod_{|a_n| \leq R} \left(1 - \frac{z}{a_n}\right)^{-1}$$

for all $z \in \mathbb{C}$. Taking $|z| = 2R$ and $|a_n| \leq R$, we have $|1 - z/a_n| \geq 1$, so that

$$|F_R(z)| \leq \left|\frac{f(z)}{f(0)}\right| = O(e^{(2R)^{\lambda+\varepsilon}}). \tag{2.13}$$

Note that as $f(0) \neq 0$, the function $F_R(z)$ is entire, so that inequality (2.13) also holds in $|z| < 2R$, and $F_R(z)$ is nonzero in $|z| \leq R$. Hence the function $G_R(z) = \log F_R(z)$ with $G_R(0) = 0$ is analytic in $|z| \leq R$. As a consequence, for such z, we can write

$$\operatorname{Re}(G_R(z)) < KR^{\lambda+\varepsilon}.$$

By applying Corollary 1.11 of the Hadamard–Borel–Carathéodory theorem we have

$$|G_R^{(\eta+1)}(z)| \leq \frac{2^{\eta+3}(\eta+1)!R}{(R-r)^{\eta+2}} KR^{\lambda+\varepsilon}$$

for $|z| = r < R$, whence for $|z| = r = R/2$,

$$G_R^{(\eta+1)}(z) = O(R^{\lambda+\varepsilon-\eta-1}).$$

Now for all $z \in \mathbb{C}$,

$$G_R(z) = \log F_R(z) = \log f(z) - \log f(0) - \sum_{|a_n| \leq R} \log\left(1 - \frac{z}{a_n}\right),$$

implying

$$G_R^{(\eta+1)}(z) = \frac{d^\eta}{dz^\eta}\left(\frac{f'(z)}{f(z)}\right) + \eta! \sum_{|a_n| \leq R} \left(\frac{1}{(a_n - z)^{\eta+1}}\right).$$

We can substitute this result back into equation (2.12) to obtain

$$q^{(\eta+1)}(z) = G_R^{(\eta+1)}(z) + \sum_{|a_n| > R} \left(\frac{1}{(a_n - z)^{\eta+1}}\right)$$

$$= O(R^{\lambda+\varepsilon-\eta-1}) + O\Big(\sum_{|a_n|>R}\Big(\frac{1}{(a_n-z)^{\eta+1}}\Big)\Big)$$
$$= O(R^{\lambda+\varepsilon-\eta-1}) + O\Big(\sum_{|a_n|>R}\Big(\frac{1}{|a_n|^{\eta+1}}\Big)\Big)$$

for $|z| = r = R/2$, and likewise for $|z| < R/2$. Then for ε sufficiently small, $\lambda + \varepsilon < \eta + 1$, and thus

$$O(R^{\lambda+\varepsilon-\eta-1}) \to 0$$

as $R \to \infty$. Since $\eta + 1 > \lambda$, Corollary 2.12 implies that $\sum_{n=1}^{\infty} 1/|a_n|^{\eta+1}$ is convergent, so that $\sum_{|a_n|>R}(\frac{1}{|a_n|^{\eta+1}}) \to 0$ as $R \to \infty$. We conclude that $q^{(\eta+1)}(z) = 0$ and that $q(z)$ is a polynomial of degree less than or equal to λ, proving the theorem. □

In general, for an *entire function $f(z)$ of order λ*, we know that regarding the exponent of convergence of the zeros, $\rho_1 \le \lambda$. If for a moment, we restrict our attention to just the canonical product term of $f(z)$, then we can say something stronger.

2.14 Theorem. *For a canonical product*

$$P(z) = \prod_{n=1}^{\infty} E_p\Big(\frac{z}{a_n}\Big)$$

of order $\kappa, \kappa = \rho_1$, the exponent of convergence of the zeros of $P(z)$.

The proof is rather technical but worthwhile, as it allows for a characterization of the order of an entire function $f(z) = e^{q(z)}P(z)$, where $q = \deg q(z)$, and the order of $e^{q(z)}$ equals q. From the theorem since ρ_1 is the order of $P(z)$, the order λ of $f(z)$ satisfies (exercise)

$$\lambda = \max(q, \rho_1).$$

Proof of the theorem. As $\rho_1 \le \kappa$, we will demonstrate that $\kappa \le \rho_1$. Since $\log|P(z)| = \sum_{n=1}^{\infty} \log|E_p(\frac{z}{a_n})|$, we can write

$$\log|P(z)| = \sum_{|a_n|\le 2r} \log\Big|E_p\Big(\frac{z}{a_n}\Big)\Big| + \sum_{|a_n|>2r} \log\Big|E_p\Big(\frac{z}{a_n}\Big)\Big|. \tag{2.14}$$

Considering the second series of (2.14), by inequality (2.5) for $|z| = r$, we have

$$\sum_{|a_n|>2r} \log\Big|E_p\Big(\frac{z}{a_n}\Big)\Big| = O\Big(\sum_{|a_n|>2r}\Big(\frac{r}{|a_n|}\Big)^{p+1}\Big)$$
$$= O\Big(r^{p+1}\sum_{|a_n|>2r}\frac{1}{|a_n|^{p+1}}\Big). \tag{2.15}$$

Since $p \le \rho_1$, if $p = \rho_1 - 1$,[10] then the right-hand side of (2.15) becomes $O(r^{\rho_1})$. Otherwise, either $p = \rho_1$ or $p = [\rho_1]$, and in both cases, we can find $\varepsilon > 0$ sufficiently small such that $\rho_1 + \varepsilon < p + 1$. Therefore, again for the right-hand side of (2.15),

$$\begin{aligned} r^{p+1} \sum_{|a_n|>2r} \frac{1}{|a_n|^{p+1}} &= r^{p+1} \sum_{|a_n|>2r} |a_n|^{\rho_1+\varepsilon-p-1} |a_n|^{-\rho_1-\varepsilon} \\ &< r^{p+1} (2r)^{\rho_1+\varepsilon-p-1} \sum_{|a_n|>2r} |a_n|^{-\rho_1-\varepsilon} = O(r^{\rho_1+\varepsilon}). \end{aligned}$$

Thus the second series of (2.14) is $O(r^{\rho_1+\varepsilon})$ in each instance. Regarding the first series of (2.14), where $|z/a_n| \ge 1/2$, each term satisfies

$$\log\left|E_p\left(\frac{z}{a_n}\right)\right| \le \log\left(1 + \left|\frac{z}{a_n}\right|\right) + \left(\left|\frac{z}{a_n}\right| + \cdots + \frac{1}{p}\left|\frac{z}{a_n}\right|^p\right) < K\left|\frac{z}{a_n}\right|^p.$$

Then for the series itself, using the ε from above, we obtain

$$\begin{aligned} \sum_{|a_n|\le 2r} \log\left|E_p\left(\frac{z}{a_n}\right)\right| &< O\left(r^p \sum_{|a_n|\le 2r} |a_n|^{-p}\right) \\ &= O\left(r^p \sum_{|a_n|\le 2r} |a_n|^{\rho_1+\varepsilon-p} |a_n|^{-\rho_1-\varepsilon}\right) \\ &= O\left(r^p (2r)^{\rho_1+\varepsilon-p} \sum_{|a_n|\le 2r} |a_n|^{-\rho_1-\varepsilon}\right) \\ &= O(r^{\rho_1+\varepsilon}). \end{aligned}$$

In conclusion, we have shown that

$$\log|P(z)| < O(r^{\rho_1+\varepsilon})$$

and $\kappa \le \rho_1$, as desired. □

We also have the following interesting consequence of the fact that $\lambda = \max(q, \rho_1)$.

2.15 Proposition. *If the order λ of an entire function $f(z) = e^{q(z)}P(z)$ is not an integer, then $\lambda = \rho_1$.*

In fact, by the Hadamard theorem, $q(z)$ is a polynomial of degree $q \le \lambda$, and as λ is not an integer but q must be, it follows that $q < \lambda$. Therefore $\lambda = \rho_1$.

This also entails that if λ is not an integer, then $P(z)$ cannot reduce to a polynomial, that is, the function $f(z)$ must have infinitely many zeros.

---

10 Recall this is the case where ρ_1 is an integer and $\sum \frac{1}{|a_n|^{\rho_1}}$ converges.

Laguerre theorem

The derivative $f'(z)$ not only shares a common order with the entire function $f(z)$, but their zeros can be intertwined if the zeros of $f(z)$ are real in a particular instance. This result is due to Edmond Laguerre (1834–1886).

2.16 Laguerre theorem. *Suppose that $f(z)$ is an entire function of order less than 2 that is real on the real axis and that has only real zeros. Then the zeros of $f'(z)$ are all real and separated from each other by the zeros of $f(z)$.*

Proof. The function $f(z)$ has the form

$$f(z) = cz^m e^{az} \prod_{n=1}^{\infty}\left(1 - \frac{z}{a_n}\right)e^{\frac{z}{a_n}},$$

where c and a are constants, m is a nonnegative integer, and $a_1, a_2, a_3, \ldots$ are the real zeros of $f(z)$. Taking the logarithmic derivative gives

$$\frac{f'(z)}{f(z)} = \frac{m}{z} + a + \sum_{n=1}^{\infty}\left(\frac{1}{z - a_n} + \frac{1}{a_n}\right), \tag{2.16}$$

and taking the imaginary parts of both sides for $z = x + iy$ yields

$$\operatorname{Im}\left(\frac{f'(z)}{f(z)}\right) = -y\left(\frac{m}{x^2 + y^2} + \sum_{n=1}^{\infty}\frac{1}{(x - a_n)^2 + y^2}\right)$$

since the a_n are real. Hence $\operatorname{Im}(\frac{f'(z)}{f(z)})$ vanishes only when $y = 0$, and we deduce that all the zeros of $f'(z)$ must be real. Furthermore, if $z = x$ is real and $z \neq a_n$ for any n, then

$$\frac{d}{dz}\left(\frac{f'(z)}{f(z)}\right) = -\frac{m}{x^2} - \sum_{n=1}^{\infty}\frac{1}{(x - a_n)^2} < 0,$$

which means that $f'(z)/f(z)$ is only decreasing as real values of z increase from a_n to a_{n+1}. Since $f'(z)/f(z)$ becomes infinite at all the points a_n, it follows that the function changes sign in the interval (a_n, a_{n+1}) producing the desired single zero of $f'(z)$ between any two zeros of $f(z)$. □

The theorem is just a generalization of the fact that if $p(x)$ is a real polynomial, then between any two consecutive roots, there lies a root of the derivative $p'(z)$ by Rolle's theorem.

To show that the order of $f(z)$ must be strictly less than two in the Laguerre theorem, consider the function $f(z) = ze^{z^2/2}$ with derivative $f'(z) = (z^2 + 1)e^{z^2/2}$, so that the zeros of $f'(z)$ are not real. Furthermore, for the function $f(z) = (z^2 - 4)e^{z^2/2}$, we have $f'(z) = (z^2 - 2)ze^{z^2/2}$, and the zeros of $f'(z)$ are all real; however, they are not separated by the zeros of $f(z)$.

2.17 Corollary. *Under the above conditions, the zeros of $f(z)$ and those of $f'(z)$ have the same exponent of convergence.*

This follows directly from the theorem since if $a_1, a_2, a_3, \ldots$ are the zeros of $f(z)$ and $a_1', a_2', a_3', \ldots$ are the zeros of $f'(z)$, then both the series

$$\sum_{n=1}^{\infty} \frac{1}{|a_n|^\rho} \quad \text{and} \quad \sum_{n=1}^{\infty} \frac{1}{|a_n'|^\rho}$$

converge or diverge together.

3 Meromorphic functions

This chapter is divided into two parts, the classical theory of meromorphic functions prior to the work of Rolf Nevanlinna (1895–1980) and the work that followed. The entire landscape changed in a profound way in 1925 with the work of the latter, and in subsequent years that of his student Lars Ahlfors (1907–1996).

Classical theory

Meromorphic functions are analytic with the exception of poles. It is certainly the case that in the domain where a function is meromorphic, the poles have no accumulation point for if they did, then the accumulation point would be a singularity, which is clearly not a pole due to the Laurent series expansion. A simple consequence of this is the following:

3.1 Theorem. *Any meromorphic function $f(z)$ in the complex plane $\mathbb{C}$ is the quotient of two entire functions having no common zeros.*

Indeed, this was just Theorem 2.3 derived as a consequence of the Weierstrass product theorem and an example of the interaction between entire and meromorphic functions.

Laurent series

Recall that an analytic function $f(z)$ on a domain $\Omega \subseteq \mathbb{C}$ with a pole of order m_j at a point b_j can be expressed by the Laurent series

$$\begin{aligned} f(z) &= \frac{k_{m_j}}{(z-b_j)^{m_j}} + \frac{k_{m_j-1}}{(z-b_j)^{m_j-1}} + \cdots + \frac{k_1}{z-b_j} + \sum_{n=1}^{\infty} c_n (z-b_j)^n \\ &= P_j\left(\frac{1}{z-b_j}\right) + \sum_{n=1}^{\infty} c_n (z-b_j)^n \end{aligned}$$

for $0 < |z-b_j| < r$, where r is the distance from b_j to the next nearest pole or to the boundary of Ω. The sum of the terms with negative powers of $z-b_j$ is the *principal part*, and k_1 is the *residue* of $f(z)$ at b_j. Here $P_j(\frac{1}{z-b_j})$ represents a polynomial in negative powers of $z-b_j$ but with no constant term.

In the general case, to create a meromorphic function in a domain Ω with poles at the points $\{b_j\}$ and $\lim_{j\to\infty} b_j = \infty$, we cannot simply sum up all the principal parts since the result will in general diverge, but as was done for the Weierstrass product theorem 2.2, a small modification to the polynomials $P_j(\frac{1}{z-b_j})$ will produce the desired convergence.

https://doi.org/10.1515/9783110757828-003

The Mittag–Leffler theorem is the analog of the Weierstrass product theorem (for the zeros of an entire function) for meromorphic functions in terms of their poles.

3.2 Mittag–Leffler theorem.[1] *For any sequence of points $\{b_j\}$ in $\mathbb{C}$ with $\lim_{j\to\infty} b_j = \infty$, suppose that $P_j(z)$ are polynomials with no constant term. Then there exists a meromorphic function $f(z)$ possessing poles at exactly the points $\{b_j\}$ with no others and having principal parts given by the terms $P_j(\frac{1}{z-b_j})$. The function is not unique, and the most general such meromorphic function is of the form*

$$f(z) = \sum_{j=1}^{\infty}\left(P_j\left(\frac{1}{z-b_j}\right) - p_j(z)\right) + g(z),$$

where $p_j(z)$ are polynomials, and $g(z)$ is an entire function.

Proof. Without loss of generality, we assume that $0 < |b_j|$ since if $b_0 = 0$, then we include the term $P_0(\frac{1}{z})$.

Since for each $j = 1, 2, 3, \ldots$ and for $|z| < b_j$, the polynomial P_j is analytic, we can write

$$P_j\left(\frac{1}{z-b_j}\right) = \sum_{n=0}^{\infty} c_n z^n$$

for $|z| < b_j$. Then the series converges uniformly in the disk $D_j : |z| \le |b_j|/2$, and thus for a partial sum $p_j(z) = \sum_{n=1}^{m_j} c_n z^n$, we obtain

$$\left|P_j\left(\frac{1}{z-b_j}\right) - p_j(z)\right| \le \frac{1}{2^j} \tag{3.1}$$

for m_j sufficiently large.

Now taking any fixed $R > 0$, the disk $D_R : |z| \le R$ will be contained in some D_j. Then on D_R, considering the function

$$f(z) = \sum_{j=1}^{m_j}\left(P_j\left(\frac{1}{z-b_j}\right) - p_j(z)\right) + \sum_{j=m_j+1}^{\infty}\left(P_j\left(\frac{1}{z-b_j}\right) - p_j(z)\right),$$

the second series converges uniformly to an analytic function in view of (3.1). The first series is analytic except for poles at the points $z = b_j$ for $j = 1, 2, \ldots, m_j$. Thus $f(z)$ is meromorphic in D_R with the desired principal parts, and since R was arbitrary, the first part of the theorem is complete. The last part follows immediately. □

1 Sur la representation analytique des fonctions monogènes uniformes d'une variable indépendante, *Acta Math.* 4 (1884), 1–79.

Often, the polynomials $p_j(z)$ used to obtain convergence can be dispensed with altogether.

3.3 Example. If we require simple poles at the points $b_j = 0, 1, 4, \dots, n^2, \dots$ with principal parts $\frac{1}{(z-n^2)}$, then the function

$$f(z) = \frac{1}{z} + \sum_{n=1}^{\infty} \frac{1}{z - n^2}$$

is sufficient. Indeed, in the disk $|z| \le R$, for any $R > 0$ and all $n > \sqrt{2R}$,

$$\left| \frac{1}{z - n^2} \right| \le \frac{1}{n^2 - R} \le \frac{2}{n^2},$$

so that the series converges uniformly.

A more practical result for determining the expansion in terms of the poles of a meromorphic function in the complex plane is the following:

3.4 Mittag–Leffler expansion theorem. *Let $f(z)$ be a meromorphic function in $\mathbb{C}$ with simple poles at the points $a_1, a_2, a_3, \dots$ with $0 \le |a_1| \le |a_2| \le |a_3| \le \cdots$ and $\operatorname{Res}(a_n) = b_n$, $n = 1, 2, 3, \dots$. Suppose that $\{C_n\}$ is a sequence of circles*[2] *of radius R_n centered at the origin that do not pass through any of the poles of $f(z)$ on which $|f(z)| \le M$ with M independent of n and $\lim_{n\to\infty} R_n = \infty$. Then*

$$f(z) = f(0) + \sum_{n=1}^{\infty} b_n \left(\frac{1}{z - a_n} + \frac{1}{a_n} \right).$$

Proof. Note that the function $f(\zeta)/(\zeta - z)$ for $z \ne a_n$ has poles at $z, a_1, a_2, a_3, \dots$ and that for this function,

$$\operatorname{Res}(a_n) = \frac{b_n}{a_n - z},$$

so that by the Cauchy residue theorem, with z interior to C_n,

$$\frac{1}{2\pi i} \int_{C_n} \frac{f(\zeta)}{\zeta - z} d\zeta = f(z) + \sum_{\nu} \frac{b_\nu}{a_\nu - z}, \tag{3.2}$$

where the sum is taken over all the poles inside the circle C_n. On the other hand,

2 Squares or rectangles centered about the origin can substitute for circles with the semidiagonal taking the place of the radius. This is evident from the proof.

$$\frac{1}{2\pi i}\int_{C_n}\frac{f(\zeta)}{\zeta - z}d\zeta = \frac{1}{2\pi i}\int_{C_n}\frac{f(\zeta)}{\zeta}d\zeta + \frac{z}{2\pi i}\int_{C_n}\frac{f(\zeta)}{\zeta(\zeta - z)}d\zeta$$

$$= f(0) + \sum_{\nu}\frac{b_\nu}{a_\nu} + \frac{z}{2\pi i}\int_{C_n}\frac{f(\zeta)}{\zeta(\zeta - z)}d\zeta \tag{3.3}$$

in view of (3.2). Furthermore,

$$\left|\frac{z}{2\pi i}\int_{C_n}\frac{f(\zeta)}{\zeta(\zeta - z)}d\zeta\right| \leq \frac{|z|M}{R_n - |z|} \to 0$$

as $n \to \infty$. The result now follows by combining (3.2) and (3.3) and letting $n \to \infty$. □

3.5 Example. Consider the function $\pi \cot \pi z = \frac{\cos \pi z}{\sin \pi z}$, which has simple poles at the points $z = 0, \pm 1, \pm 2, \ldots$. Then the function $f(z) = \pi \cot \pi z - \frac{1}{z}$ has a removable singularity at the origin, since

$$\lim_{z\to 0} f(z) = \lim_{z\to 0}\left(\frac{z\cos \pi z - \sin \pi z}{z \sin \pi z}\right) = 0,$$

and so we take $f(0) = 0$. Moreover, at the poles $z = \pm n$,

$$\operatorname{Res}(\pm n) = \left.\frac{\pi \cos \pi z}{(\sin \pi z)'}\right|_{\pm n} = 1.$$

To apply the preceding expansion theorem, let us consider the closed square contours C_n bounded by the lines $x = \pm(n + \frac{1}{2})$ and $y = \pm(n + \frac{1}{2})$ as in Figure 3.1.

Therefore, for $z = x + iy$,

$$|\cot \pi z|^2 = \left|\frac{e^{i\pi(x+iy)} + e^{-i\pi(x+iy)}}{e^{i\pi(x+iy)} - e^{-i\pi(x+iy)}}\right|^2 = \frac{e^{2\pi y} + e^{-2\pi y} + 2\cos(2\pi x)}{e^{2\pi y} + e^{-2\pi y} - 2\cos(2\pi x)}.$$

On the vertical sides of C_n with $x = \pm(n + \frac{1}{2})$, we have $\cos(2\pi x) = -1$, and hence $|\cot \pi z|^2 \leq 1$.

On the horizontal sides of C_n with $y = \pm(n + \frac{1}{2})$, we have $\cos 2\pi x \leq 1$, implying

$$|\cot \pi z|^2 = \frac{e^{2\pi y} + e^{-2\pi y} + 2\cos(2\pi x)}{e^{2\pi y} + e^{-2\pi y} - 2\cos(2\pi x)} \leq \frac{e^{2\pi y} + e^{-2\pi y} + 2}{e^{2\pi y} + e^{-2\pi y} - 2} \leq 2.$$

We conclude that the function $f(z) = \pi \cot \pi z - \frac{1}{z}$ is uniformly bounded on all the contours C_n, $n = 1, 2, 3, \ldots$. It follows from the Mittag–Leffler expansion theorem that

$$f(z) = f(0) + \sum_{n\neq 0}\left(\frac{1}{z - n} + \frac{1}{n}\right),$$

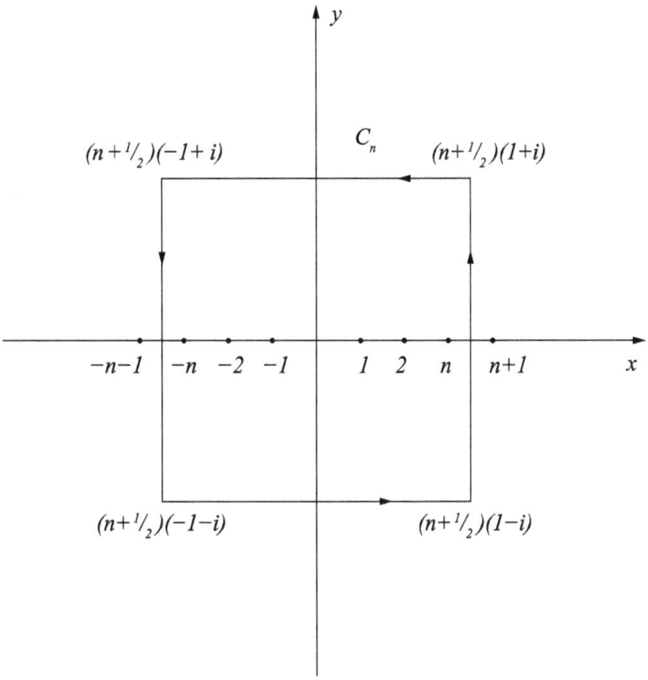

Figure 3.1: The contour of integration C_n described in the text. Courtesy Katy Metcalf.

that is,

$$\pi \cot \pi z = \frac{1}{z} + \sum_{n\neq 0}\left(\frac{1}{z-n} + \frac{1}{n}\right). \tag{3.4}$$

Note that the infinite sum in (3.4) converges uniformly on compact subsets that avoid the poles since for any closed disk $|z| \leq R$,

$$\left|\frac{1}{z-n} + \frac{1}{n}\right| \leq \frac{2R}{n^2}$$

for all $n \geq 2R$. Thus the sum converges uniformly on any compact set that contains no poles.

Expansion (3.4) can also be expressed as

$$\pi \cot \pi z = \lim_{k\to\infty}\left(\sum_{n=-k}^{k} \frac{1}{z-n}\right) = \frac{1}{z} + \sum_{n=1}^{\infty} \frac{2z}{z^2 - n^2},$$

simply by grouping the terms involving n and $-n$, allowed by the local uniform convergence.

Elliptic functions

The study of elliptic functions began in the works of Gauss, Abel, and Jacobi and has played a fundamental role in complex function theory. Here we examine only the basic highlights.

3.6 Definition. A meromorphic function $f(z)$ in $\mathbb{C}$ is said to be *elliptic* if there are nonzero complex numbers ω, ω' satisfying $\operatorname{Im}(\frac{\omega'}{\omega}) > 0$ with

$$f(z + \omega) = f(z)$$

and

$$f(z + \omega') = f(z)$$

for all $z \in \mathbb{C}$. The points ω, ω' are called *periods* of the function $f(z)$, and the function is said to be *doubly periodic*.

Note that if ω, ω' are periods, then so is any linear combination $n\omega + m\omega'$ for integer values of n and m (exercise). Given any two periods ω, ω', if any other period is of the form $n\omega + m\omega'$ for integers n and m, then ω and ω' are said to be *primitive* or *fundamental periods*. Taking the four points $0, \omega, \omega', \omega + \omega'$,[3] we can form the *fundamental parallelogram*

$$\Pi = \{s\omega + t\omega' : 0 \leq s, t < 1\},$$

which includes the lines from 0 to ω and from 0 to ω', as well as the point at the origin, but not either of the points ω or ω'. The other regions formed by the points $n\omega + m\omega'$ are called *period parallelograms* (Figure 3.2) and just translates of Π, also called *cells* whose disjoint union forms a tiling of $\mathbb{C}$.

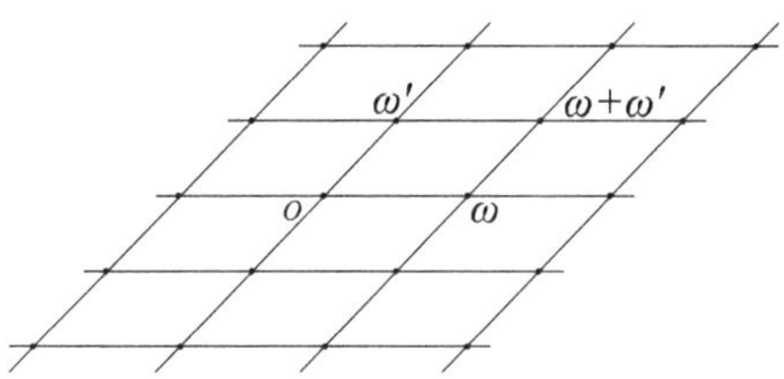

Figure 3.2: A fundamental parallelogram formed by the primitive periods ω and ω'. Courtesy Katy Metcalf.

3 Sometimes, an arbitrary point $a \in \mathbb{C}$ is taken instead of the origin, but we lose nothing by taking $a = 0$, which simplifies the notation.

Two points z and w in $\mathbb{C}$ are *congruent* with respect to the lattice

$$\Lambda = \{n\omega + m\omega' : n, m \in \mathbb{Z}\}$$

if $w = z + n\omega + m\omega'$ for some $n, m \in \mathbb{Z}$, and we write $w \equiv z$ and observe that $f(w) = f(z)$. For a particular given lattice, we will also use the term Λ-*periodic*.

It is clear then that the values of an elliptic function are completely determined by its values in a fundamental parallelogram. Therefore an elliptic function cannot be strictly analytic at all points of the fundamental parallelogram, for otherwise it would be bounded on $\overline{\Pi}$ and therefore in $\mathbb{C}$, hence reduce to a constant by Liouville's theorem.

Note further that the sum, difference, product, and quotient of two elliptic functions is also an elliptic function, and so is the derivative of an elliptic function with the observation that the resulting function may be a constant. Since there can be only finitely many poles in any cell (why?), we have:

3.7 Definition. If $f(z)$ is an elliptic function, then the sum of the orders of the poles in a cell is the *order* of $f(z)$.

It is interesting that the order must be greater than 1.

3.8 Theorem. *If $f(z)$ is a nonconstant elliptic function, then its order is equal to or greater than* 2.

Proof. Let Π be the fundamental parallelogram. We may assume that there are no poles on the boundary of Π, since if they were, we could consider a suitably small translation in the direction of a diagonal so that the boundary of the parallelogram avoids the poles. Computing the integral in the positive sense over the four sides of the boundary of the parallelogram, we have

$$\int_{\partial\Pi} f(z)dz = 0,$$

since by periodicity the integrals over opposite sides of the parallelogram cancel each other. If there was only a single simple pole in Π with residue r, then we would have $\int_{\partial\Pi} f(z)dz = 2\pi i r \neq 0$, a contradiction. Therefore the sum of the residues equaling zero implies that the order of $f(z)$ must be at least 2. □

The case of an elliptic function having order 2 can arise by having a single pole of order 2 (with residue 0) or by having two simple poles (whose residues sum to 0) in the period parallelogram.

The order of an elliptic function is significant for the following reason.

3.9 Theorem. *A nonconstant elliptic function $f(z)$ of order n takes every value in the complex plane exactly n times in a period parallelogram.*

Proof. For any $a \in \mathbb{C}$, the function $f'(z)/(f(z)-a)$ is also elliptic, so that by the preceding theorem and the argument principle

$$0 = \int_{\partial\Pi} \frac{f'(z)}{f(z)-a}\,dz = N - P,$$

where N and P are the numbers of zeros and poles of $f(z)-a$ in Π counted according to multiplicity, respectively. Therefore $N = P = n$. □

Something further can be said about the zeros and poles of an elliptic function. Consider the fundamental period parallelogram Π, and this time we integrate

$$\frac{1}{2\pi i}\int_{\partial\Pi} \frac{zf'(z)}{f(z)}\,dz,$$

where as above we may suppose that there are no zeros or poles on $\partial\Pi$. Denoting the zeros $a_1,\dots,a_n$ and poles $b_1,\dots,b_n$ that lie inside Π, by the residue theorem for the function in the integrand, a direct calculation gives $\mathrm{Res}(a_i) = a_i$, and $\mathrm{Res}(b_i) = -b_i$, $i = 1,\dots,n$. Thus the integral equals $a_1 + \cdots + a_n - b_1 - \cdots - b_n$. Integrating over the two sides of $\partial\Pi$ from 0 to ω and from to $\omega+\omega'$ to ω' as in Figure 3.2, we have

$$\frac{1}{2\pi i}\int_0^{\omega} \frac{zf'(z)}{f(z)}\,dz - \frac{1}{2\pi i}\int_{\omega'}^{\omega+\omega'} \frac{zf'(z)}{f(z)}\,dz = \frac{-\omega'}{2\pi i}\int_0^{\omega} \frac{f'(z)}{f(z)}\,dz$$

upon substituting $z+\omega'$ for z in the second integral. Writing $w = f(z)$, as z varies from 0 to ω, the function $f(z)$ traces out a closed curve C so that

$$-\omega'\frac{1}{2\pi i}\int_C \frac{dw}{w-0} = -\omega'\cdot \mathrm{Ind}(C,0),$$

where $\mathrm{Ind}(C,0)$ is the integer-valued winding number of C with respect to the origin. Similarly, the integral along the two remaining sides is an integer multiple of ω (exercise), and consequently,

$$a_1 + \cdots + a_n - b_1 - \cdots - b_n = \frac{1}{2\pi i}\int_{\partial\Pi} \frac{zf'(z)}{f(z)}\,dz = n\omega + m\omega',$$

or, in other words, $a_1 + \cdots + a_n \equiv b_1 + \cdots + b_n \bmod(\omega,\omega')$, which is our desired result.

Weierstrass $\wp$-function

The smallest order of an elliptic function being 2, we consider the case of the function having a double pole with residue zero and the classical function of Weierstrass.[4]

For two nonzero points ω, ω' in $\mathbb{C}$ with $\operatorname{Im}(\frac{\omega'}{\omega}) > 0$, we enumerate the lattice of points $n\omega + m\omega', n, m = 0, \pm 1, \pm 2, \ldots$, as a sequence $0 = \omega_0, \omega_{1,}, \ldots, \omega_k, \ldots$ and define the Weierstrass $\wp$-function by

$$\wp(z) = \wp(z, \omega, \omega') = \frac{1}{z^2} + \sum_{k=1}^{\infty}\left(\frac{1}{(z-\omega_k)^2} - \frac{1}{\omega_k^2}\right). \tag{3.5}$$

To verify that the series converges, let $|z| = r$ and observe that

$$\left|\frac{1}{(z-\omega_k)^2} - \frac{1}{\omega_k^2}\right| = \left|\frac{z(2\omega_k - z)}{(z-\omega_k)^2\omega_k^2}\right| = O\left(\frac{1}{\omega_k^3}\right)$$

as $\omega_k \to \infty$. We must now show that $\sum \frac{1}{\omega_k^3} < \infty$.

Indeed, taking $n^2 + m^2 > 0$, we can write

$$\frac{n\omega + m\omega'}{\sqrt{n^2+m^2}} = \frac{n}{\sqrt{n^2+m^2}}\omega + \frac{m}{\sqrt{n^2+m^2}}\omega'$$

and set

$$\cos\theta = \frac{n}{\sqrt{n^2+m^2}}, \qquad \sin\theta = \frac{m}{\sqrt{n^2+m^2}}.$$

Now the function

$$f(\theta) = \omega\cos\theta + \omega'\sin\theta$$

yields a nondegenerate ellipse (exercise), so that $|f(\theta)| \geq b > 0$, that is,

$$|n\omega + m\omega'| \geq b\sqrt{n^2+m^2} \geq \frac{b}{\sqrt{2}}(|n| + |m|) = a(|n| + |m|)$$

for all n, m. As we can express the number $k = |n| + |m|$ in $4k$ different ways, it follows that

$$\sum_{k=1}^{\infty}\frac{1}{\omega_k^3} \leq \frac{4}{a^3}\sum_{k=1}^{\infty}\frac{1}{k^2} < \infty.$$

4 There is of course the alternative of the function having two simple poles whose respective residues are opposite in sign which is the basis for the Jacobian theory that will not be pursued here.

We conclude that the series in (3.5) converges uniformly on every compact set which avoids poles, to a meromorphic function in $\mathbb{C}$ with poles of order 2 at the lattice points $\omega_k, k = 0, 1, 2, \ldots$.

Before showing that $\wp(z)$ is Λ-periodic, note that it is an even function. Indeed, for $z \neq \omega_k$, the infinite series converges absolutely, and thus the sum is independent of the order of its terms. Since the sequence $\{\omega_k\}$ is the same as $\{-\omega_k\}$,

$$\begin{aligned}\wp(z) &= \frac{1}{z^2} + \sum_{k=1}^{\infty}\left(\frac{1}{(z-(-\omega_k))^2} - \frac{1}{(-\omega_k)^2}\right) \\ &= \frac{1}{z^2} + \sum_{k=1}^{\infty}\left(\frac{1}{(z+\omega_k)^2} - \frac{1}{(\omega_k)^2}\right) \\ &= \frac{1}{z^2} + \sum_{k=1}^{\infty}\left(\frac{1}{(-z-\omega_k)^2} - \frac{1}{(\omega_k)^2}\right). \\ &= \wp(-z).\end{aligned}$$

Moreover, we can differentiate (3.5) term-by-term, so that

$$\wp'(z) = \frac{-2}{z^3} - \sum_{k=1}^{\infty}\frac{2}{(z-\omega_k)^3} = -2\sum_{k=0}^{\infty}\frac{1}{(z-\omega_k)^3}. \tag{3.6}$$

It is clear from this expression that

$$\wp'(z) = \wp'(z+\omega), \quad \wp'(z) = \wp'(z+\omega'),$$

that is, $\wp'(z)$ is Λ-periodic. Integrating the first equation yields

$$\wp(z) - \wp(z+\omega) = c,$$

and setting $z = -\omega/2$ shows that $c = 0$. Hence $\wp(z) = \wp(z+\omega)$, and likewise $\wp(z) = \wp(z+\omega')$, so that $\wp(z)$ is an elliptic function with periods ω and ω'.

Furthermore, since $\wp(z)$ has a pole at the origin, if there were any periods other than $n\omega + m\omega'$, then $\wp(z)$ would have poles at points other than $z = \omega_k$. As this is not the case, ω and ω' are the fundamental periods, and $\wp(z)$ is an elliptic function of order 2.

We can easily glean some further information about $\wp(z)$. In the first instance, equality (3.6) implies that $\wp'(z)$ is an odd function having fundamental periods ω and ω'. As $\wp'(z)$ is also Λ-periodic, we have (with $z = -\omega/2$)

$$\wp'\left(\frac{\omega}{2}\right) = \wp'\left(-\frac{\omega}{2}\right),$$

and as it is an odd function, $\wp'(-\frac{\omega}{2}) = -\wp'(\frac{\omega}{2})$, implying that $\wp'(\frac{\omega}{2}) = 0$. Likewise,

$$\wp'\left(\frac{\omega'}{2}\right) = 0 = \wp'\left(\frac{\omega+\omega'}{2}\right).$$

Since $\wp'(z)$ has order 3, by Theorem 3.9 the *half-periods* $\frac{\omega}{2}, \frac{\omega'}{2}, \frac{\omega+\omega'}{2}$ are simple zeros of $\wp'(z)$, as well as the only zeros in the fundamental parallelogram. From this we can conclude that for $a \in \mathbb{C}$, the equation $\wp(z) = a$ only has the points $\frac{\omega}{2}, \frac{\omega'}{2}, \frac{\omega+\omega'}{2}$ as roots of multiplicity 2 in the fundamental parallelogram. For all other points $a \in \mathbb{C}$, $\wp(z) = a$ has two distinct simple roots in the fundamental parallelogram. Denote

$$\wp\left(\frac{\omega}{2}\right) = e_1, \quad \wp\left(\frac{\omega'}{2}\right) = e_2, \quad \wp\left(\frac{\omega+\omega'}{2}\right) = e_3.$$

It is clear that the numbers e_1, e_2, e_3 are distinct for if not and two are the same, then that common value would be taken four times, a contradiction since the order of $\wp(z)$ is 2.

Thus we have arrived at the following result.

3.10 Theorem. *For $a \in \hat{\mathbb{C}}$, the equation $\wp(z) = a$ has roots of multiplicity 2 whenever a is any one of the points e_1, e_2, e_3, ∞ and only simple roots for any other value of a.*

One of the consequences of the Nevanlinna theory is that if $f(z)$ is a nonconstant meromorphic function such that all the roots of $f(z) = a_\nu$ have multiplicity at least 2, then there are at most four such values a_ν. The Weierstrass $\wp$-function is such an example.

As we might expect, the numbers e_1, e_2, e_3 provide an intimate relationship between $\wp'(z)$ and $\wp(z)$.

3.11 Corollary. *The derivative $\wp'$ satisfies the cubic equation*

$$(\wp')^2 = 4(\wp - e_1)(\wp - e_2)(\wp - e_3). \tag{3.7}$$

Proof. Define

$$\mathcal{E}(z) = (\wp(z) - e_1)(\wp(z) - e_2)(\wp(z) - e_3).$$

The only roots of $\mathcal{E}(z) = 0$ in the fundamental parallelogram are the double roots at the half-periods $\frac{\omega}{2}, \frac{\omega'}{2}, \frac{\omega+\omega'}{2}$, which, in view of our preceding considerations regarding $\wp'(z)$, are also double roots of $(\wp'(z))^2 = 0$. Moreover, $\mathcal{E}(z)$ has poles of order 6 at each of the lattice points, and the same holds true for $(\wp'(z))^2$.[5] Hence the quotient

$$\mathcal{G}(z) = \frac{(\wp'(z))^2}{\mathcal{E}(z)}$$

5 This is so because $\wp'(z)$ has a pole of order 3 at each lattice point.

has removable singularities and is thus analytic and Λ-periodic in a period parallelogram and therefore constant by Liouville's theorem.

Now in a neighborhood of the origin, we have

$$\wp(z) = \frac{1}{z^2} + \cdots, \quad \wp'(z) = \frac{-2}{z^3} + \cdots,$$

and we infer that the constant of the corollary equals 4. □

Differential equation

Let us consider the first few terms of the Laurent series for $\wp(z)$ about $z = 0$. Note that for $|z| < |\omega_k|$,

$$\frac{1}{(z-\omega_k)^2} - \frac{1}{\omega_k^2} = \frac{1}{\omega_k^2(1-\frac{z}{\omega_k})^2} - \frac{1}{\omega_k^2} = \frac{2z}{\omega_k^3} + \frac{3z^2}{\omega_k^4} + \frac{4z^3}{\omega_k^5} + \frac{5z^4}{\omega_k^6} + \cdots .$$ [6]

Next, for $n \geq 3$, define

$$G_n = \sum_{k=1}^{\infty} \frac{1}{\omega_k^n}$$

and note that if n is odd, then $G_n = 0$ by symmetry. In view of the above and interchanging the order of summation, the Weierstrass $\wp$-function can be expressed as

$$\wp(z) = \frac{1}{z^2} + 3G_4 z^2 + 5G_6 z^4 + O(z^6),$$ [7]

so that

$$\begin{aligned}
\wp'(z) &= \frac{-2}{z^3} + 6G_4 z + 20G_6 z^3 + O(z^5),\\
\wp'(z)^2 &= \frac{4}{z^6} - \frac{24G_4}{z^2} - 80G_6 + O(z^2),\\
\wp(z)^3 &= \frac{1}{z^6} + \frac{9G_4}{z^2} + 15G_6 + O(z^2).
\end{aligned}$$

As a consequence, we have the relationship

$$\wp'(z)^2 - 4\wp(z)^3 - 60G_4\wp(z) - 140G_6 = O(z^2).$$

---

6 For $|x| < 1$, $\frac{1}{(1-x)^2} = (1 + x + x^2 + x^3 + \cdots)(1 + x + x^2 + x^3 + \cdots) = 1 + 2x + 3x^2 + 4x^3 + 5x^4 + \cdots$.

7 More specifically,

$$\wp(z) = \frac{1}{z^2} + \sum_{n=1}^{\infty}(2n+1)G_{2n+2}z^{2n}.$$

It is customary to set $g_2 = 60G_4 = 60\sum_{k=1}^{\infty}\frac{1}{\omega_k^4}$ and $g_3 = 140G_6 = 140\sum_{k=1}^{\infty}\frac{1}{\omega_k^6}$. Since $O(z^2) \to 0$ as $z \to 0$, we obtain the equality

$$\wp'(z)^2 = 4\wp(z)^3 - g_2\wp(z) - g_3. \tag{3.8}$$

We conclude from equations (3.7) and (3.8) that e_1, e_2, e_3 are zeros of the cubic polynomial $4w^3 - 4g_1w^2 - g_2w - g_3$, where

$$\begin{aligned} g_1 &= e_1 + e_2 + e_3 = 0, \\ g_2 &= -4(e_1e_2 + e_2e_3 + e_1e_3), \\ g_3 &= 4e_1e_2e_3. \end{aligned}$$

From equation (3.8) we can write

$$z = \int_{\wp(z_0)}^{\wp(z)} \frac{dw}{\sqrt{4w^3 - g_2w - g_3}} + z_0$$

for $z = z(w)$, which demonstrates that $\wp(z)$ is an inverse of an elliptic integral. Of course, care must be taken that the path of integration avoids the zeros and poles of $\wp'(z)$ and the sign of the square root is such that the function equals the value of $\wp'(z)$.

The expression

$$y^2 = 4x^3 - g_2x - g_3$$

defines an elliptic curve E in $\mathbb{C}^2$, and we have shown that the points $(\wp(z), \wp'(z))$ lie on E. Real elliptic curves have now found extensive use in cryptography as part of public-key cryptosystems.[8]

A completely different approach to the Weierstrass elliptic $\wp$-function was initiated by K. Venkatachaliengar and developed further by S. Cooper in *Ramanujan's Theta Functions*, Springer 2017.

Historically, the inverses defined by elliptic integrals led Abel and Jacobi to the notion of doubly periodic functions, which they called "elliptic functions."

The $\wp$-function may be regarded in some sense as the "simplest" elliptic function, since there are no elliptic functions having order less than 2. In addition, the function has only a single pole with the simplest principal part $1/z^2$ in the period parallelogram. Furthermore, if there were another elliptic function of order 2 having the same pole and principal part, then it would differ from the $\wp$-function by a constant as their difference would be an elliptic function of order zero.

8 See https://cryptobook.nakov.com/asymmetric-key-ciphers/elliptic-curve-cryptography-ecc

Finally, to complete the characterization of elliptic functions in terms of any other elliptic function, we have the following results, the first of which is a particular case of the second more general result.

3.12 Proposition. *If $\varepsilon(z)$ is an even elliptic function having periods ω, ω' with $\mathrm{Im}(\frac{\omega'}{\omega}) > 0$, then $\varepsilon(z)$ can be expressed as a rational function of $\wp(z, \omega, \omega')$.*

Proof. Since $\varepsilon(z)$ is an even function, any zero or pole it may have at $z = 0$ has even order. Then the function $\varepsilon(z)\wp(z)^k$ for some integer k has no zero or pole, respectively, at the four corner points of the lattice. Hence we will assume that there is no zero or pole of $\varepsilon(z)$ at any of the four corner lattice points.

We will first consider the zeros of the function $\varepsilon(z)$ and attempt to recreate them with the $\wp$-function. For any zero $z = a$ in a cell, the point z that is congruent to $-a$ is also a zero in the cell since $\varepsilon(z)$ is an even function. Therefore we can enumerate all the zeros of $\varepsilon(z)$ in any cell as $a_1, a_2, \ldots, a_n$ and the points that are congruent to $-a_1, -a_2, \ldots, -a_n$. These are repeated according to multiplicity with the proviso that if a_i is not a half-period, then a_i and $-a_i$ are repeated according to the multiplicity of $\varepsilon(z)$, and if a_i is a half-period (in which case a_i is congruent to $-a_i$[9]), then the multiplicity of a_i and $-a_i$ are taken to be one-half that of $\varepsilon(z)$. There are two possibilities:

(i) If the zero a_i is not a half-period of $\varepsilon(z)$, then the function

$$\wp_i(z) = \wp(z) - \wp(a_i)$$

has two simple zeros in any cell, one at a_i and one at the point congruent to $-a_i$.

(ii) If the zero a_i is a half-period, in which case $-a_i$ is congruent to a_i itself, then $\wp_i(z)$ has zero of order 2 at a_i.

In both cases the zeros of $\varepsilon(z)$ in any cell coincide with the zeros of the term

$$\prod_{i=1}^{n} (\wp(z) - \wp(a_i)).$$

As well, the poles of $\varepsilon(z)$ in each cell are the same in number and can be listed and counted according to multiplicity in the same manner as above: $b_1, b_2, \ldots, b_n$ and the points congruent to $-b_1, -b_2, \ldots, -b_n$. Hence the rational function

$$\mathcal{R}(z) = \prod_{i=1}^{n} \frac{(\wp(z) - \wp(a_i))}{(\wp(z) - \wp(b_i))}$$

is elliptic possessing the same zeros and poles as the function $\varepsilon(z)$, counted according to multiplicity. Then the quotient $\varepsilon(z)/\mathcal{R}(z)$ is analytic and elliptic, hence constant, thereby concluding the proof. □

9 Note that points a and $-a$ are congruent if and only if a is a half-period.

The full significance of the Weierstrass $\wp$-function for elliptic functions culminates in the following:

3.13 Theorem. *Every elliptic function $f(z)$ having periods ω, ω' with $\operatorname{Im}(\frac{\omega'}{\omega}) > 0$ can be represented as a rational function of $\wp(z, \omega, \omega')$ and $\wp'(z, \omega, \omega')$.*

Proof. Writing

$$f(z) = \left(\frac{f(z) + f(-z)}{2}\right) + \left(\frac{f(z) - f(-z)}{2}\right),$$

note that the first expression is an even function, so that by the preceding proposition it can be expressed as a rational function $\mathcal{R}_1(z)$ of $\wp(z)$. The second expression is an odd function, and since $\wp'(z)$ is also an odd function, the quotient

$$\left(\frac{f(z) - f(-z)}{2}\right) / \wp'(z)$$

is an even function, which can be expressed as a rational function $\mathcal{R}_2(z)$ of $\wp(z)$. Then $f(z) = \mathcal{R}_1(z) + \mathcal{R}_2(z)\wp'(z)$, proving the theorem. □

Addition formula

Consider two equations

$$\wp'(z) = a\wp(z) + b, \quad \wp'(y) = a\wp(y) + b,$$

which have a unique solution for a and b whenever $\wp(z) - \wp(y) \neq 0$, which is the case with the exception when z is congruent to $\pm y$ mod (ω, ω'). This is so as the function $\mathcal{Q}(z) = \wp(z) - \wp(y)$ has a zero at the points $z = y$ and $z \equiv -y$ in any cell, which determines all the zeros of $\mathcal{Q}(z)$.

Consider the function $\wp'(\zeta) = a\wp(\zeta) + b$, which has a pole of order 3 at $\zeta = 0$, and hence if $\zeta = z$ and $\zeta = y$ are two of the zeros, then the third zero w satisfies $z + y + w \equiv 0 \bmod (\omega, \omega')$ in view of the discussion subsequent to Theorem 3.9. Hence $w = -z - y$, and

$$\wp'(-z - y) = a\wp(-z - y) + b,$$

implying $-\wp'(z + y) = a\wp(z + y) + b$. The combination of the first two equations with the last one can be expressed as

$$\begin{pmatrix} \wp(z) & \wp'(z) & 1 \\ \wp(y) & \wp'(y) & 1 \\ \wp(z+y) & -\wp'(z+y) & 1 \end{pmatrix} \begin{pmatrix} a \\ -1 \\ b \end{pmatrix} = \begin{pmatrix} 0 \\ 0 \\ 0 \end{pmatrix},$$

and thus

$$\begin{vmatrix} \wp(z) & \wp'(z) & 1 \\ \wp(y) & \wp'(y) & 1 \\ \wp(z+y) & -\wp'(z+y) & 1 \end{vmatrix} = 0,$$

which is our *addition formula*. This relation allows for the expression of $\wp(z+y)$ in terms of $\wp(z)$ and $\wp(y)$ since the differentiated terms can be expressed in terms of $\wp(z), \wp(y)$, and $\wp(z+y)$ as above.

Elliptic modular function

Modular functions play an important role in complex analysis, and we have already encountered one instance regarding Picard's little theorem.

With $\tau = \frac{\omega'}{\omega}$, the λ-*modular function* defined by

$$\lambda(\tau) = \frac{e_3 - e_2}{e_1 - e_2}$$

is so called as it is invariant under the action of the modular group of Chapter 1. However, we will merely outline an equivalent approach to a modular function as it involves hyperbolic triangles that will be discussed in Chapter 5.

The formidable looking Schwarzian function constructed from the Schwarz–Christoffel formula[10]

$$w = f(z) = \frac{\int_0^1 t^{-\frac{1}{2}(1+a+b+c)}(1-t)^{-\frac{1}{2}(1+a-b-c)}(1-zt)^{-\frac{1}{2}(1-a+b-c)}dt}{\int_0^1 t^{-\frac{1}{2}(1+a+b+c)}(1-t)^{-\frac{1}{2}(1-a-b+c)}(1-t+zt)^{-\frac{1}{2}(1-a+b-c)}dt}$$

represents the conformal mapping of the upper half-plane $\operatorname{Im}(z) > 0$ onto a curvilinear triangle having angles $\pi a, \pi b, \pi c$, where $a+b+c < 1$. In our case, we are interested when $a = b = c = 0$ (see Figure 3.3), which simplifies the above expression considerably and, together with a change of variable $t = s^2$ and reverting back to the variable t, gives

$$f(z) = \frac{K(z)}{K(1-z)},$$

where

$$K(z) = \int_0^1 \frac{dt}{\sqrt{(1-t^2)(1-zt^2)}}.$$

10 Cf. Nehari (1952, eq. (75)).

Figure 3.3: By Saric – Own work, Public Domain, https://commons.wikimedia.org/w/index.php?curid=9627793.

Together with a Möbius transformation, we have a conformal mapping of the upper half-plane onto the central ideal triangle of Figure 3.3 contained in $|w| < 1$. Then the inverse mapping $\mu(w)$ maps the shaded region to $\text{Im}(z) > 0$ with the three curvilinear sides mapped to the real line.

Since $\mu(w)$ is real on each of the three boundary arcs of the initial curvilinear triangle, it may be extended analytically by the Schwarz reflection principle to the three adjacent white regions, which are individually mapped analytically to the lower half-plane $\text{Im}(z) < 0$. Continuing to reflect each new curvilinear triangle across its three boundaries in this manner, we get an analytic mapping of the disk $U : |w| < 1$ that takes every value in $\mathbb{C}$ with the exception of $0, 1, \infty$.

The function that is specifically involved in the proof of Picard's little theorem in Chapter 2 is the inverse mapping

$$\nu(z) = \mu^{-1}(z),$$

which of course is multiple-valued. However, we can and do select a particular branch at a point z_0 which will have an analytic continuation along any curve not passing through the points $0, 1, \infty$.

Now, if we assume that $z = f(\zeta)$ is a nonconstant entire function that omits two distinct values a and b, then the function

$$g(\zeta) = \frac{f(\zeta) - a}{b - a}$$

is an entire function omitting the values $0, 1$ in the z-plane. Then $\nu(z)$ is analytic in $\mathbb{C} - \{0, 1\}$, and $|\nu(z)| < 1$. Hence the composition $h(\zeta) = \nu(g(\zeta))$ becomes a bounded

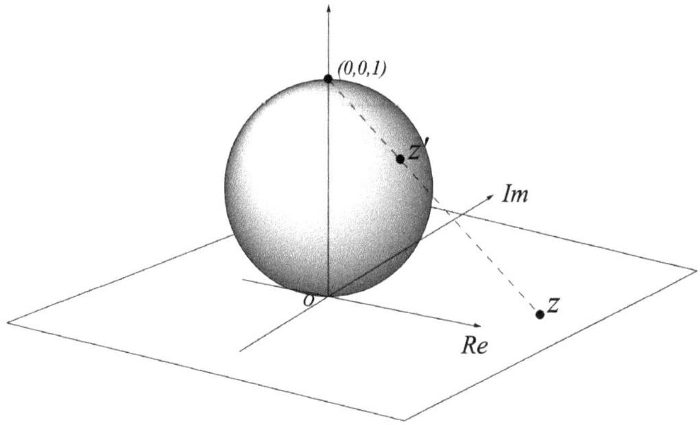

Figure 3.4: The Riemann sphere $\mathbb{S}$ where the North pole representing infinity is at the point (0, 0, 1). Points in the complex plane have a unique representation with points on the sphere. Courtesy Katy Metcalf.

entire function, and hence $h(\zeta)$ must be constant with the result that $g(\zeta)$ must be constant. This contradiction proves Picard's little theorem.

To discuss meromorphic functions to their full extent in a modern context, we need to expand the domain to the extended complex plane.

Riemann sphere

When considering meromorphic functions, it is useful to treat ∞ as any other point, and to do this, we consider the Riemann sphere $\mathbb{S}$, a sphere of diameter = 1 sitting on the complex plane at the origin (Figure 3.4). All points z in the complex plane $\mathbb{C}$ are in a one-to-one correspondence with points z' on the sphere by connecting the point $z \in \mathbb{C}$ with a line to $(0, 0, 1)$ on the sphere. This intersects the sphere at a unique point z' on the sphere.

We further associate the point $(0, 0, 1)$ with ∞ and identify $\mathbb{S}$ with the extended complex plane $\hat{\mathbb{C}} = \mathbb{C} \cup \{\infty\}$ via the stereographic projection above. To extend our function theory to the Riemann sphere, we say that a function f has a pole/removable singularity/essential singularity at $z = \infty$ if the function $F(z) = f(\frac{1}{z})$ has a pole/removable singularity/essential singularity (resp.) at $z = 0$.

Spherical metric

The *chordal* (*spherical*) distance $\chi(z_1, z_2)$ between two points z_1 and z_2 in $\mathbb{C}$ is given by the length of the chord connecting the corresponding points z_1' and z_2' on the sphere $\mathbb{S}$. With a bit of geometry, this works out to be

$$\chi(z_1, z_2) = \frac{|z_1 - z_2|}{\sqrt{1 + |z_1|^2}\sqrt{1 + |z_2|^2}}.$$

Moreover, for $z_2 = \infty$,

$$\chi(z_1, \infty) = \frac{1}{\sqrt{1 + |z_1|^2}}.$$

Furthermore, $\chi(z_1, z_2)$ satisfies all the properties to be a metric on $\hat{\mathbb{C}}$, known as the *spherical metric*, and clearly $\chi(z_1, z_2) \leq |z_1 - z_2|$ on $\mathbb{C}$. A useful property for points in $\mathbb{C}$ is

$$\chi(z_1, z_2) = \chi\left(\frac{1}{z_1}, \frac{1}{z_2}\right),$$

and if $|z_1| \leq |z_2|$, then $\chi(0, z_1) \leq \chi(0, z_2)$.

3.14 Definition. A function $f(z)$ is *spherically continuous at a point* $z_0 \in \mathbb{C}$ if for any $\varepsilon > 0$, there exists $\delta > 0$ such that

$$\chi(f(z), f(z_0)) < \varepsilon$$

whenever $|z - z_0| < \delta$.

This is just a direct analogue of continuity in the Euclidean metric, but, interestingly, we now obtain the property of spherical continuity for meromorphic functions.

Proposition. *A meromorphic function $f(z)$ defined in a domain $\Omega \subseteq \mathbb{C}$ is spherically continuous in Ω.*

Proof. Suppose that $f(z)$ is analytic at $z_0 \in \Omega$. Then

$$\chi(f(z), f(z_0)) \leq |f(z) - f(z_0)|. \tag{3.9}$$

If $f(z)$ has a pole at z_0, then $\frac{1}{f(z)}$ is continuous at z_0, and the fact that

$$\chi(f(z), f(z_0)) = \chi\left(\frac{1}{f(z)}, \frac{1}{f(z_0)}\right)$$

establishes the result. □

The notion of equicontinuity also carries over to meromorphic functions, and we see by inequality (3.9) that in the case of analytic functions, equicontinuity implies spherical equicontinuity (the converse is false). Another natural carryover to meromorphic functions is the notion of a sequence of functions $\{f_n\}$ *converging spherically uniformly on compact subsets to a function* f, that is, $\chi(f_n, f) \to 0$ as $n \to \infty$ on compact subsets. It is clear from the above that ordinary uniform convergence implies spherical uniform convergence, and in general the converse is not true except when the limit function is bounded.

3.15 Theorem. *A sequence $\{f_n\}$ of functions that converge spherically uniformly to a bounded function f on a set E also converges uniformly to f on E.*

Proof. Let $|f(z)| \leq M < \infty$ for $z \in E$, so that

$$\chi(0, f(z)) \leq \chi(0, M) = \frac{M}{\sqrt{1+M^2}} < 1.$$

Taking $0 < \varepsilon < 1 - \frac{M}{\sqrt{1+M^2}}$, the spherical convergence implies that there is $n_0 > 0$ such that

$$\chi(f(z), f_n(z)) < \varepsilon$$

for all $n \geq n_0$ and $z \in E$. In view of the triangle inequality,

$$\frac{|f_n(z)|}{\sqrt{1+|f_n(z)|^2}} = \chi(0, f_n(z)) \leq \chi(0, f(z)) + \chi(f(z), f_n(z)) < \frac{M}{\sqrt{1+M^2}} + \varepsilon = A < 1.$$

Unwinding the left-hand side leads to

$$|f_n(z)| < \frac{A}{\sqrt{1-A^2}} = B$$

for $n \geq n_0$. If follows that

$$\begin{aligned} |f(z) - f_n(z)| &= \sqrt{1+|f(z)|^2}\sqrt{1+|f_n(z)|^2} \cdot \chi(f(z), f_n(z)) \\ &< \sqrt{1+M^2}\sqrt{1+B^2} \cdot \chi(f(z), f_n(z)) \end{aligned}$$

for $n \geq n_0$ and $z \in E$, establishing the desired uniform convergence of the theorem. □

In general, for a sequence of continuous functions that converge uniformly on a compact set K to a function f, the latter is uniformly continuous on K. In exactly the same manner, we can establish the following:

3.16 Theorem. *Let $\{f_n\}$ be a sequence of spherically continuous functions that converge spherically uniformly on a compact set $K \subset \mathbb{C}$ to a function f. Then f is uniformly spherically continuous on K, and the family $\{f_n\}$ is spherically equicontinuous on K.*

Proof. Exercise. The only new aspect here is the spherical equicontinuity of $\{f_n\}$, but this fact falls out as a consequence of the details of the proof. □

A question that will be of importance in the chapter on normal families is what is the consequence of a sequence of meromorphic functions that converge *spherically uniformly* to a function f on compact subsets of a domain.

3.17 Theorem. *A sequence $\{f_n\}$ of meromorphic functions on a domain Ω converges spherically uniformly to f on compact subsets of Ω if and only if about each point $z_0 \in \Omega$, there is a closed disk $\overline{D}(z_0, r)$ on which either*

$$|f_n - f| \to 0$$

or

$$\left|\frac{1}{f_n} - \frac{1}{f}\right| \to 0$$

uniformly as $n \to \infty$.

Proof. One direction is clear since $\chi(w_n, w) \leq |w_n - w|$ and $\chi(w_n, w) \leq |\frac{1}{w_n} - \frac{1}{w}|$, so that if either $|f_n - f| \to 0$ or $|\frac{1}{f_n} - \frac{1}{f}| \to 0$ uniformly on each $\overline{D}(z_0, r)$, then $\chi(f_n, f) \to 0$ uniformly on compact subsets of Ω.

Conversely, assuming that $\chi(f_n, f) \to 0$ uniformly on compact subsets of Ω, we must consider the value of the function $f(z)$ at the point z_0 in two separate cases:

(i) $f(z_0) \neq \infty$: The function $f(z)$ is spherically continuous on Ω by the preceding theorem. Therefore in some closed disk $\overline{D}(z_0, r)$ the function $f(z)$ is bounded, and by Theorem 3.15 the convergence $|f_n - f| \to 0$ is uniform on $\overline{D}(z_0, r)$. Note that the function f is in fact analytic in $D(z_0, r)$.

(ii) $f(z_0) = \infty$: Again, there is a closed disk $\overline{D}(z_0, r)$ in which $\frac{1}{f(z)}$ is bounded, and as $\chi(\frac{1}{f_n}, \frac{1}{f}) \to 0$ uniformly in this disk, another application of Theorem 3.15 implies that $|\frac{1}{f_n} - \frac{1}{f}| \to 0$ uniformly in $\overline{D}(z_0, r)$. Moreover, $\frac{1}{f}$ is analytic in $D(z_0, r)$. □

The significance of the preceding result is that it plays an essential role when it comes to normal families of meromorphic functions.

3.18 Corollary. *If $\{f_n\}$ is a sequence of meromorphic (analytic) functions on a domain Ω that converges spherically uniformly on compact sets to a function f, then f is either meromorphic (analytic) in Ω or identically ∞.*

Proof. We first treat the meromorphic case. Assuming that $f \not\equiv \infty$, we will show that f is analytic except for possible poles. At a point $z_0 \in \Omega$ such that $f(z_0) \neq \infty$, by case (i) in the theorem, f is analytic in a neighborhood of z_0.

If $f(z_0) = \infty$, then suppose that it is *not* an isolated singularity. Then there exists a sequence of points $z_n \to z_0$ with $f(z_n) = \infty$. As in case (ii), $\frac{1}{f}$ is analytic in some disk $D(z_0, r)$, but $\frac{1}{f(z_n)} = 0$ for each n implies that $\frac{1}{f} \equiv 0$. Thus $f \equiv \infty$ in $D(z_0, r)$.

Consequently, if there is some point with $f(z_0) = \infty$, then let

$$P = \{z \in \Omega : f(z) = \infty\}.$$

We have just shown that P is an open set, but so is the complement. Since Ω is connected, we are forced to conclude that $P \equiv \Omega$ and $f \equiv \infty$. As this is not the case, the singularity z_0 must indeed be isolated, that is, f is a meromorphic function.

In the analytic case, if $f(z_0) = \infty$ for some $z_0 \in \Omega$, then as in case (ii) of the theorem, $\frac{1}{f_n} \to \frac{1}{f}$ uniformly in some $D(z_0, r)$. It follows that for all n sufficiently large, the functions $\frac{1}{f_n}$ are analytic and nonzero in $D(z_0, r)$. As a consequence of the Hurwitz theorem 1.28, $\frac{1}{f} \equiv 0$ in $D(z_0, r)$, so that the preceding connectedness argument implies that $f \equiv \infty$ in Ω. We conclude that if $f \not\equiv \infty$, then it has no poles, and $f(z)$ is analytic in a neighborhood of every point of Ω. □

Spherical derivative

In our spherical setting, we also need a derivative, which we will define analogously as the derivative in the Euclidean metric. If $f(z)$ is meromorphic in a domain Ω and $z_0 \in \Omega$, then we define the *spherical derivative* $f^{\#}(z_0)$ as

$$\begin{aligned} f^{\#}(z_0) &= \lim_{z \to z_0} \frac{\chi(f(z), f(z_0))}{|z - z_0|} \\ &= \lim_{z \to z_0} \frac{|f(z) - f(z_0)|}{|z - z_0|} \cdot \frac{1}{\sqrt{1 + |f(z_0)|^2}\sqrt{1 + |f(z)|^2}} \\ &= \frac{|f'(z_0)|}{1 + |f(z_0)|^2}, \end{aligned}$$

and for a pole z_0 of $f(z)$, since the poles are isolated,

$$f^{\#}(z_0) = \lim_{z \to z_0} \frac{|f'(z)|}{1 + |f(z)|^2},$$

whence $f^{\#}(z)$ is continuous, and, moreover, $f^{\#}(z) = \left(\frac{1}{f(z)}\right)^{\#}$ (exercise).

For any contour C in Ω, the image of C on the Riemann sphere under the mapping $f(z)$ has *spherical arc length element* given by

$$ds = f^{\#}(z)|dz|,$$

so that the spherical length of $f(C)$ on the Riemann sphere becomes

$$\int_C f^{\#}(z)|dz|.$$

Moreover, the spherical area element is given by

$$dA = [f^{\#}(z)]^2 dxdy, \quad z = x + iy.$$

Then for a function $f(z)$ meromorphic in $|z| \le r$, we have the quantities

$$L(r) = \int_{|z|=r} f^{\#}(z)|dz| = \int_0^{2\pi} \frac{|f'(re^{i\theta})|}{1 + |f(re^{i\theta})|^2} r d\theta, \tag{3.10}$$

$$S(r) = \frac{1}{\pi} \iint_{|z|<r} [f^{\#}(z)]^2 dxdy = \frac{1}{\pi} \int_0^{2\pi} \int_0^r \frac{|f'(\rho e^{i\theta})|^2 \rho d\rho d\theta}{(1 + |f(\rho e^{i\theta})|^2)^2}. \tag{3.11}$$

Here $L(r)$ is the length of $f(\{|z| = r\})$ on the Riemann sphere, and $\pi S(r)$ is the area of $f(\{|z| < r\})$ on the Riemann sphere, both determined with due regard for multiplicity.

For example, taking the mapping $f(z) = z$ for $z = z(t) = t$, $-\infty < t < \infty$, we find that the image on the Riemann sphere of the real axis (an infinite circle) has the length

$$\int_{-\infty}^{\infty} \frac{dt}{1 + t^2} = \tan^{-1} t\Big|_{-\infty}^{\infty} = \pi,$$

which is the circumference of the Riemann sphere.

Nevanlinna theory

The seeds of the deep theory of meromorphic functions due to Rolf Nevanlinna (1895–1980) lie in the Poisson–Jensen formula, which has many applications in its own right. The formula tells us how the zeros and poles constrain the values of the function.

Poisson–Jensen formula

Suppose that an analytic function $f(z)$ defined in $|z| < R < \infty$ has *no* zeros there. Then a single-valued analytic branch of $\log f(z)$ can be defined, and $\log |f(z)| = \mathrm{Re}(\log f(z))$ is a harmonic function in $|z| < R$. The Poisson formula for a harmonic function $u(z)$ for $|z| \le r < R$, which we establish in Chapter 7, gives a weighted average to the values of $u(z)$ for $0 \le |z| = \rho < r < R$ according to the formula

$$u(z) = \frac{1}{2\pi} \int_0^{2\pi} K(z, \zeta) f(re^{i\phi}) d\phi,$$

where

$$K(z,\zeta) = \frac{r^2 - \rho^2}{r^2 - 2r\rho\cos(\theta - \phi) + \rho^2},$$

$z = \rho e^{i\theta}$, $\zeta = re^{i\phi}$. Thus

$$\log|f(z)| = \frac{1}{2\pi}\int_0^{2\pi} K(z,\zeta)\log|f(re^{i\phi})|d\phi \tag{3.12}$$

and

$$\log|f(0)| = \frac{1}{2\pi}\int_0^{2\pi} \log|f(re^{i\phi})|d\phi. \tag{3.12$'$}$$

Although we have stipulated that formulations (3.12) and (3.12′) required that $f(z)$ has no zeros (or poles) on the circle $|z| = r$, the formulas remain valid even with (finitely many) zeros (or poles) on the circle itself. To this end, take for each zero (or pole) on $|z| = r$, a small semicircular path γ_δ of radius δ from each zero/pole (avoiding any other zero/pole) so that γ_δ extends into the domain $|z| < r$. Thus our new boundary consists of arcs of $|z| = r$ together with the small semicircular indentations of radius δ. The length of each γ_δ is less than $\pi\delta$, and each integrand in (3.12) or (3.12′) is uniformly $O(\log(1/\delta))$ (why?). Since $\delta\log(1/\delta) \to 0$ as $\delta \to 0$, we obtain (3.12) and (3.12′) *whether or not there are zeros or poles on* $|z| = r$. The value of the integral mean is still $\log|f(z)|$.

If $f(z)$ has finitely many zeros at the points $a_1, a_2, \ldots, a_n \neq 0$ and poles $b_1, b_2, \ldots, b_m \neq 0$ ($f(z) \neq 0, \infty$ on $|z| = r$), both listed according to multiplicity in $|z| < r < R$, then we consider the function

$$F(z) = f(z)\prod_{i=1}^{n}\frac{r^2 - \overline{a_i}z}{r(z - a_i)}\prod_{j=1}^{m}\frac{r(z - b_j)}{r^2 - \overline{b_j}z}.$$

It follows that $F(z)$ is analytic in $|z| \leq r$ and $|F(z)| = |f(z)|$ on $|z| = r$. Replacing $f(z)$ by $F(z)$ in (3.12), we obtain the *Poisson–Jensen formula* for $|z| < r$,

$$\frac{1}{2\pi}\int_0^{2\pi} K(z,\zeta)\log|f(re^{i\phi})|d\phi = \log|f(z)| + \sum_{i=1}^{n}\log\left|\frac{r^2 - \overline{a_i}z}{r(z - a_i)}\right| + \sum_{j=1}^{m}\left|\frac{r(z - b_j)}{r^2 - \overline{b_j}z}\right|. \tag{3.13}$$

Upon setting $z = 0$ we arrive at the important *Jensen formula*

$$\frac{1}{2\pi}\int_0^{2\pi} \log|f(re^{i\phi})|d\phi = \sum_{i=1}^{n}\log\frac{r}{|a_i|} - \sum_{j=1}^{m}\log\frac{r}{|b_j|} + \log|f(0)|. \tag{3.14}$$

In view of our preceding remarks, we can infer that formulas (3.13) and (3.14) *remain valid for all values of $r < R$*.[11]

Let us now remove the restriction that $f(0) \neq 0, \infty$ and suppose that in a neighborhood of the origin, $f(z)$ has the Laurent series expansion

$$f(z) = c_k z^k + c_{k+1} z^{k+1} + \cdots$$

for some integer $k \neq 0$. To apply the Jensen formula, we consider the function $F(z) = r^k z^{-k} f(z)$, which is analytic, $F(0) \neq 0$, $|F(z)| = |f(z)|$ on $|z| = r$, and $|F(0)| = r^k c_k$. Since $F(z)$ also possesses the same zeros and poles except at the origin, we find that

$$\frac{1}{2\pi} \int_0^{2\pi} \log|f(re^{i\phi})|d\phi = \sum_{i=1}^{n} \log \frac{r}{|a_i|} - \sum_{i=1}^{m} \log \frac{r}{|b_j|} + k \log r + \log |c_k|. \tag{3.15}$$

Note that k is either positive, whereby it represents the multiplicity of the zero at the origin, or negative and represents the multiplicity of the pole at the origin.

Counting function

All of this allows us introduce some new notation with regard to a meromorphic function $f(z)$ defined in $|z| < R < \infty$:

$n(t, 0)$ = number of zeros counted according to multiplicity in $|z| \le t$;
$n(t, a)$ = number of zeros of $f(z) - a$ counted according to multiplicity in $|z| \le t$;
$n(t, \infty)$ = number of poles counted according to multiplicity in $|z| \le t$.

By taking into account the possible zeros of $f(z)$ at the origin, we can write subject to the proviso as in footnote 10 of Chapter 1,

$$\sum_{i=1}^{n} \log \frac{r}{|a_i|} = \int_0^r \frac{n(t,0) - n(0,0)}{t} dt,$$

$$\sum_{j=1}^{m} \log \frac{r}{|b_j|} = \int_0^r \frac{n(t,\infty) - n(0,\infty)}{t} dt.$$

Therefore equality (3.15) now reads

11 Jensen actually proved this general case where $f(z)$ has zeros and poles by another means. He even states that the formula remains valid when the circle $|z| = r$ passes through zeros or poles: Sur un nouvel et important théorème de la théorie des fonctions, *Acta Math.* (22) (1899), 359–364. From the title of his article, Jensen knew that this was an important result, and he was absolutely correct!

$$\frac{1}{2\pi}\int_0^{2\pi} \log|f(re^{i\phi})|d\phi = \int_0^r \frac{n(t,0)-n(0,0)}{t}dt - \int_0^r \frac{n(t,\infty)-n(0,\infty)}{t}dt$$
$$+ \left(n(0,0)-n(0,\infty)\right)\log r + \log|c_k|. \tag{3.16}$$

Admittedly, equation formula (3.16) is rather cumbersome. So let us denote

$$N(r,f) = N(r,\infty) = \int_0^r \frac{n(t,\infty)-n(0,\infty)}{t}dt + n(0,\infty)\log r$$

and

$$N\left(r,\frac{1}{f}\right) = \int_0^r \frac{n(t,0)-n(0,0)}{t}dt + n(0,0)\log r.$$

Formula (3.16) now reads as

$$\frac{1}{2\pi}\int_0^{2\pi} \log|f(re^{i\phi})|d\phi = N\left(r,\frac{1}{f}\right) - N(r,f) + \log|c_k|. \tag{3.17}$$

We can think of $N(r,f)$ as a "pole counting average" and $N(r,\frac{1}{f})$ as a "zero counting average" in $|z| \le r$.

Proximity function

We next proceed to deconstruct the integral mean on the left-hand side of formula (3.17). Consider the truncated logarithm function $\log^+ x = \max(\log x, 0)$. Note that for $x > 0$,

$$\log x = \log^+ x - \log^+ \frac{1}{x},$$

which we now apply to $|f(re^{i\phi})|$ in the preceding. Setting

$$m(r,f) = m(r,\infty) = \frac{1}{2\pi}\int_0^{2\pi} \log^+|f(re^{i\phi})|d\phi,$$

which is the *proximity function*,[12] our representation of Jensen's formula of (3.17) can now be expressed in the form

$$m(r,f) + N(r,f) = m\left(r,\frac{1}{f}\right) + N\left(r,\frac{1}{f}\right) + \log|c_k|.$$

12 The basis for the name is explained in the sequel.

Characteristic function

Another quantity of significance in the study of meromorphic functions is the *characteristic function*

$$T(r,f) = m(r,f) + N(r,f),$$

which plays a fundamental role in the study of both entire and meromorphic functions. Therefore the Jensen formula now takes the form

$$T(r,f) = T\left(r, \frac{1}{f}\right) + \log|c_k|.^{13} \tag{3.18}$$

Let us now mention a few elementary facts regarding $\log^+ x$:

$$\log^+(x_1 + x_2) \le \log^+(x_1) + \log^+(x_2) + \log 2,$$

so that for any $a \in \mathbb{C}$,

$$\log^+|x - a| \le \log^+|x| + \log^+|a| + \log 2,$$

and

$$\log^+|x| \le \log^+|x - a| + \log^+|a| + \log 2,$$

which results in

$$\left|\log^+|x - a| - \log^+|x|\right| \le \log^+|a| + \log 2.$$

Applying this last result to the proximity function $m(r,f)$, we obtain

$$\left|m(r,f - a) - m(r,f)\right| \le \log^+|a| + \log 2,$$

as any point z_0 is a pole of f if and only if it is a pole of $f - a$, which means that

$$N(r,f - a) = N(r,f),$$

and hence we deduce that

$$\left|T(r,f - a) - T(r,f)\right| \le \log^+|a| + \log 2. \tag{3.19}$$

13 Under the assumption that $f(0) \ne 0, \infty$, Jensen's formula (3.18) reads

$$T(r,f) = T\left(r, \frac{1}{f}\right) + \log\left|f(0)\right|. \tag{3.18$'$}$$

Finally, we denote

$$m(r,a) = m\left(r, \frac{1}{f-a}\right) = \frac{1}{2\pi}\int_0^{2\pi} \log^+\left|\frac{1}{f(re^{i\theta})-a}\right| d\theta,$$

$$N(r,a) = N\left(r, \frac{1}{f-a}\right) = \int_0^r \frac{n(t,a)-n(0,a)}{t}\,dt + n(0,a)\log r,$$

where the latter term is a measure of the mean density of the a-points[14] in $|z| \le r$, and the former term represents a measure of the average proximity of $f(z)$ to a on $|z| = r$. Only on those arcs where $f(re^{i\theta})$ is in reasonable proximity[15] to the value a will there be any significant contribution to the value of $m(r,a)$, and therefore the proximity function can be considered a measure of the mean deviation on $|z| = r$ of $f(z)$ from the value a. Thus the sum

$$m(r,a) + N(r,a)$$

in some sense represents a measure of the "total affinity" of the function $f(z)$ for the value a in the disk $|z| \le r$.

Therefore, in view of Jensen's formulation of formula (3.18), for the function $f - a$, we have

$$T(r, f-a) = T\left(r, \frac{1}{f-a}\right) + \log|c_a|, \tag{3.20}$$

where c_a is the first nonzero coefficient of the Laurent series of $f(z) - a$ in a neighborhood of the origin. Then inequality (3.19) coupled with equality (3.20) gives Nevanlinna's first main theorem.

3.19 First fundamental theorem.[16] *If $f(z)$ is meromorphic in $|z| < R \le \infty$, then for any $a \in \mathbb{C}$ and $0 < r < R$,*

$$m(r,a) + N(r,a) = T\left(r, \frac{1}{f-a}\right) = T(r,f) + \varepsilon(r,a),$$

where $|\varepsilon(r,a)| \le \log^+|a| + \log 2 + |\log|c_a||$ with c_a as above.

3.20 Remark. The result of the theorem can be rephrased as

$$m(r,a) + N(r,a) = T(r,f) + O(1)$$

14 That is, points where $f(z) = a$.

15 If $|f(re^{i\theta}) - a| \ge 1$, then $m(r,a) = 0$. If $|f(re^{i\theta}) - a| < 1$, then Jensen's formula is applicable.

16 Zur Theorie der meromorphen Funktionen, *Acta Math.* 46 (1925), 1–99.

as $r \to R$. Moreover, the result also holds for $a = \infty$ using the definitions of $m(r, \infty)$ and $N(r, \infty)$ as above, since $T(r,f) = m(r, \infty) + N(r, \infty)$, i. e.,

$$m(r, a) + N(r, a) = m(r, \infty) + N(r, \infty) + \varepsilon(r, a).$$

In general terms, in the words of the master, Nevanlinna, "The sum $m(r, a)+N(r, a)$ for different values of a maintains a total, given by the quantity $T(r,f)$, which is invariant, up to additive terms that are bounded for $r < R$."

If $f(0) \neq 0, \infty$, then in view of equality (3.18′), we can rephrase the first fundamental theorem as

$$\begin{aligned} m(r, a) + N(r, a) &= T\left(r, \frac{1}{f - a}\right) = T(r,f - a) - \log|f(0) - a| \\ &= T(r,f) - \log|f(0) - a| + \varepsilon(r, a), \end{aligned}$$

where $|\varepsilon(r, a)| \leq \log^+ |a| + \log 2$ for any $a \in \mathbb{C}$.

The first fundamental theorem in this form dispenses with the constant c_a, which is often a nuisance in the theory and diminishes the aesthetic appeal of some results. However, without the restriction $f(0) \neq 0, \infty$, the theory still holds with minor modifications.

3.21 Example. Suppose that $P(z)$ and $Q(z)$ are two relatively prime polynomials of degrees p and q, respectively, and let

$$f(z) = \frac{P(z)}{Q(z)},$$

so that there no zeros or poles at the origin.

(i) $p > q$ and $a \neq \infty$:
Then $m(r, a) = 0$ for all r sufficiently large since $f(z) \to \infty$ as $z \to \infty$. As well, $P(z) - aQ(z) = 0$ has $n(t, a) = p$ for all t sufficiently large and $n(t, a) = 0$ for all t sufficiently small, so that for r sufficiently large,

$$N(r, a) = \int_0^r \frac{n(t, a)}{t} dt = p \log r + O(1),$$

and therefore $T(r,f) = p \log r + O(1)$ as $r \to \infty$.

(ii) $p < q$ and $a \neq 0$:
Here $n(t, a) = q$ for all t sufficiently large regardless if a is finite nor not, so that

$$N(r, a) = \int_0^r \frac{n(t, a)}{t} dt = q \log r + O(1),$$

and since $f(z) \to 0$ as $z \to \infty$, again $m(r,a) = O(1)$ as $r \to \infty$. Therefore $T(r,f) = q\log r + O(1)$ as $r \to \infty$.

(iii) $p = q$ and $a \neq 1$:
Again, $T(r,f) = p\log r + O(1)$ as $r \to \infty$.

3.22 Exercise. For the function $f(z) = e^z$, note that if z_0 is a zero of $e^z - a$, then the other roots are given by $z_0 + 2k\pi i$, k an integer.

(i) Show that $m(r,0) = \frac{r}{\pi} = m(r,\infty)$ and $N(r,0) = N(r,\infty) = 0$, so that $T(r,f) = \frac{r}{\pi}$.
(ii) Find a relationship between $n(t,a)$ and the radius t.
(iii) For $a \neq 0, \infty$, show that

$$N(r,a) = \frac{r}{\pi} + O(1), \quad m(r,a) = O(1).$$

Thus all cases are in accordance with the first fundamental theorem, $m(r,a)+N(r,a) = T(r,f) + O(1)$, where $T(r,f) = \frac{r}{\pi}$.

We can now obtain a growth estimate of the characteristic function. For $f(z)$ analytic in $|z| \leq R$, denote

$$M(r,f) = \max_{|z|=r} |f(z)|.$$

3.23 Theorem. *The characteristic function for $f(z)$ analytic in $|z| \leq R$ satisfies*

$$T(r,f) \leq \log^+ |M(r,f)| \leq \frac{R+r}{R-r} T(R,f)$$

for $0 \leq r < R$.

Proof. As $f(z)$ is analytic, $N(r,f) = 0$, implying

$$T(r,f) = m(r,f) = \frac{1}{2\pi} \int_0^{2\pi} \log^+ |f(re^{i\phi})| d\phi \leq \log^+ M(r,f).$$

Furthermore, for $z = re^{i\theta}$ for $0 \leq r < R$, the Poisson–Jensen formula (3.13) gives

$$\log|f(z)| \leq \frac{1}{2\pi} \int_0^{2\pi} K(z,\zeta) \log^+ |f(Re^{i\phi})| d\phi \leq \frac{R+r}{R-r} m(R,f).$$

Consequently,

$$\log^+ |f(z)| \leq \frac{R+r}{R-r} T(R,f),$$

establishing the result. In the specific case for $r = 2R$, we have the result

$$T(r,f) \leq 3T(2r,f).$$

□

Order

As for entire functions, we have the notion of order.

3.24 Definition. The *order* of a meromorphic function $f(z)$ is given by

$$\overline{\lim_{r\to\infty}}\frac{\log T(r,f)}{\log r}=\rho,$$

or, in other words, for any $\varepsilon>0$,

$$T(r,f)=O(r^{\rho+\varepsilon})$$

as $r\to\infty$.

If $f(z)$ happens to be an entire function, since the proof of the above theorem implies $\log M(r)\le 3T(2r,f)$, we see that the definition of the order of $f(z)$ given above is in agreement with that given in Chapter 2.

Since

$$m(r,a)\le T(r,f)+O(1),\quad N(r,a)\le T(r,f)+O(1),$$

we maintain the following:

3.25 Corollary. *If $f(z)$ is a meromorphic function of order ρ, then for every $\varepsilon>0$*

$$m(r,a)=O(r^{\rho+\varepsilon}),\quad N(r,a)=O(r^{\rho+\varepsilon})$$

as $r\to\infty$.

Cartan theorem

There is a noteworthy identity proved by Henri Cartan (1904–2008)[17] (although the theorem was originally proved by Nevanlinna) regarding the characteristic function

3.26 Theorem. *If $f(z)$ is meromorphic in $|z|<R$, then the characteristic function is given by*

$$T(r,f)=\frac{1}{2\pi}\int_0^{2\pi}N(r,e^{i\theta})d\theta+\log^+|f(0)|\quad (f(0)\ne\infty)$$

for $0<r<R$.

17 Sur la function de croissance attachée à une function méromorphic de deux variable, et ses applications aux fonctions méromorphes d'une variable, *C. R. Acad. Sci. Paris*, 1 (1929), 521–523.

Proof. Jensen's formula (3.14) applied to the function $f(z) - e^{i\theta}$ yields

$$\log|f(0) - e^{i\theta}| = \frac{1}{2\pi}\int_0^{2\pi} \log|f(re^{i\phi}) - e^{i\theta}|d\phi - N(r, e^{i\theta}) + N(r, \infty) \tag{3.21}$$

for $0 < r < R$. A separate application of Jensen's formula to the function $g(z) = a - z$ this time with $r = 1$ yields

$$\frac{1}{2\pi}\int_0^{2\pi} \log|a - e^{i\theta}|d\theta = \begin{cases} \log|a| & \text{if } |a| \geq 1, \\ -\log|a| + \log|a| = 0 & \text{if } |a| < 1. \end{cases}$$

Hence, in either case, we have

$$\frac{1}{2\pi}\int_0^{2\pi} \log|a - e^{i\theta}|d\theta = \log^+|a|. \tag{3.22}$$

Integration of both sides of (3.21) with respect to θ gives

$$\begin{aligned}
\log^+|f(0)| &= \frac{1}{2\pi}\int_0^{2\pi} \log|f(0) - e^{i\theta}|d\theta \\
&= \frac{1}{2\pi}\int_0^{2\pi}\left(\frac{1}{2\pi}\int_0^{2\pi} \log|f(re^{i\phi}) - e^{i\theta}|d\phi\right)d\theta - \frac{1}{2\pi}\int_0^{2\pi} N(r, e^{i\theta})d\theta + \frac{1}{2\pi}\int_0^{2\pi} N(r, \infty)d\theta \\
&= \frac{1}{2\pi}\int_0^{2\pi}\left(\frac{1}{2\pi}\int_0^{2\pi} \log|f(re^{i\phi}) - e^{i\theta}|d\theta\right)d\phi - \frac{1}{2\pi}\int_0^{2\pi} N(r, e^{i\theta})d\theta + N(r, \infty) \\
&= \frac{1}{2\pi}\int_0^{2\pi} \log^+|f(re^{i\phi})|d\phi - \frac{1}{2\pi}\int_0^{2\pi} N(r, e^{i\theta})d\theta + N(r, \infty) \\
&= T(r,f) - \frac{1}{2\pi}\int_0^{2\pi} N(r, e^{i\theta})d\theta,
\end{aligned}$$

where the interchange of the order of integration is justified by the absolute convergence of the double integral. Hence

$$T(r,f) = \frac{1}{2\pi}\int_0^{2\pi} N(r, e^{i\theta})d\theta + \log^+|f(0)|,$$

proving the theorem.[18] □

---

18 If $f(0) = \infty$, then the constant term is replaced by $\log^+|c|$, where c is the first nonvanishing term of the Laurent series.

As a consequence, we next show that the mean value of the quantity $m(r,a)$ is in fact bounded on each circle.

3.27 Corollary. *For* $0 < r < R$ *(*$f(0) \neq 0, \infty$*),*

$$\frac{1}{2\pi}\int_0^{2\pi} m(r,e^{i\theta})d\theta \leq \log 2.$$

Proof. From the rephrased first fundamental theorem of Remark 3.20 for $a = e^{i\theta}$, we have

$$T(r,f) = m(r,e^{i\theta}) + N(r,e^{i\theta}) + \log|f(0) - e^{i\theta}| - \varepsilon(r,\theta),$$

where $|\varepsilon(r,\theta)| \leq \log 2$. Integrating both sides with respect to θ implies by the theorem and eq. (3.22)

$$T(r,f) = \frac{1}{2\pi}\int_0^{2\pi} m(r,e^{i\theta})d\theta + T(r,f) - \frac{1}{2\pi}\int_0^{2\pi} \varepsilon(r,\theta)d\theta.$$

Consequently,

$$\frac{1}{2\pi}\int_0^{2\pi} m(r,e^{i\theta})d\theta = \frac{1}{2\pi}\int_0^{2\pi} \varepsilon(r,\theta)d\theta \leq \log 2,$$

as desired. □

Similarly, the integral mean value of $m(r,a)$ over any circle will also have a suitable bound. Thus, for large values of $T(r,f)$, most of the contribution will come from $N(r,a)$.

Following Nevanlinna, an interesting geometric property arises as a consequence of Cartan's theorem 3.26 involving the integral on the right, namely

$$\frac{1}{2\pi}\int_0^{2\pi} N(r,e^{i\theta})d\theta = \frac{1}{2\pi}\int_0^{2\pi}\int_0^{r} \frac{n(t,e^{i\theta})}{t}dtd\theta$$
$$= \frac{1}{2\pi}\int_0^{r}\frac{dt}{t}\int_0^{2\pi} n(t,e^{i\theta})d\theta.$$

Differentiating the Cartan result, we then obtain

$$\frac{d}{d\log r}T(r,f) = r\frac{d}{dr}T(r,f) = \frac{1}{2\pi}\int_0^{2\pi} n(r,e^{i\theta})d\theta,$$

which shows that $T(r,f)$ is an increasing convex function of $\log r$. Furthermore, the integral $\int_0^{2\pi} n(r, e^{i\theta})d\theta$ represents the total $L(r)$ of all the lengths of the arcs on the unit circle $|w| = 1$ that are covered by the values $w = f(z)$ for $|z| \le r$, with each arc counted according to its multiplicity. As a consequence,

$$\frac{d}{d\log r}T(r,f) = \frac{1}{2\pi}L(r).$$

As an example, taking $f(z) = e^z$, by Exercise 3.22 we have (i) $T(r,f) = \frac{r}{\pi}$, so that $\frac{d}{d\log r}T(r,f) = \frac{r}{\pi}$. The mapping $w = e^z = e^x e^{iy}$ maps each line segment $z = iy$ for $k2\pi \le y < (k+1)2\pi$, k an integer, onto the unit circle $|w| = 1$. Therefore the disk $|z| \le r$ contains two line segments $[0, r]$ and $[0, -r]$ on the y-axis, each of which is mapped once onto an arc on $|w| = 1$ having length r. Then $L(r) = 2r$ and $\frac{1}{2\pi}L(r) = \frac{r}{\pi}$ as per the formula.

Ahlfors–Shimizu characteristic

A more geometric interpretation of the characteristic function was independently given by Ahlfors (1929) and Shimizu (1929). Indeed, we can define the *Ahlfors–Shimizu characteristic* for a meromorphic function $f(z)$ in $|z| \le r$ as

$$T_0(r,f) = \int_0^r \frac{S(t)}{t}dt,$$

where $S(t) = \frac{1}{\pi}A(t)$ is the area function (3.11). It can be viewed as the average of the spherical area of the image disk under $w = f(z)$. The relation to the Nevanlinna characteristic is

$$|T(r,f) - T_0(r,f)| \le \log\sqrt{2} + \log^+|f(0)|,$$

so that the two characteristic functions can be treated as essentially the same.

We next demonstrate an application of the characteristic $T_0(r,f)$ to a nonconstant meromorphic function defined on $\mathbb{C}$. Clearly, $S(r)$ is positive increasing for all $r > 0$. Choosing $0 < r_0 < r$ implies $S(r_0) < S(r)$, so that

$$T_0(r,f) = \int_0^r \frac{S(t)}{t}dt > \int_{r_0}^r \frac{S(t)}{t}dt > S(r_0)\log\frac{r}{r_0},$$

which implies

$$\varliminf_{r\to\infty} \frac{T_0(r,f)}{\log r} > 0.$$

3.28 Theorem. *If $f(z)$ is meromorphic in $\mathbb{C}$ and the characteristic $T_0(r,f)$ satisfies*

$$\varliminf_{r\to\infty} \frac{T_0(r,f)}{\log r} = 0,$$

then $f \equiv$ constant in $\mathbb{C}$.

Note that this means that if $T_0(r,f)$ remains bounded in $\mathbb{C}$, then $f(z)$ reduces to a constant, an analogue of Liouville's theorem. Therefore in $\mathbb{C}$, for a nonconstant meromorphic function, $T(r,f) \to \infty$ as $r \to \infty$.

Functions of bounded characteristic

However, the functions $f(z)$ that are meromorphic in $|z| < 1$ and satisfy

$$\lim_{r\to 1} T(r,f) < \infty$$

are said to have *bounded characteristic*. Since $T(r,f)$ is an increasing function, the limit as $r \to 1$ exists (possibly, infinite). Note that

$$\log^+(x_1 x_2) \le \log^+(x_1) + \log^+(x_2),$$

so that

$$m(r, f_1 f_2) \le m(r, f_1) + m(r, f_2).$$

Furthermore, since the order of a pole of $f = f_1 f_2$ at a point b is at most the sum of the orders of the poles of f_1 and f_2 at b, it follows that

$$N(r, f_1 f_2) \le N(r, f_1) + N(r, f_2),$$

and, consequently,

$$T(r, f_1 f_2) \le T(r, f_1) + T(r, f_2).$$

We then conclude that if $f_1(z)$ and $f_2(z)$ are meromorphic in $|z| < 1$ and have bounded characteristic, then the product has bounded characteristic, and if $f_2(z) \not\equiv 0$, then by formula (3.18)

$$T\left(r, \frac{f_1}{f_2}\right) \le T(r, f_1) + T\left(r, \frac{1}{f_2}\right) \le T(r, f_1) + T(r, f_2) + O(1),$$

so that the quotient has bounded characteristic.

In this context, we can say even more.

3.29 Theorem. *If $f(z)$ has bounded characteristic, then it is the quotient of two bounded analytic functions, and, conversely, the quotient of two bounded analytic functions has bounded characteristic.*

Although the proof due to Nevanlinna[19] is too technical to reproduce here, we can take some preliminary steps, which are of interest in their own right.

Suppose that $f(z)$ is meromorphic in the unit disk U and has bounded characteristic, $f(0) \neq 0, \infty$. What can be said about how the zeros $\{a_k\}$ and poles $\{b_j\}$ (counted according to multiplicity) of $f(z)$ converge to $|z| = 1$? Indeed, invoking the Stieltjes integral and integration by parts, we find

$$\sum_j (1 - |b_j|) = \sum_j (1 - t_j) = \int_0^1 (1 - t)dn(t, \infty) = \int_0^1 n(t, \infty)dt$$
$$= \lim_{r \to 1} \int_0^r n(t, \infty)dt \leq \lim_{r \to 1} N(r, \infty) \leq T(1, f) < \infty.$$

Likewise, $\sum_k (1 - |a_k|) < \infty$. As a consequence, we can construct the Blaschke products for both the zeros and poles:

$$B_1(z) = \prod_k \frac{|a_k|}{a_k} \left(\frac{a_k - z}{1 - \overline{a_k} z} \right), \quad B_2(z) = \prod_j \frac{|b_j|}{b_j} \left(\frac{b_j - z}{1 - \overline{b_j} z} \right),$$

so that

$$g(z) = \frac{B_2(z)}{B_1(z)} f(z)$$

is a nonzero analytic function in U having bounded characteristic since both B_1, B_2 are bounded. If $f(0) = 0, \infty$ and $f(z)$ has a zero or pole of order m at the origin, then we accordingly consider the function $z^{-m} f(z)$ as above. We conclude that in all cases,

$$g(z) = z^{-m} \frac{B_2(z)}{B_1(z)} f(z)$$

is a nonzero analytic function in U, and hence we may write $g(z) = e^{h(z)}$ for $h(z)$ analytic in U.

3.30 Proposition. *If $f(z)$ is a meromorphic function of bounded characteristic, then*

$$f(z) = z^m \frac{B_1(z)}{B_2(z)} e^{h(z)},$$

where m is an integer, B_1 and B_2 are Blaschke products, and $h(z)$ is analytic in U.

19 Le théoreme de Picard–Borel et la théorie des fonctions méromorphes, Gauthier-Villars, Paris, 1929.

3.31 Exercise. If $h = u + iv$ is the function of the preceding proposition, then show that $u \in h^1(U)$. See Definition 7.41.

Second fundamental theorem

To proceed further, we need one additional quantity to take into account the multiple roots of our function $f(z)$. Let $\overline{n}(t,a)$ = number of distinct roots of $f(z) - a = 0$ in $|z| \le t$ and define

$$\overline{N}(r,a) = \overline{N}(r,a,f) = \int_0^r \frac{\overline{n}(t,a) - \overline{n}(0,a)}{t} dt + \overline{n}(0,a) \log r.$$

We will consider a meromorphic function $f(z)$ in $|z| < R \le \infty$ with its characteristic $T(r,f) \to \infty$ as $r \to R$. Denote the normalized quantities

$$\delta(a) = \delta(a,f) = \varliminf_{r\to R} \frac{m(r,a)}{T(r,f)} = 1 - \varlimsup_{r\to R} \frac{N(r,a)}{T(r,f)},$$

$$\theta(a) = \theta(a,f) = \varliminf_{r\to R} \frac{N(r,a) - \overline{N}(r,a)}{T(r,f)},$$

$$\Theta(a) = \Theta(a,f) = 1 - \varlimsup_{r\to R} \frac{\overline{N}(r,a)}{T(r,f)},$$

all of which satisfy $0 \le \delta(a), \theta(a), \Theta(a) \le 1$. The quantity $\delta(a)$ is called the *deficiency* of the value a and gives a measure of the density of the points where $f(z) = a$. If the number of such a-points is small, then the deficiency $\delta(a)$ is relatively large. The quantity $\theta(a)$ is the *index of multiplicity* (*ramification index*) and is a measure of the density of the multiple roots of $f(z) = a$.

We can easily show that $\delta(a) + \theta(a) \le \Theta(a)$. In fact, let $\varepsilon > 0$ be such that for r sufficiently near to R,

$$N(r,a) < (1 - \delta(a) + \varepsilon)T(r,f),$$
$$N(r,a) - \overline{N}(r,a) > (\theta(a) - \varepsilon)T(r,f),$$

implying that

$$\overline{N}(r,a) < (1 - \delta(a) - \theta(a) + 2\varepsilon)T(r,f).$$

We conclude that

$$\delta(a) + \theta(a) \le 1 - \varlimsup_{r\to R} \frac{\overline{N}(r,a)}{T(r,f)} = \Theta(a).$$

This leads us to the renowned result.[20]

3.32 Second fundamental theorem. *If $f \not\equiv$ constant and $R = \infty$, then the set of values $a \in \hat{\mathbb{C}}$ for which $\Theta(a) > 0$ is countable, and*

$$\sum_a (\delta(a) + \theta(a)) \le \sum_a \Theta(a) \le 2. \tag{3.23}$$

The same holds if $R < \infty$ and

$$\overline{\lim_{r\to R}} \frac{T(r,f)}{\log \frac{1}{R-r}} = \infty.$$

This version of the second fundamental theorem can actually be proved from a much more technical version, which bears the same name and which will not be discussed.[21] See Hayman (1964), Nevanlinna (1970), or Rubel and Colliander (1996) for the intricate details. The number 2 is sharp.

As an example, we have a simple proof of Picard's little theorem for meromorphic functions. Indeed, if $f(z)$ is nonconstant meromorphic in $\mathbb{C}$ and $f(z) \neq a \in \hat{\mathbb{C}}$, then $N(r,a) = 0$ and $\delta(a) = 1$. Since $\theta(a) \ge 0$, there can be no more than two omitted values. More generally, the same conclusion holds if $N(r,a) = o(T(r,f))$ as $r \to R$ for $a \in \hat{\mathbb{C}}$.

3.33 Corollary. *Suppose that $f(z)$ is a nonconstant meromorphic function in $\mathbb{C}$ such that all the roots of $f(z) = a_\nu$ have multiplicities at least $m_\nu \ge 2$. Then*

$$\sum_\nu \left(1 - \frac{1}{m_\nu}\right) \le 2. \tag{3.24}$$

Proof. We maintain that

$$\overline{N}(r,a_\nu) \le \frac{1}{m_\nu} N(r,a_\nu) \le \frac{1}{m_\nu} T(r,f) + O(1),$$

and therefore $\Theta(a_\nu) \ge 1 - \frac{1}{m_\nu}$ $(\ge \frac{1}{2})$. Then the result is a consequence of (3.23). □

---

20 The eminent mathematical physicist Herman Weyl has stated (1943) on Nevanlinna's publication in 1925: "The appearance of this paper has been one of the few great mathematical events of our century."

21 **Second fundamental theorem:** If *$f(z)$ is a nonconstant meromorphic function defined in $|z| < R \le \infty$ and $a_1, a_2, \dots, a_q$, $q > 2$, are mutually distinct points (finite or infinite), then for $0 \le r < R$,*

$$m(r,f) + \sum_{i=1}^{q} m(r,a_i) \le 2T(r,f) - N_1(r) + S(r),$$

where $N_1(r)$ is a term related to the number of multiple roots, and $S(r)$ is an inconsequential error term. Generally speaking, any sum $\sum m(r,a_i)$ cannot be much larger than $2T(r,f)$.

3.34 Example. (i) Consider $f(z) = e^z$. Then $\delta(0) = \delta(\infty) = 1$ and $\delta(a) = 0$ for all $a \neq 0, \infty$, since $m(r, a) = O(1)$. As $\theta(a) = 0$ for all $a \in \hat{\mathbb{C}}$, we have $\sum_a (\delta(a) + \theta(a)) = 2$.

(ii) If $f(z) = \cos z$, then $\delta(\infty) = 1$, but $\delta(a) = 0$ for $a \neq \infty$. As well, $\theta(a) = 0$ except for $a = 1, -1$, in which case $\theta(1) = \theta(-1) = \frac{1}{2}$. More generally, for an entire function $f(z)$, there can be at most two points that are roots of $f(z) = a$ having multiplicity greater than 1. All other roots must be simple.

3.35 Remark. As in the corollary, if all the roots of $f(z) = a_\nu$ are multiple roots, then a_ν is called *totally ramified*, and $\Theta(a_\nu) \geq \frac{1}{2}$. In view of inequality (3.24), *a meromorphic function can have at most four totally ramified values* a_ν, which is quite an extraordinary result. We have already encountered a function with exactly four values, each of multiplicity two, namely the Weierstrass $\wp$-function.

3.36 Example. Let a, b, c be distinct real numbers, and let m, n, p be three positive integers such that

$$\frac{1}{m} + \frac{1}{n} + \frac{1}{p} = 1.$$

Then the Schwarz–Christoffel mapping[22]

$$z = z(w) = \int_0^w (t-a)^{\frac{1}{m}-1}(t-b)^{\frac{1}{n}-1}(t-c)^{\frac{1}{p}-1}dt \tag{3.25}$$

maps the upper half-plane conformally onto a rectilinear triangle having angles $\frac{\pi}{m}, \frac{\pi}{n}, \frac{\pi}{p}$. By continuing reflecting the inverse function $w = f(z)$ by the Schwarz reflection principle in the sides of triangles the result is a doubly periodic meromorphic function in the plane.[23] Here a fundamental region in the z-plane consists of a triangle, and its reflection over one side and each value in the w-plane is taken only once in a fundamental region.

Following Hayman (1964), differentiating the function (3.25) and letting $z \to z(a)$, we get

$$\frac{dz}{dw} \sim c(w-a)^{\frac{1}{m}-1},$$

so that

$$(z(w) - z(a)) \sim cm(w-a)^{\frac{1}{m}},$$

which implies

22 Cf. Ahlfors (1979, p. 233).

23 Any triangle can tessellate the plane.

$$(w - a) \sim \left(\frac{(z(w) - z(a))}{cm} \right)^m$$

as $z \to z(a)$. We deduce that the function $f(z) - a = 0$ has a root of multiplicity m. Note that the smaller the angle at the point $\frac{\pi}{m}$, the higher the multiplicity m at the vertex $z(a)$, which is geometrically as what one expects. Similarly, the points b, c are taken with multiplicities n, p, respectively.

Regarding the function $w = f(z)$, we have $\overline{N}(r, a) = \frac{1}{m} N(r, a)$, implying that $\Theta(a) \geq 1 - \frac{1}{m}$, and, likewise, $\Theta(b) \geq 1 - \frac{1}{n}$, $\Theta(c) \geq 1 - \frac{1}{p}$. In view of the constraint on m, n, p, we deduce that $\Theta(a) + \Theta(b) + \Theta(c) = 2$ and $\Theta(a) = 1 - \frac{1}{m}$, $\Theta(b) = 1 - \frac{1}{n}$, $\Theta(c) = 1 - \frac{1}{p}$, whence

$$\frac{1}{m} = \overline{\lim_{r \to \infty}} \frac{\overline{N}(r, a)}{T(r, f)} = \frac{1}{m} \overline{\lim_{r \to \infty}} \frac{N(r, a)}{T(r, f)},$$

which implies that $\delta(a) = 0$, and similarly $\delta(b) = 0 = \delta(c)$. Furthermore, if $\omega \neq a, b, c$, then $\Theta(\omega) = 0 = \delta(\omega) = \theta(\omega)$.

There are three cases: $m = n = p = 3$; $m = 2$, $n = p = 4$; $m = 2$, $n = 3$, $p = 6$. So, for example, in the second case, $\Theta(a) = \frac{1}{2}$ and $\Theta(b) = \Theta(c) = 3/4$, and their sum equals 2. Similarly for the other cases.

We conclude this chapter with a most elegant result.

3.37 Corollary. *Let $f(z)$ be meromorphic in $\mathbb{C}$ such that (i) all the zeros of $f(z)$ have multiplicity $\geq h$, (ii) all the poles have multiplicity $\geq k$, and (iii) all the zeros of $f(z) - 1$ have multiplicity $\geq \ell$. If*

$$\frac{1}{h} + \frac{1}{k} + \frac{1}{\ell} < 1,$$

then $f(z) \equiv$ constant.

Indeed, if $f(z)$ is nonconstant, then

$$\left(1 - \frac{1}{h}\right) + \left(1 - \frac{1}{k}\right) + \left(1 - \frac{1}{\ell}\right) > 2,$$

violating inequality (3.24).

4 Normal families

Analytic functions

The famous Bolzano–Weierstrass theorem states that every bounded sequence of real numbers has a convergent subsequence. In the late 19th century, attempts were made to apply this concept to sequences of continuous functions. One version of the work arising from this period can be formulated as follows.

4.1 Arzelà–Ascoli theorem.[1] *Let X be a compact metric space, and let $\{f_n\}$ be a sequence of continuous functions defined on X. Then there exists a subsequence of $\{f_n\}$ that converges uniformly on X to a continuous function f if and only if $\{f_n\}$ is uniformly bounded and equicontinuous on X.*

Normality/compactness

The roots of the theory of normal families developed in the early 20th century by Paul Montel (1876–1975) lies within this theorem. From his seminal paper in 1907 Montel over several decades developed the notion of a normal family into a formidable tool in complex function theory (Montel 1927).

4.2 Definition. A family of analytic functions $\mathcal{F}$ defined on a domain Ω in the complex plane is said to be *normal* in Ω, or a *normal family in* Ω, if every sequence $\{f_n\} \subseteq \mathcal{F}$ contains a subsequence that converges uniformly on compact subsets of Ω either to a limit function f or to ∞. Furthermore, a family $\mathcal{F}$ is *normal at a point* $z_0 \in \Omega$ if it is normal in some neighborhood of z_0.

It is clear by the Weierstrass Theorem 1.27 that any such limit function $f \not\equiv \infty$ is analytic in Ω. In the latter case, if $f_n \to \infty$ uniformly on compact subsets, it is understood that for any compact subset $K \subset \Omega$ and any constant $M > 0$, there is a number $n_0 = n_0(K, M)$ such that $n \geq n_0$ implies that $|f_n(z)| > M$ for all $z \in K$. Note that there are sequences $\{f_n\}$ for which both alternatives occur.

4.3 Remark. It is evident that if a family of analytic functions $\mathcal{F}$ is normal in a domain Ω, then it is normal at each of its points. The converse is subsumed under the same result for meromorphic functions (Theorem 4.29).

Again, consistent with sequences of real numbers, we say that a normal family $\mathcal{F}$ is *compact* if the limit of every convergent subsequence belongs to $\mathcal{F}$.

---

1 C. Arzelà, Sulle funzioni di line, *Mem. Accad. Sci. Bologna* (5) 5 (1895), 225–244, and Ascoli *op. cit.*

https://doi.org/10.1515/9783110757828-004

Montel theorem

With regard to the Arzelà–Ascoli theorem, it was Montel's inspiration that regarding analytic functions, local boundedness was in itself sufficient to imply that any sequence had a uniformly convergent subsequence since by Theorem 1.36 locally bounded sequences of analytic functions are already equicontinuous on compact subsets.

4.4 Theorem (Montel).[2] *A locally bounded family $\mathcal{F}$ defined on a domain Ω is normal in Ω.*

Proof. Since the complex plane is separable, let $\{z_n\}$ be a countable dense subset of Ω.[3] For any sequence of functions $\{f_n\}$ in $\mathcal{F}$ the sequence of complex numbers $\{f_n(z_1)\}$, $n = 1, 2, 3, \ldots$, is bounded by hypothesis. Therefore by the Bolzano–Weierstrass theorem this bounded sequence has a convergent subsequence

$$f_{n_1}^{(1)}(z_1), f_{n_2}^{(1)}(z_1), f_{n_3}^{(1)}(z_1), \ldots;$$

in other words, the subsequence of functions $\{f_{n_k}^{(1)}\}$ converges at the point z_1. Considering *this* subsequence evaluated at the point z_2, which is again bounded, we see that, as before, $\{f_{n_k}^{(1)}(z_2)\}$, $k = 1, 2, 3, \ldots$, has a convergent subsequence

$$f_{n_1}^{(2)}(z_2), f_{n_2}^{(2)}(z_2), f_{n_3}^{(2)}(z_2), \ldots,$$

that is, the subsequence $\{f_{n_k}^{(2)}\}$ converges at the point z_2 as well as the point z_1 being a subsequence of the preceding one.

We continue in this fashion, each time extracting a subsequence $\{f_{n_k}^{(m)}\}$ from the preceding one, which converges at the points $z_1, z_2, \ldots, z_m$ for each positive integer m. Displaying all the subsequences in the array

$$\begin{array}{c} f_{n_1}^{(1)} \, f_{n_2}^{(1)} \, f_{n_3}^{(1)} \cdots \\ f_{n_1}^{(2)} \, f_{n_2}^{(2)} \, f_{n_3}^{(2)} \cdots \\ \vdots \\ f_{n_1}^{(m)} \, f_{n_2}^{(m)} \, f_{n_3}^{(m)} \cdots \\ \vdots \end{array}$$

2 Sur les suites infinies de fonctions, *Ann. École Norm. Sup.* 24 (1907), 233–234.

3 For example, the set of all $z_n = x_n + iy_n$ in Ω where x_n and y_n are rational numbers is a countable dense subset of Ω.

it becomes clear that the diagonal sequence $\{f_{n_k}^{(k)}\}$ converges at every point of the countable dense set $\{z_n\}$. To complete the proof, we need to show that this diagonal sequence in fact converges uniformly on compact subsets of Ω.

Simplifying the notation, set $F_k = f_{n_k}^{(k)}$, let $K \subset \Omega$ be compact, and let $\varepsilon > 0$. Given that the family $\mathcal{F}$ is equicontinuous, by Theorem 1.36 there exists $\delta = \delta(K, \varepsilon) > 0$ such that

$$|F_n(z) - F_n(z')| < \frac{\varepsilon}{3} \tag{4.1}$$

for all $z, z' \in K$ such that $|z - z'| < \delta$, $n = 1, 2, 3, \ldots$. With this value of $\delta > 0$, we consider the family of open disks centered about the points of our countable dense subset $\{D(z_k, \delta)\}$ for $k = 1, 2, 3, \ldots$, which forms an open cover of K. Hence there is a finite open subcover $\bigcup_{k=1}^{k_0} D(z_k, \delta) \supset K$, where, if necessary, we have suitably relabeled the points $z_1, z_2, \ldots, z_{k_0}$.

Furthermore, since the diagonal functions $\{F_k\}$ converge at each point of the countable dense subset, there is a positive integer n_0 such that for $n, m \geq n_0$, we have

$$|F_n(z_k) - F_m(z_k)| < \frac{\varepsilon}{3} \tag{4.2}$$

for $k = 1, 2, \ldots, k_0$. Finally, we conclude that for any point $z \in K$, it belongs to some $D(z_j, \delta)$, and so by (4.1) and (4.2)

$$|F_n(z) - F_m(z)| \leq |F_n(z) - F_n(z_j)| + |F_n(z_j) - F_m(z_j)| + |F_m(z_j) - F_m(z)| < \varepsilon.$$

This establishes the uniform convergence of the diagonal sequence on the compact set K, and since the limit function must be analytic by the Weierstrass Theorem 1.27, the theorem is proved. □

Of course, if $\mathcal{F}$ is a family of analytic functions defined on a domain Ω and $|f(z)| \leq M < \infty$ for all $f \in \mathcal{F}$ and $z \in \Omega$, then $\mathcal{F}$ is obviously a normal family in Ω and compact as well.

4.5 Examples.

(i) The family of linear fractional transformations from the open unit disk U onto itself

$$\mathcal{F} = \left\{T(z) = e^{i\gamma}\frac{z - \alpha}{1 - \bar{\alpha}z} : \gamma \in \mathbb{R}, |\alpha| < 1\right\}$$

is a normal family in U, and whenever $|\alpha| \leq \beta < 1$, it forms a compact subfamily depending on β.

(ii) $\mathcal{F} = \{f_n(z) = z^n : n = 1, 2, 3, \ldots\}$ is a normal family in the domain $\{|z| > 1\}$ converging uniformly on compact subsets to ∞.

(iii) $\mathcal{F} = \{f_y\}$ for f_y analytic in U, $f_y(z) = \sum_{n=1}^{\infty} c_n^{(y)} z^n$, and $|c_n^{(y)}| \le M < \infty$ for all y and $n = 1, 2, 3, \ldots$. Then for $|z| < 1$,

$$|f_y(z)| \le \frac{M}{1 - |z|},$$

implying that the family $\mathcal{F}$ is locally bounded in U and hence normal and compact.

Although for a locally bounded family $\mathcal{F}$ of analytic functions, the corresponding family $\mathcal{F}'$ of derivatives is also locally bounded by Theorem 1.32, the same implication does not apply to the property of normality.

(iv) In the unit disk U the family of functions

$$f_n(z) = n + \frac{nz^2}{2}, \quad n = 1, 2, 3, \ldots,$$

satisfies $|f_n(z)| \ge n - \frac{n}{2} = \frac{n}{2}$ and thus forms a normal family in U, but the family of derivatives $\{nz : n = 1, 2, 3, \ldots\}$ is normal neither in U nor in fact in any neighborhood of the origin.

(v) Consider the family of analytic functions $\mathcal{F}$ in U whose characteristic function is bounded, that is, $T(r, f) = \frac{1}{2\pi} \int_0^{2\pi} \log^+ |f(re^{i\phi})| d\phi \le M < \infty$, $0 < r < 1$, for all $f \in \mathcal{F}$. Then $\mathcal{F}$ is locally bounded by Theorem 3.23 and hence normal and compact.

The other implication of the Arzelà–Ascoli theorem can also be dealt with in our present context of analytic functions.

4.6 Theorem. *Let $\mathcal{F}$ be a family of analytic functions defined on a domain Ω. If every sequence in $\mathcal{F}$ has a convergent subsequence converging uniformly on compact subsets to an analytic function f, then $\mathcal{F}$ is locally bounded and therefore equicontinuous on compact subsets of Ω.*

Proof. On the contrary, suppose that $\mathcal{F}$ is not locally bounded. Then there is a closed disk $\overline{D}$ in Ω and a point $z_n \in \overline{D}$ such that for some function $f_n \in \mathcal{F}$,

$$|f_n(z_n)| > n, \quad n = 1, 2, 3 \ldots.$$

Then for some analytic function f defined on Ω, a subsequence $f_{n_k} \to f$ uniformly on $\overline{D}$. In particular,

$$|f_{n_k}(z) - f(z)| < 1$$

for all $z \in \overline{D}$, so that if $M = \max_{z \in \overline{D}} |f(z)|$, then

$$|f_{n_k}(z)| < 1 + M.$$

However, this contradicts the fact that $|f_{n_k}(z_{n_k})| \to \infty$ as $k \to \infty$. □

4.7 Exercise. Let $f(z)$ be a nonvanishing analytic function defined on a domain Ω, and let $\mathcal{F} = \{\alpha f(z) : \alpha \in \mathbb{R}^+\}$. Prove that $\mathcal{F}$ is a normal family but neither locally bounded nor equicontinuous in Ω.

The issue arising in Exercise 4.7 can be circumvented by an additional condition.

4.8 Corollary. *Let $\mathcal{F}$ be a normal family of analytic functions defined on a domain Ω such that $|f(z_0)| \leq M < \infty$ for some point $z_0 \in \Omega$ and for all $f \in \mathcal{F}$. Then $\mathcal{F}$ is locally bounded and equicontinuous on compact subsets of Ω.*

One of the cornerstones of complex function theory is the Riemann mapping theorem. The original flawed proof given by Riemann in his inaugural dissertation of 1851 was based on the so-called Dirichlet principle, which is not always valid and which will be discussed further in Chapter 7. The first valid proof was given by W. F. Osgood in 1900, who overcame the shortcomings of Riemann's proof.[4] The proof given here is based on a normal family argument and uses the compactness of a certain family that admits an extremal function (cf. the discussion of a continuous functional in Chapter 7). The basic ideas for the proof are due to Paul Koebe.

4.9 Riemann mapping theorem. *Let Ω be a simply connected domain that is not the whole plane. Then for each $z_0 \in \Omega$ there is a unique analytic univalent function $w = f(z)$ mapping Ω onto the open unit disk U such that $f(z_0) = 0$ and $f'(z_0) > 0$.*

Proof. Let us consider the family of analytic univalent functions $\mathcal{F}$ defined on Ω satisfying the following conditions:

$$|f(z)| \leq 1,$$
$$f(z_0) = 0,$$
$$f'(z_0) \geq \alpha > 0,$$

where α is yet to be determined. The first question we must address is whether $\mathcal{F} \neq \emptyset$. By the hypothesis there is a finite point $a \notin \Omega$. Therefore Ω being simply connected implies that we can define (in view of the monodromy theorem) a single-valued analytic branch $g(z)$ of $\sqrt{z-a}$ in Ω. In this instance, $g(z)$ is univalent and maps Ω onto a domain $\tilde{\Omega}$ with $g(z_0) = w_0 \in \tilde{\Omega}$. Since $g(z)$ cannot take any value $w \in \tilde{\Omega}$ as well as $-w$ (why not?), then for some disk $D(w_0, \rho_0) \subseteq \tilde{\Omega}$, the disk $D(-w_0, \rho_0)$ lies exterior to $\tilde{\Omega}$, which implies that

$$|g(z) + w_0| \geq \rho_0$$

for all $z \in \Omega$. Since $g(z_0) = w_0$, we obtain $|w_0| \geq \rho_0/2$.

---

4 See J. L. Walsh, History of the Riemann Mapping Theorem, *Amer. Math. Monthly*, 80, (1973), 270–276; also R.E. Greene and K.-T. Kim, The Riemann mapping theorem from Riemann's viewpoint, Springer Open Access DOI 10.1186/s40627-016-0009-7, where Riemann's original idea has been rehibilitated.

Next, we consider the function that exactly fulfills our requirements to belong to the family $\mathcal{F}$, namely

$$h(z) = \frac{\rho_0}{4}\frac{|g'(z_0)|}{g'(z_0)}\frac{w_0}{|w_0|^2}\frac{g(z)-w_0}{g(z)+w_0}.$$

To see this, by our two preceding inequalities involving w_0 we have

$$\left|h(z)\right| = \frac{\rho_0}{4|w_0|}\left|\frac{g(z)-w_0}{g(z)+w_0}\right| = \frac{\rho_0}{4}\left|\frac{1}{w_0} - \frac{2}{g(z)+w_0}\right| \le 1$$

for $z \in \Omega$, $h(z_0) = 0$, and $h'(z_0) = \frac{\rho_0}{8}\frac{|g'(z_0)|}{|w_0|^2} > 0$. Therefore $h \in \mathcal{F}$ by taking $\alpha = \frac{\rho_0}{8}\frac{|g'(z_0)|}{|w_0|^2}$ since $h(z)$ is also analytic and univalent in Ω, establishing that the family $\mathcal{F}$ is nonempty.

The Riemann mapping function will be the function in $\mathcal{F}$ with maximal derivative at the point z_0 (which is mapped to the origin). To show that such a function exists, let[5]

$$\beta = \sup_{h\in\mathcal{F}} h'(z_0) \le \infty.$$

Then there is a sequence of functions $h_n \in \mathcal{F}$ with $\lim_{n\to\infty} h_n'(z_0) = \beta$. As each $h_n \in \mathcal{F}$ satisfies $|h_n(z)| \le 1$, $\mathcal{F}$ is normal in Ω, so that we can extract a subsequence h_{n_k} that converges uniformly on compact subsets to an analytic function $f(z)$. Hence $|f(z)| \le 1$, $f(z_0) = 0$, and by Corollary 1.29 the function $f(z)$ is univalent. By the Cauchy integral formula for derivatives, $\lim_{k\to\infty} h_{n_k}'(z_0) = f'(z_0)$, and, consequently, $f'(z_0) \ge \alpha > 0$. Therefore $f \in \mathcal{F}$, and most importantly, $f'(z_0) = \beta = \sup_{h\in\mathcal{F}} h'(z_0) < \infty$, illustrating the elegance of this normal family argument.

Finally, to complete the proof, it remains to show that $f(z) : \Omega \to U$ surjectively. So let us suppose that there is a point $\zeta_0 \in U$ such that $f(z) \ne \zeta_0$ for all $z \in \Omega$. As above, we can find a single-valued analytic branch in Ω of the function

$$G(z) = \sqrt{\frac{f(z)-\zeta_0}{1-\overline{\zeta_0}f(z)}},$$

where $G(z)$ is univalent, and $|G(z)| \le 1$. Then the function

$$H(z) = \frac{|G'(z_0)|}{G(z_0)}\frac{G(z)-G(z_0)}{1-\overline{G(z_0)}G(z)}$$

is also analytic univalent in Ω satisfying $|H(z)| \le 1$ and $H(z_0) = 0$. In addition,

5 See the analogous discussion on continuous functionals in Chapter 7.

$$H'(z_0) = \frac{|G'(z_0)|}{1-|G(z_0)|^2} = \frac{1+|\zeta_0|}{2\sqrt{|\zeta_0|}} f'(z_0)$$

since $f \in \mathcal{F}$. On the other hand, as $|\zeta_0| < 1$,

$$1+|\zeta_0| = \left(1-\sqrt{|\zeta_0|}\right)^2 + 2\sqrt{|\zeta_0|} > 2\sqrt{|\zeta_0|},$$

implying that $H'(z_0) > f'(z_0)$, in violation of the maximal property of $f'(z_0)$. Thus the function $f(z) : \Omega \to U$ is the desired Riemann mapping function.

To establish the uniqueness, if there were two such functions $f_1(z)$ and $f_2(z)$, then the function

$$g(w) = f_1 \circ f_2^{-1}(w)$$

would be an analytic univalent mapping of U onto U with $g(0) = 0$. Applying the Schwarz lemma, we find that $|f_1(z)| \leq |f_2(z)|$, and reversing $f_1(z)$ and $f_2(z)$, we obtain $|f_2(z)| \leq |f_1(z)|$, that is, $|f_1(z)| \equiv |f_2(z)|$, $z \in U$. Since the function $k(z) = f_1(z)/f_2(z)$ is analytic in U with constant modulus, we have $k(z) \equiv c$ with $|c| = 1$. By the positivity of the derivatives, $c = 1$, and $f_1(z) \equiv f_2(z)$, concluding the proof of the Riemann mapping theorem. □

Of course, there is nothing in the theorem to say if there is any correspondence via the Riemann mapping function between the boundary of Ω and the boundary of U, and we should not expect a continuous extension of the mapping function to all of the extreme cases possible for $\partial\Omega$. On the other hand, *if $\partial\Omega$ is a simple closed contour, then the mapping $f(z) : \Omega \to U$ has a continuous one-to-one extension to the boundary.*[6]

Moreover, we cannot expect to obtain a similar result for multiply connected domains, but there is the following:[7]

4.10 Theorem. *For a domain Ω bounded by a finite number of analytic contours $\Gamma_1, \ldots, \Gamma_n$, there exists a univalent function ϕ_k mapping Ω onto the unit disk U with $n-1$ circular slits such that Γ_k is mapped into $|w| = 1$.*

In general, a sequence of functions that converge pointwise in a domain does not converge uniformly on compact subsets. For example, the functions $f_n(z) = nz$, $n = 1, 2, 3, \ldots$, do not converge uniformly on any disk containing the origin. But in the case of a locally bounded family of analytic functions, we have a very strong conclusion in the following scenario.

6 Cf. Nehari (1952), p. 179 or Collingwood and Lohwater (1966), p. 49.

7 Nehari (1952), p. 336.

4.11 Vitali–Porter theorem.[8] *Suppose that* $\{f_n\}$ *is a locally bounded sequence of analytic functions on a domain* Ω. *If for each point* z *of a set* $S \subseteq \Omega$, $\lim_{n\to\infty} f_n(z) = f(z)$ *exists and* S *has a point of accumulation in* Ω, *then the sequence* $\{f_n\}$ *converges uniformly on compact subsets of* Ω *to an analytic function.*

Our elementary proof uses a normal family argument. By the hypotheses we can select a normally convergent subsequence $\{f_{n_k}\}$ such that $\lim_{k\to\infty} f_{n_k}(z) = f(z)$ for all $z \in S$. Assuming that the original sequence $\{f_n\}$ does not converge uniformly on compact subsets would mean that for some compact set $K \subset \Omega$ and some $\varepsilon > 0$, there is a subsequence $\{f_{m_i}\}$ and points $z_i \in K$ satisfying

$$|f_{m_i}(z_i) - f(z_i)| \geq \varepsilon \tag{4.3}$$

for $i = 1, 2, 3, \ldots$. Since this subsequence is locally bounded, it also has a subsequence that converges uniformly on compact subsets of Ω to an analytic function, say g. By (4.3) $f \not\equiv g$, but, on the other hand, f and g agree on the set S, so that $f \equiv g$ on Ω by the identity theorem for analytic functions. This contradiction proves the theorem.

Note that the normality of the sequence $\{f_n\}$ is all that is required to obtain the consequence of the Vitali–Porter theorem.

In the same spirit of the Vitali–Porter theorem, there is an interesting result established by Blaschke (1915).

4.12 Theorem. *Let* $\{f_n\}$ *be a uniformly bounded sequence of analytic functions in* U *such that* $\{f_n\}$ *converges at each of the points* $a_n \in U$, $n = 1, 2, 3, \ldots$, *satisfying* $\sum_{n=1}^{\infty}(1 - |a_n|) = \infty$. *Then* $\{f_n\}$ *converges uniformly on compact subsets of* U *to an analytic function.*

Proof. Since $\{f_n\}$ is bounded and hence normal, there is a subsequence that converges to a bounded limit function, say f. If we assume that the sequence itself does not converge uniformly on compact subsets to f, then there must be another bounded analytic limit function, say g, such that $f \not\equiv g$. However, $f(a_n) - g(a_n) = 0$, and Corollary 1.25 implies that $f - g \equiv 0$. □

Fundamental normality test

One of the very core results in the theory of normal families is that a family $\mathcal{F}$ of analytic functions is normal if it omits two distinct fixed values a and b in $\mathbb{C}$. Of course, they can be transferred to the points 0 and 1 by simply considering the associated family

---

8 G. Vitali, Sopra le serie di funzioni analitiche, *Rend. Della R. Inst. Lombardo di Sci. Lett.* 36 (1903), 772–774.
M. B. Porter, Concerning series of analytic functions, *Ann. Math.* 6 (1904–05), 190–192.

$$\mathcal{G} = \left\{ g(z) = \frac{f(z) - a}{b - a} : f \in \mathcal{F} \right\},$$

which omits the values 0 and 1, and $\mathcal{G}$ is normal if and only if $\mathcal{F}$ is normal.

The original proof by Montel (1912) used the elliptic modular function and was quite technical. Our deferred proof (Theorem 4.34) relies on more modern developments.

4.13 Fundamental normality test. *Let $\mathcal{F}$ be a family of analytic functions defined on a domain Ω such that for two distinct finite values a and b, $f(z) \neq a$ and $f(z) \neq b$ for all $f \in \mathcal{F}$. Then $\mathcal{F}$ is normal in Ω.*

At this juncture, let us just mention that there is something of deep significance going on here recalling that an entire function that omits two values is a constant. We will explore this somewhat distinctive connection in the next section.

The normal family notion can be employed to make light work of proving a number of difficult theorems. Montel's key idea was to replace a specified property of a single function by the same property of a family of functions, which then become a normal family. The following result, sometimes referred to as Montel's theorem,[9] exemplifies this idea and is in the same spirit as the Phragmén–Lindelöf theorem.

4.14 Theorem. *Let $f(z)$ be a bounded analytic function in the half-strip $S : \{a < x < b, y > 0\}$. If $\lim_{y\to\infty} f(\xi + iy) = \ell$ for some $a < \xi < b$, then $f \to \ell$ uniformly as $y \to \infty$ in the strip $a + \delta \le x \le b - \delta$.*

Proof. For $z = x + iy \in S$, define the sequence of functions

$$f_n(z) = f(z + in)$$

for $n = 1, 2, 3, \ldots$ in the open rectangle $R_0 : a < x < b, 0 < y < 2$. Since $f(z)$ is bounded, so is the family $\{f_n\}$, and hence it is normal in R_0. Moreover, $\lim_{n\to\infty} f_n(z) = \ell$ at each point $z = \xi + iy$ means that we can apply the Vitali–Porter theorem to conclude that $f_n \to \ell$ uniformly on the closed rectangle

$$R : a + \delta \le x \le b - \delta, \quad \frac{1}{3} \le y \le \frac{5}{3}.$$

Consequently, given $\varepsilon > 0$, there exists $n_0 = n_0(\varepsilon, R) > 0$ such that

$$|f_n(z) - \ell| < \varepsilon$$

for all $n \ge n_0$ and $z \in R$. It follows that

9 Sur les familles de fonctions analytiques qui admettent des valuers exceptionelles dans un domaine, *Ann. École Norm. Sup.* 23 (1912), 487–435.

$$\left|f(z+in)-\ell\right|<\varepsilon$$

for all $n \geq n_0$ and $a+\delta \leq x \leq b-\delta$, proving the theorem. □

The cornerstone of the theory of entire functions as stated in Chapter 2 is the following:

4.15 Picard little (first) theorem.[10] *A nonconstant entire function takes every complex value with one possible exception.*

Note that the function $w = e^z$ is entire and omits the value $w = 0$. A proof along the lines of Picard's original was given in Chapter 3, but it uses the elliptic modular function, so in some sense the proof that follows is considered more "elementary."

Proof of the theorem. Let us assume that an entire function $f(z)$ omits two distinct values, say a and b in the complex plane. Define a sequence of open disks $D_n : |z| < 2^n$, $n = 0, 1, 2, \ldots$, and define the sequence of functions

$$f_n(z) = f(2^n z),$$

which are necessarily entire. Observe that $f_n(D_1) = f(D_{n+1})$ for $n = 0, 1, 2\ldots$, and, consequently, the sequence $\{f_n\}$ omits the values a and b in the disk D_1. By the fundamental normality test the sequence $\{f_n\}$ is a normal family. Moreover, as $f_n(0) = f(0)$, $n = 0, 1, 2\ldots$, Corollary 4.8 implies that the family $\{f_n\}$ is uniformly bounded on the compact subset set $\overline{D_0}$ of D_1. Therefore the function $f(z)$ is bounded in $\mathbb{C}$, and from Liouville's theorem we conclude that $f(z)$ is constant. This proves the theorem.

4.16 Exercise. Let $\mathcal{F}$ be a family of analytic functions on a domain Ω such that $|f(z)| < M < \infty$ for all $f \in \mathcal{F}$ and $z \in \Omega$. Let $\{f_n\}$ be a sequence in $\mathcal{F}$ such that $f_n(z_0) \to M$ as $n \to \infty$ for some $z_0 \in \Omega$. Show that $\{f_n\}$ has a subsequence that converges to a constant uniformly on compact subsets of Ω.

The year 1879 was an *annus mirabilis* for Émile Picard (1856–1941)[11] when he proved his remarkable second theorem.

4.17 Picard great (second) theorem.[12] *If an analytic function $f(z)$ has an essential singularity at a point z_0, then in a punctured disk about z_0, $f(z)$ infinitely often takes every complex value with at most one possible exception.*

---

10 Sur une propriété des fonctions entières, *C. R. Acad. Sci. Paris* 88 (1879), 1024–1027.

11 Picard also authored one of the first textbooks on the theory of relativity: *La théorie de la relativité et ses applications à l'astronomie*, Gauthier-Villars, Paris, 1922.

12 Sur les fonctions analytique uniformes dans le voisinage d'un point singulier essential, *C. R. Acad. Sci. Paris* 89 (1879), 745–747.

The proof is again by Montel (1912). We can take $z_0 = 0$ (why?) and suppose that $f(z)$ omits two values a and b in the disk $D = \{0 < |z| < r\}$. In this instance, we consider the family $\mathcal{F}$ consisting of functions defined by

$$f_n(z) = f\left(\frac{z}{2^n}\right), \quad n = 1, 2, 3, \ldots .$$

Then each $f_n \in \mathcal{F}$ is analytic in the annulus $\mathcal{A} : \{\frac{r}{2} < |z| < r\}$ and omits the values a and b in $\mathcal{A}$. As $\mathcal{F}$ is normal in $\mathcal{A}$ by the fundamental normality test, we can extract a subsequence $\{f_{n_k}\}$ that converges uniformly on the compact set $|z| = \rho$ for $\frac{r}{2} < \rho < r$ either to an analytic function $g(z)$ or to $g(z) \equiv \infty$ in $\mathcal{A}$.

In the case that $g(z)$ is analytic and hence bounded on $|z| = \rho$, it follows that the subsequence $\{f_{n_k}\}$ is uniformly bounded there, that is,

$$|f_{n_k}(z)| \le M < \infty$$

for all $k = 1, 2, 3, \ldots$ and $|z| = \rho$. As a consequence, $|f(z)| \le M$ on each of the circles $|z| = \rho/2^{n_k}$, $k = 1, 2, 3, \ldots$, so that by the maximum modulus theorem, $|f(z)| \le M$ in each region between every pair of these circles. *A fortiori*, $|f(z)| \le M$ in the punctured disk $\{0 < |z| < \frac{\rho}{2^{n_1}}\}$, but this contradicts the fact that $z = 0$ is an essential singularity and hence f is unbounded in any neighborhood of the origin.

In the case that $g(z) \equiv \infty$ in $\mathcal{A}$, note that the subsequence $\{\frac{1}{f_{n_k} - a}\}$ converges to zero uniformly on compact subsets of $\mathcal{A}$. The same argument as above implies that the function $\{\frac{1}{f-a}\}$ is bounded in a deleted neighborhood of the origin and therefore $z = 0$ is either a removable singularity or a pole of the function f. Again, we have contradiction.

To conclude the proof, suppose that there are two values that are only attained a finite number of times by $f(z)$. Then the same two values would be completely omitted in some sufficiently small neighborhood of the origin by $f(z)$, but we have already established that this cannot happen, proving the theorem.

The result is best possible since the function $e^{1/z}$ has an isolated essential singularity at the origin but omits the value zero.

Our preceding analysis in fact allows us to say something more about the sequence of analytic functions defined above. Again, without loss of generality, we take the essential singularity at $z_0 = 0$. First, we require a lemma whose applicability is not particularly obvious.

4.18 Lemma. *If $f(z)$ is analytic in a punctured neighborhood of an essential singularity at $z_0 = 0$, then the sequence of functions given by $f_n(z) = f(\frac{z}{2^n})$ is not normal in some annular region about the origin.*

Proof. Assume that the sequence $\{f_n\}$ constitutes a normal family in the annulus $A = \{\frac{r}{2^3} < |z| < r\}$. Then we can extract a subsequence $\{f_{n_k}\}$ that converges uniformly on the closed annulus

$$\overline{A_0} = \left\{ \frac{r}{2^2} \le |z| \le \frac{r}{2} \right\}$$

to a limit function $g(z)$. As in the preceding proof of Picard's theorem, the function $g(z) \equiv \infty$.

Furthermore, we claim that the original sequence $\{f_n\}$ must converge uniformly on $\overline{A_0}$ to ∞. If this were not the case, then there would exist a sequence of points $z_i \in \overline{A_0}$, a positive number M, and a subsequence $\{f_{m_i}\}$ of $\{f_n\}$ satisfying

$$|f_{m_i}(z_i)| \le M$$

for $i = 1, 2, 3, \ldots$. Now the sequence $\{f_{m_i}\}$ is itself normal in A and therefore contains a subsequence that converges uniformly on $\overline{A_0}$ to ∞ as above, which is a contradiction. We conclude that $\{f_n\}$ converges uniformly to ∞ on $\overline{A_0}$; in other words, for any $M > 0$, there exists n_0 depending on M such that

$$|f_n(z)| > M, \quad z \in \overline{A_0}, \tag{4.4}$$

whenever $n \ge n_0$.

At this juncture, consider the sequence of closed annuli

$$\overline{A_n} = \left\{ \frac{r}{2^{n+2}} \le |z| \le \frac{r}{2^{n+1}} \right\}$$

for $n = 1, 2, 3, \ldots$ and note that for any $z \in \overline{A_0}$, since $f_n(z) = f(\frac{z}{2^n})$, we can write $f_n(\overline{A_0}) = f(\overline{A_n})$. Hence, in view of (4.4),

$$|f(z)| > M$$

for $0 < |z| \le \frac{r}{2^{n_0+1}}$. This means that $\lim_{z\to 0} f(z) = \infty$, and since $z = 0$ is an essential singularity, this is a contradiction. We are forced to conclude that the sequence $\{f_n\}$ as defined is not normal in the annulus A, as desired. □

This leads us to an interesting extension of Picard's great theorem due to Gaston Julia (1893–1978) that he deduced from the preceding lemma.[13]

4.19 Julia theorem. *If z_0 is an essential singularity of an analytic function $f(z)$, then there exists at least one ray emanating from z_0 given by $\arg(z - z_0) = \alpha$ such that in every sector $\alpha - \varepsilon < \arg(z - z_0) < \alpha + \varepsilon$, the function $f(z)$ takes every complex value infinitely often with at most one possible exception.*

13 Leçons sur les fonctions uniformes à point singulier essential isolé, 104, *Gauthier-Villars*, Paris, 1924. While serving in the French army, Gaston Julia lost his nose in a serious World War I injury and thereafter wore a patch over it. Nonetheless, he had a long productive life.

Proof. As in the proof of Picard's great theorem, we take the point z_0 to be the origin and for $n = 1, 2, 3, \dots$, define

$$f_n(z) = f\left(\frac{z}{2^n}\right) \tag{4.5}$$

in a punctured disk $D : \{0 < |z| < r\}$. By the preceding lemma the sequence $\{f_n\}$ is not normal at some point $z' \in D$ and hence not normal in some sufficiently small disk $D_0 : |z - z'| < \rho \subset D$. We then form the family of disks

$$D_n : \left|z - \frac{z'}{2^n}\right| < \frac{\rho}{2^n}, \quad n = 1, 2, 3, \dots,$$

so that $f_n(D_0) = f(D_n)$ by (4.5).

Suppose now that there are two values $a, b \in \mathbb{C}$ such that it is not true that at least one of them is taken in infinitely many disks D_n; that is, there is a positive integer N such that both a and b are not taken by the function $f(z)$ in D_n for $n \geq N$ or, equivalently, that a and b are not attained by the functions f_n in D_0 for $n \geq N$. This implies that the sequence $\{f_n\}$ is normal in D_0 by the fundamental normality test, and this contradiction proves the theorem. □

The ray emanating from the essential singularity z_0 that passes through the point z' is known as the *line (direction, ray) of Julia*. The theorem has an interesting counterpart for harmonic functions (Theorem 7.56) even without the presence of an essential singularity.

4.20 Corollary.[14] *If $f(z)$ is a nonpolynomial entire function, then $f(z)$ takes every complex value infinitely often with at most one exception.*

Indeed, suppose to the contrary that $f(z)$ takes the values $a, b, a \neq b$, finitely many times. Then for $r > 0$ sufficiently small, $f(z)$ does not take the values a, b at all in the complement of the disk $D(0, 1/r)$. Then the theorem implies that in the disk $0 < |z| < r$ the function $g(z) = f(1/z)$ cannot have an essential singularity at $z = 0$ by Julia's theorem. As a consequence, the Laurent series expansion for $g(z)$ can have only finitely many terms having negative powers of z. Since $f(z)$ is entire, it reduces to a polynomial, a contradiction, which establishes the result.

Another classical result, which can be established from the fundamental normality test, is the following:

4.21 Schottky theorem.[15] *Suppose that $f(z)$ is analytic in the disk $|z| < R$ such that $f(z) \neq 0$, $f(z) \neq 1$, and $f(0) = a_0$. Then for each value λ with $0 < \lambda < 1$, there is a constant $M(a_0, \lambda) > 0$ such that*

---

14 É. Picard, Memoire sur les fonctions entières, *Ann. Sci. École Norm. Sup.* (2) 9 (1880), 145–166.

15 Über den Picardschen Satz und die Borelschen Ungleichungen, *Sitz. der Preussischen Akad. der Wiss. Berlin* (1904), 1244–1263.

$$|f(z)| \le M(a_0, \lambda)$$

for $|z| \le \lambda R$.

Proof. Consider the family of functions

$$\mathcal{G} = \{g(z) = f(Rz) : z \in U\}$$

where U is the open unit disk. Then every function $g \in \mathcal{G}$ omits 0 and 1, and hence $\mathcal{G}$ is normal in U. Since $g(0) = a_0$ for each $g \in \mathcal{G}$, Corollary 4.8 implies that $\mathcal{G}$ is locally bounded, that is, $|g(z)| \le M = M(a_0, \lambda)$ on any compact subset $|z| \le \lambda < 1$. As a consequence, $|f(z)| \le M(a_0, \lambda)$ for $|z| \le \lambda R$, proving the theorem. □

More explicit bounds for Schottky's theorem have been found over the years by various authors.[16]

We can apply the Schottky theorem to prove a theorem of E. Landau (1904).[17]

4.22 Landau theorem. *For any two complex numbers* $a_0, a_1 \ne 0$*, there is a constant* $M(a_0, a_1) > 0$ *such that if* $f(z)$ *is analytic in* $|z| < R$ *with* $f(0) = a_0$ *and* $f'(0) = a_1$ *and if* $f(z)$ *omits the values* 0 *and* 1*, then*

$$R \le M(a_0, a_1).$$

Proof. Taking $\lambda = \frac{1}{2}$ in Schottky's theorem gives $|f(z)| \le M(a_0)$ in the disk $|z| \le R/2$. In view of the Cauchy integral formula,

$$a_1 = \frac{1}{2\pi i} \int_{|\zeta|=R/2} \frac{f(\zeta)}{\zeta^2} d\zeta,$$

so that

$$|a_1| \le \frac{2M(a_0)}{R}.$$

In other words, $R \le \frac{2M(a_0)}{|a_1|}$, as desired. □

4.23 Corollary. *For any two complex numbers* $a_0, a_1 \ne 0$*, there is a constant* $M(a_0, a_1) > 0$ *such that for the family of all functions* $f(z) = a_0 + a_1 z + \cdots$ *that are analytic in* $|z| < R$*, where* $R > M(a_0, a_1)$*,* $f(z)$ *must assume one of the values* 0 *or* 1 *in* $|z| < R$.

---

16 See J. A. Hempel, Precise bounds in the theorems of Schottky and Picard, *J. Lond. Math. Soc.* (2), 21 (1980), 279–286.

17 Über eine Verallgemeinerung der Picardschen Satzes, *Sitz. Kön. Preuss Akad. Wiss. Berlin* 38 (1904), 1118–1133.

4.24 Exercise. Show that for analytic functions $f(z) = a_0 + a_n z^n + a_{n+1} z^{n+1} + \cdots$ as in Landau's theorem with $a_n \neq 0, n \geq 1$, we have

$$R \leq 2\sqrt[n]{\frac{M(a_0)}{|a_n|}}.$$

Meromorphic functions

We study normal families of meromorphic functions in terms of the spherical metric. Some new normality criteria will appear, such as Marty's theorem, and some criteria will be generalizations of what has already been encountered with analytic functions, although the pivotal notion of local boundedness for a family of analytic functions is now not so relevant.

4.25 Definition. A family $\mathcal{F}$ of meromorphic functions defined on a domain Ω is *normal* in Ω if from every sequence $\{f_n\}$ in $\mathcal{F}$ we can extract a subsequence that converges uniformly on compact subsets of Ω in the spherical metric.

The limit function is either meromorphic or identically ∞ by Corollary 3.18. Now it is required to show that this definition when applied to analytic functions coincides with the previously given one, that is,

> *A sequence of analytic functions $\{f_n\}$ converges uniformly to a function f (which may be $\equiv \infty$) on compact subsets of a domain Ω if an only if $\{f_n\}$ converges spherically uniformly to f on compact subsets.*

In fact, whenever the limit function is analytic, spherical uniform convergence is a consequence of uniform convergence, and the converse also holds by Theorem 3.15 on any compact set. If $f \equiv \infty$, then it is evident that

$$\chi(f_n, \infty) = \frac{1}{\sqrt{1 + |f_n|^2}}$$

gives the desired convergence. As a consequence, the two definitions of a normal family coincide in the case of families of analytic functions, which is as it should be.

Equicontinuity

Normal families of meromorphic functions also have a very direct relation with the notion of equicontinuity since the notion of local boundedness is not a factor.

4.26 Theorem (A. Ostrowski).[18] *A family $\mathcal{F}$ of meromorphic functions defined on a domain Ω is normal in Ω if and only $\mathcal{F}$ is spherically equicontinuous on Ω.*

Proof. In view of the compactness of the Riemann sphere in the spherical metric (why?), if the family $\mathcal{F}$ is equicontinuous, then the result follows via a proof directly analogous to that of the Montel Theorem 4.4 since a key component of that proof was the requirement of the equicontinuity of the family.

Conversely, suppose that $\mathcal{F}$ is a normal family in Ω but not spherically equicontinuous. As a consequence, there exist a point $z_0 \in \Omega$, a sequence $\{z_n\}$ in Ω with $z_n \to z_0$, and some $\varepsilon > 0$ such that for some sequence $\{f_n\} \subseteq \mathcal{F}$,

$$\chi(f_n(z_n), f_n(z_0)) > \varepsilon \tag{4.6}$$

for $n = 1, 2, 3, \ldots$. As $\mathcal{F}$ is normal, we can find a subsequence $\{f_{n_k}\}$ that converges spherically uniformly on compact subsets of Ω. Just such a compact subset is the sequence $E = \{z_n\} \cup \{z_0\}$, and by Theorem 3.16 the sequence $\{f_{n_k}\}$ is spherically equicontinuous on E. This contradicts (4.6), so that $\mathcal{F}$ is spherically equicontinuous on Ω. □

Marty theorem

In the normal family theory of analytic functions, a key ingredient was the condition of the functions themselves being local bounded. In the meromorphic setting, this condition is replaced by the local boundedness of the spherical derivative discussed in the previous chapter, which is equivalent to the spherical derivative being uniformly bounded on compact subsets.

4.27 Marty theorem.[19] *A family $\mathcal{F}$ of meromorphic functions defined on a domain Ω is normal if and only if for each compact subset $K \subset \Omega$, there exists a constant $M = M(K)$ such that*

$$f^{\#}(z) = \frac{|f'(z)|}{1 + |f(z)|^2} \leq M$$

for all $z \in K$ and $f \in \mathcal{F}$.

Proof. Assuming that $f^{\#}(z)$ is uniformly bounded on compact subsets, take some point $z_0 \in \Omega$ and a closed disk $\overline{D}(z_0, r) \subset \Omega$. For any point $z \in \overline{D}(z_0, r)$, consider a straight-line path γ from z_0 to z and note that the chordal distance between the points $f(z_0)$

18 Über Folgen analytischer Funktionen und einige Verschärfungen des Picardschen Satzes, *Math. Zeit.* 24 (1926), 215–258.

19 Recherches sur la repartition des valeurs d'une function méromorphe, *Ann. Fac. Sci. Univ. Toulouse* 23, (1931).

and $f(z)$ is less than or equal to the spherical length of $f(\gamma)$ on the Riemann sphere. Mathematically,

$$\chi(f(z_0), f(z)) \le \int_\gamma f^{\#}(\zeta)|d\zeta|.$$

By our assumption there is a constant M depending on our closed disk which yields

$$\chi(f(z_0), f(z)) \le M|z - z_0|$$

for all $f \in \mathcal{F}$ and $z \in \overline{D}(z_0, r)$. Therefore $\mathcal{F}$ is spherically equicontinuous in Ω and hence normal by Ostrowski's theorem.

On the other hand, assuming that $\mathcal{F}$ is a normal family, suppose that for some compact subset K in Ω and sequence of points $\{z_n\} \in K$, we have a sequence of functions $\{f_n\}$ in $\mathcal{F}$ such that $f_n^{\#}(z_n) \to \infty$ as $n \to \infty$. We then extract a subsequence $\{f_{n_k}\}$ that converges spherically uniformly on K to some limit function f. Then by Theorem 3.17, for each point $z_0 \in K$, there is a closed disk $\overline{D}(z_0, r) \subset \Omega$ in which either $|f_{n_k} - f| \to 0$ or $|\frac{1}{f_{n_k}} - \frac{1}{f}| \to 0$ uniformly as $k \to \infty$.

Considering the first case, since f is a bounded analytic function in $\overline{D}(z_0, r)$, for k sufficiently large, the functions f_{n_k} are also analytic in $\overline{D}(z_0, r)$. Hence by the Weierstrass Theorem 1.27 the functions $f_{n_k}^{\#} \to f^{\#}$ uniformly in $\overline{D}(z_0, r)$, and since $f^{\#}$ is bounded in $D(z_0, r)$, for k sufficiently large, the functions $f_{n_k}^{\#}$ are bounded there as well.

Similarly, in the second case, for $|\frac{1}{f_{n_k}} - \frac{1}{f}| \to 0$ as $k \to \infty$, we can apply the same argument as above replacing f_{n_k} by $\frac{1}{f_{n_k}}$ and f by $\frac{1}{f}$ and noting that $g^{\#} = (\frac{1}{g})^{\#}$. This gives the same conclusion that the functions $f_{n_k}^{\#}$ for k sufficiently large are bounded in $D(z_0, r)$.

Covering the compact set K with a finite number of disks in each of which the functions $f_{n_k}^{\#}$ are bounded implies that the functions $f_{n_k}^{\#}$ are bounded on the compact set K, which contradicts our assumption and thus proves the theorem. □

As a trivial consequence of Marty's theorem, we remark that if a family $\mathcal{F}$ of analytic functions has the property that the family of derivatives $\mathcal{F}'$ is locally bounded, then $\mathcal{F}$ is a normal family.

We can also characterize normality as a local property.

4.28 Definition. A family of meromorphic functions $\mathcal{F}$ is *normal at a point* z_0 if $\mathcal{F}$ is normal in some neighborhood of z_0.

4.29 Theorem. *A family of meromorphic function is normal in a domain $\Omega \subseteq \mathbb{C}$ if and only if it is normal at each of its points.*

Indeed, if the family is normal at each point of Ω, then by Marty's theorem and a standard compactness argument the family is normal in Ω, and the converse is obvious.

There is an obvious extension to the Riemann sphere. We say that a family of meromorphic functions $\{f(z)\}$ is *normal at* ∞ if the corresponding family of functions $g(z) = f(1/z)$ is normal at $z = 0$. So, for example, the family $\{nz : n = 1, 2, 3, \dots\}$ is normal at ∞ since the family $\{n/z : n = 1, 2, 3, \dots\}$ is normal at the origin.

Furthermore, for a domain Ω on the Riemann sphere that contains the point $z = \infty$, a family is *normal* in Ω if it is normal at $z = \infty$ and normal in $\Omega - \{\infty\}$ in the usual sense. Since $\{nz : n = 1, 2, 3, \dots\}$ is normal in the domain $\{|z| > 1\}$, we can say that the family is normal in $\Omega = \{|z| > 1\} \cup \{\infty\}$ on the Riemann sphere.

In other words, *a family $\mathcal{F}$ of meromorphic functions is normal in a domain Ω on the Riemann sphere if an only if for every sequence of functions $\{f_n\}$ belonging to $\mathcal{F}$, there is a subsequence that converges spherically uniformally on compact subsets* of Ω. Thus, for example, the fundamental normality test is valid for domains on the Riemann sphere.

Regarding poles, something of interest can be said about them for a normal family of meromorphic functions.

4.30 Theorem. *Let $\mathcal{F}$ be a normal family of meromorphic functions in a domain Ω such that $|f(z_0)| \leq m$ for all $f \in \mathcal{F}$. Then there is some $r > 0$ and disk $D(z_0, r) \subseteq \Omega$, in which every $f \in \mathcal{F}$ is analytic.*

Proof. Assuming that no such disk exists, there is a sequence of functions $\{f_n\}$ belonging to $\mathcal{F}$ with poles at the points z_n such that $z_n \to z_0$ as $n \to \infty$. By the normality of $\mathcal{F}$ there is a subsequence $\{f_{n_k}\}$ that converges spherically uniformly on compact subsets to a meromorphic function g. Since $|g(z_0)| \leq m$, there is a disk $D(z_0, r) \subseteq \Omega$ in which $|g(z_0)| \leq M < \infty$. However, now we can choose k sufficiently large so that the poles $z_{n_k} \in D(z_0, r)$ and also $|f_{n_k}(z) - g(z)| < 1$ for $z \in D(z_0, r)$. It follows that

$$\left|f_{n_k}(z)\right| < 1 + M$$

for $z \in D(z_0, r)$, a contradiction, establishing the result. □

Similarly, for a normal family of meromorphic functions such that

$$\left|f(z_0) - a\right| > \rho > 0,$$

we have that $f(z) - a = 0$ has no roots in a neighborhood of z_0 for all functions f in the family. In fact, the corresponding functions $1/(f(z) - a)$ satisfy the conditions of the theorem.

Bloch principle

The Bloch principle is in fact a heuristic principle dating back to a statement in the 1926 monograph[20] of André Bloch (1893–1948):[21] *if $\mathcal{P}$ is a property that reduces an analytic or meromorphic function defined in $\mathbb{C}$ to a constant, then a family of analytic or meromorphic functions all possessing the property $\mathcal{P}$ in a domain Ω will likely be a normal family in Ω.* For example, an entire function omitting two distinct values reduces to a constant and a family of analytic functions on a domain Ω each of which omits the same two distinct values is also normal in Ω by the fundamental normality test. There are many such examples, but there are also some counterexamples, as we will see further. Nevertheless, the idea is fascinating and has led some researchers to formalize this notion.

In a 1973 address by Abraham Robinson,[22] he gave a formulation of the Bloch principle and established it within the context of nonstandard analysis. An analytical version was given by Lawrence Zalcman,[23] which we present here.

4.31 Zalcman lemma. *Let $\mathcal{F}$ be a family of analytic (meromorphic) functions in U. Then $\mathcal{F}$ is not normal in U if and only if there exist:*
(i) *a number r with $0 < r < 1$,*
(ii) *points z_n with $|z_n| < r$,*
(iii) *functions $f_n \in \mathcal{F}$,*
(iv) *positive numbers $\rho_n \to 0$ as $n \to \infty$,*

such that

$$f_n(z_n + \rho_n\zeta) \to g(\zeta) \quad \text{as } n \to \infty \tag{4.7}$$

spherically uniformally on compact subsets of $\mathbb{C}$, where $g(\zeta)$ is a nonconstant entire (meromorphic) function in $\mathbb{C}$.

Proof. Assume that $\mathcal{F}$ is not normal in U. Then by Marty's Theorem 4.27 there is some r_0 with $0 < r_0 < 1$, a sequence of points z_n' in $\{|z| \le r_0\}$, and a sequence of functions $f_n \in \mathcal{F}$ such that $f_n^{\#}(z_n') \to \infty$ as $n \to \infty$. By the continuity of the spherical derivative in U we can define for $r_0 < r < 1$, where r is fixed,

20 Les fonctions holomorphes et méromorphes dans le circle unité, Gauthier-Villars, Paris, 1926.

21 André Bloch suffered severe injuries as a soldier during World War I and could not return to duties. Subsequently, at a dinner party in 1917 he murdered his brother, uncle, and aunt. Bloch was assigned to a lunatic asylum, where for the next 30 years, he quite contentedly worked on mathematics.

22 Metamathematical problems, *J. Symbolic Logic*, 38 (1973), 500–516.

23 A heuristic principle in complex function theory, *Amer. Math. Monthly* 82 (1975), 813–817.

$$M_n = \max_{|z|\le r}\left[\left(1-\frac{|z|^2}{r^2}\right)f_n^{\#}(z)\right] = \left(1-\frac{|z_n|^2}{r^2}\right)f_n^{\#}(z_n), \tag{4.8}$$

which define the points z_n. Since $f_n^{\#}(z_n') \to \infty$, it follows that $M_n \to \infty$ and hence

$$\rho_n = \frac{1}{M_n}\left(1-\frac{|z_n|^2}{r^2}\right) = \frac{1}{f_n^{\#}(z_n)} \to 0 \tag{4.9}$$

as $n \to \infty$. Rearranging this last expression, we obtain

$$\frac{\rho_n}{r-|z_n|} = \frac{r+|z_n|}{r^2 M_n} \le \frac{2}{rM_n} \to 0 \tag{4.10}$$

as $n \to \infty$.

Next, consider the function

$$g_n(\zeta) = f_n(z_n + \rho_n\zeta)$$

defined for $|\zeta| < (\frac{r-|z_n|}{\rho_n}) = R_n$ with $R_n \to \infty$ as $n \to \infty$. Setting $\zeta = 0$, we have

$$g_n^{\#}(0) = \rho_n f_n^{\#}(z_n) = 1$$

for each n by (4.9). Our immediate aim is to show that the functions $g_n^{\#}(\zeta)$ are uniformly bounded on compact subsets of $\mathbb{C}$ in order to apply Marty's theorem. To this end, take $|\zeta| \le R < R_n$, so that $|z_n + \rho_n\zeta| < r$, and by (4.8)

$$g_n^{\#}(\zeta) = \rho_n f_n^{\#}(z_n + \rho_n\zeta) \le \frac{\rho_n M_n}{1-\frac{|z_n+\rho_n\zeta|^2}{r^2}} \le \frac{r+|z_n|}{r+|z_n|-\rho_n R}\cdot\frac{r-|z_n|}{r-|z_n|-\rho_n R} \to 1$$

as $n \to \infty$ by (4.9) and by (4.10) applied respectively to the terms on the right.

Hence the spherical derivatives $\{g_n^{\#}\}$ are uniformly bounded on compact subsets, so that by Marty's theorem there exists a subsequence $\{g_{n_k}\}$ that converges spherically uniformly to a function g on compact subsets of $\mathbb{C}$. Moreover,

$$g^{\#}(0) = \lim_{k\to\infty} g_{n_k}^{\#}(0) = 1,$$

implying that g is either a nonconstant entire or meromorphic function by Corollary 3.18.

To prove the converse, suppose that all of the hypotheses are satisfied but $\mathcal{F}$ is normal in U. Then for some positive constant M,

$$\max_{|z|\le\frac{1+r}{2}} f^{\#}(z) \le M$$

for all $f \in \mathcal{F}$ by Marty's theorem. Taking a point $\zeta \in \mathbb{C}$ satisfying $|z_n + \rho_n\zeta| < \frac{1+r}{2}$ for all n sufficiently large gives $\rho_n f_n^{\#}(z_n + \rho_n\zeta) \le \rho_n M$. As a consequence, (4.7) implies that

$$g^{\#}(\zeta) = \lim_{n\to\infty} \rho_n f_n^{\#}(z_n + \rho_n\zeta) = 0.$$

It follows that g is identically constant since ζ was arbitrary, which contradicts our assumption on g. We conclude that $\mathcal{F}$ is not a normal family in U, as desired. □

The notion of a normal family is very dependent on the domain the family is defined on. For example, the family $\mathcal{F} = \{nz : n = 1, 2, 3, \dots\}$ is not normal in the unit disk U but is normal in the punctured disk $\{0 < |z| < 1\}$. So, to make the Bloch principle rigorous in some sense, we must consider functions together with their domains and distinguish between function elements $\langle f, \Omega\rangle$ and $\langle f, \Omega'\rangle$ whenever $\Omega \neq \Omega'$.

Moreover, for a given property $\mathcal{P}$, the notation $\langle f, \Omega\rangle \in \mathcal{P}$ means that the function f has the property $\mathcal{P}$ on the domain Ω. The following conditions stem from Robinson's formalization of the Bloch principle and strengthened by Zalcman.

4.32 Definition. A property $\mathcal{P}$ of analytic or meromorphic functions is called *normal with respect to a domain* Ω if it satisfies the following conditions:
(a) If $\langle f, \Omega\rangle \in \mathcal{P}$ and $\Omega' \subseteq \Omega$, then $\langle f, \Omega'\rangle \in \mathcal{P}$;
(b) If $\langle f, \Omega\rangle \in \mathcal{P}$ and $t(z) = az + b$, then $\langle f \circ t, t^{-1}(\Omega)\rangle \in \mathcal{P}$;
(c) Let $\langle f_n, \Omega_n\rangle \in \mathcal{P}$ for $\Omega_1 \subseteq \Omega_2 \subseteq \dots$ and $\mathbb{C} = \bigcup_{n=1}^{\infty} \Omega_n$. If $f_n \to f$ uniformly (spherically uniformly) on compact subsets of $\mathbb{C}$, then $\langle f, \mathbb{C}\rangle \in \mathcal{P}$;
(d) If $\langle f, \mathbb{C}\rangle \in \mathcal{P}$, then f is identically constant.

The preceding considerations lead to a very expeditious test for normality and puts the Bloch principle on a solid mathematical foundation.

4.33 Robinson–Zalcman principle. *If a property $\mathcal{P}$ is normal with respect to a domain Ω, then the family of functions*

$$\mathcal{F} = \{f : \langle f, \Omega\rangle \in \mathcal{P}\}$$

is normal in Ω.

Proof. Since normality is a local property by Theorem 4.29, assuming that $\mathcal{F}$ is not a normal family in Ω, it would not be normal in some disk $D \subseteq \Omega$ and $\langle f, D\rangle \in \mathcal{P}$ by property (a) for each $f \in \mathcal{F}$. In view of (ii), we may take $D = U$ the open unit disk. We now wish to apply the Zalcman lemma 4.31 to the nonnormal family $\mathcal{F}$ with respect to U taking the same functions f_n and parameters r, z_n, ϱ_n, R_n. Since $R_n = \frac{r-|z_n|}{\rho_n} \to \infty$ as $n \to \infty$, we may assume that it is an increasing sequence by possibly considering a subsequence. Thus the functions

$$g_n(\zeta) = f_n(z_n + \rho_n\zeta)$$

of the lemma are defined on $\Omega_n : |\zeta| < R_n$, so that for each n, since $|z_n + \rho_n\zeta| < r < 1$, we have $\langle g_n, \Omega_n\rangle \in \mathcal{P}$ by property (b). Since $g_n \to g$ uniformly on compact subsets of

$\mathbb{C}$ and $\bigcup_{n=1}^{\infty} \Omega_n = \mathbb{C}$, $\langle g, \mathbb{C}\rangle \in \mathcal{P}$ by (c). This means that $g \equiv$ constant by property (d), a contradiction. This establishes that $\mathcal{F}$ is a normal in Ω, concluding the proof. □

The R-Z principle has been extended in a number of different directions, but these developments will not be pursued here. There are now several classical results that become a simple routine to verify. One of the most significant is the following:

4.34 Fundamental normality test (Montel).[24] *Let $\mathcal{F}$ be a family of analytic functions defined on a domain Ω such that there are two distinct fixed values a and b in $\mathbb{C}$ such that $f(z) \neq a$ and $f(z) \neq b$ for all $f \in \mathcal{F}$. Then $\mathcal{F}$ is a normal family in Ω.*

Indeed, let $\mathcal{P}$ be the property on a domain Ω: *f is either constant or omits the values a and b.* Conditions (a) and (b) are evident, and (c) is a consequence of the Hurwitz theorem 1.28. Property (d) is just Picard's first theorem, and therefore $\mathcal{F} = \{f : \langle f, \Omega\rangle \in \mathcal{P}\}$ is a normal family in Ω.

The astute reader will have noticed that we used the FNT to prove Picard's first theorem! Fortunately, there are multiple proofs of both theorems that are independent of each other, and the above proof is presented merely to illustrate one of them.[25] It also illustrates the marvelous interconnectedness between the two notions. See Schiff (1993) for four other proofs of the FNT.

4.35 Corollary. *Let $\mathcal{F}$ be a family of meromorphic functions defined on a domain Ω such that there are three distinct fixed values a, b, and c in $\mathbb{C}$ such that $f(z) \neq a, f(z) \neq b$, and $f(z) \neq c$ for all $f \in \mathcal{F}$. Then $\mathcal{F}$ is a normal family in Ω.*

Indeed, the corresponding functions

$$g(z) = \frac{(c-b)(f(z)-a)}{(c-a)(f(z)-b)}$$

for each $f \in \mathcal{F}$ form a family of analytic functions $\mathcal{G}$ that omits the values 0 and 1. Then $\mathcal{G}$ is normal in the Euclidean metric and hence normal in the spherical metric, and likewise for the family $\mathcal{F}$.

4.36 Remark. By the same token the function $g(z)$ shows that a nonconstant meromorphic function in $\mathbb{C}$ can omit at most two values by Picard's first theorem.

The Robinson–Zalcman principle also allows generalizations of the fundamental normality test. One such is due to Constantin Carathéodory (1873–1950), whose proof is different from ours.[26]

24 Sur les familles de fonctions analytique qui admettent des valeurs exceptionnelles dans un domaine, *Ann. École Norm. Sup.* 29 (1912), 487–535.

25 In fact, one other proof of Picard's first theorem was given in Chapter 3 via the elliptic modular function.

26 *Theory of Functions of a Complex Variable*, vol. II, Chelsea Publ. Co., New York, 1960, p. 202.

4.37 Theorem. *Let $\mathcal{F}$ be a family of meromorphic functions on a domain Ω and suppose that each $f \in \mathcal{F}$ omits three distinct values a_f, b_f, c_f with $\min(\chi(a_f,b_f),\chi(b_f,c_f),\chi(c_f,a_f)) \geq \alpha > 0$. Then $\mathcal{F}$ is a normal family in Ω.*

Proof. Define the property $\mathcal{P}$: f omits three values a_f, b_f, c_f such that

$$\min(\chi(a_f,b_f),\chi(b_f,c_f),\chi(c_f,a_f)) \geq \alpha > 0.$$

Clearly, properties (a) and (b) of the normal property definition are satisfied. Now, for property (c), suppose that $\chi(f_n,f) \to 0$ as $n \to \infty$ uniformly on compacts subsets of Ω and each f_n omits the points a_n, b_n, c_n satisfying $\min(\chi(a_n,b_n),\chi(b_n,c_n),\chi(c_n,a_n)) \geq \alpha > 0$ for $n = 1,2,3,\ldots$. Clearly, we may assume that f is a nonconstant function since a constant function obviously satisfies condition $\mathcal{P}$. By the compactness of the Riemann sphere there is a convergent subsequence of $\{a_n\}$, say $\{a_k\}$, such that $\chi(a_k,a) \to 0$. Considering the corresponding subsequence $\{b_k\}$ it also has a convergent subsequence, say $\{b_l\}$, such that $\chi(b_l,b) \to 0$ and $\chi(a_l,a) \to 0$. Taking the corresponding subsequence $\{c_l\}$, it has a convergent subsequence $\{c_m\}$ with $\chi(c_m,c) \to 0$, $\chi(b_m,b) \to 0$, and $\chi(a_m,a) \to 0$, as well as $\chi(f_m,f) \to 0$ as $m \to \infty$. Moreover,

$$\min(\chi(a_m,b_m),\chi(b_m,c_m),\chi(c_m,a_m)) \geq \alpha > 0$$

for each m, and by continuity, $\min(\chi(a,b),\chi(b,c),\chi(c,a)) \geq \alpha > 0$.

To show that the limit function f omits the values a, b, c, suppose on the contrary that $f(z_0) = a$. In the case $a \neq \infty$, we know by Theorem 3.17 that $f(z)$ is analytic and bounded in some closed disk $\overline{D}(z_0,r)$, and, moreover, $f_m - a_m \to f - a$ uniformly on $\overline{D}(z_0,r)$. Since f is nonconstant, Hurwitz's theorem implies that for all m sufficiently large, $f_m - a_m$ has a zero in $\overline{D}(z_0,r)$, which is a contradiction. On the other hand, if $a = \infty$, then $\frac{1}{f_m} \to \frac{1}{f}$ uniformly in some $\overline{D}(z_0,r)$, again producing a contradiction as in the previous case. Similarly, f omits the values b and c, establishing property (c). Finally, property (d) follows by Picard's theorem (Remark 4.36), so that $\mathcal{P}$ is a normal property, and, consequently, $\mathcal{F}$ is a normal family in Ω by the Robinson–Zalcman principle.

A beautiful generalization of the fundamental normality test, which was originally obtained by Montel in a slightly weaker form, is the following association with a key result from the Nevanlinna theory. □

4.38 Theorem. *Let $\mathcal{F}$ be a family of meromorphic in a domain Ω satisfying the following conditions for each $f \in \mathcal{F}$: (i) all the zeros of $f(z)$ have multiplicity $\geq h$, (ii) all the poles have multiplicity $\geq k$, and (iii) all the zeros of $f(z) - 1$ have multiplicity $\geq \ell$. If*

$$\frac{1}{h} + \frac{1}{k} + \frac{1}{\ell} < 1,$$

then $\mathcal{F}$ is normal in Ω.

Proof. The conditions on each $f \in \mathcal{F}$ mean that conditions (a) and (b) are evident. Condition (c) follows from the Hurwitz theorem in the case of the zeros and in the case of the poles from Theorem 3.17, and (d) is a consequence of Corollary 3.37 in the Nevanlinna theory. □

The formalized R-Z principle allows the following extension of Landau's theorem to meromorphic functions.

4.39 Theorem. *Suppose that $\mathcal{F}$ is a family of meromorphic functions $f(z)$ in a disk $|z| < R$ such that for complex numbers $a_0, a_1 \neq 0$,*

$$f(z) = a_0 + a_1 z + \cdots$$

in a neighborhood of the origin for each $f \in \mathcal{F}$. Assume that all the roots of the equation $f(z) - a_\nu = 0$ for $a_\nu \in \hat{\mathbb{C}}$ have multiplicity $m_\nu (\geq 2)$, $\nu = 1, 2, \ldots, q$, and

$$\sum_{\nu=1}^{q} \left(1 - \frac{1}{m_\nu}\right) > 2.$$

Then $R \leq M(a_0, a_1, m_1, \ldots, m_q)$.

Proof. The last condition merely serves to make the family constant in regards to the entire complex plane in order to invoke Corollary 3.33 and, subsequently, the Robinson–Zalcman principle. □

Of course, the general Bloch principle as enunciated at the start of this section is not always valid, and there are counterexamples where the conditions of the formalized version are not met.

4.40 Example.[27] Consider the property $\mathcal{P}$ of analytic functions f such that

$$\phi(f)(z) = (f'(z) - 1)(f'(z) - 2)(f'(z) - f(z))$$

omits the value zero.

If $f(z)$ has property $\mathcal{P}$ on $\mathbb{C}$, then by Picard's first theorem $f'(z) = a$ (constant) implying $f(z) = az + b$. Since $f'(z) - f(z) \neq 0$ for all $z \in \mathbb{C}$, it follows that $a = 0$, and therefore $f(z) \equiv$ constant.

On the other hand, we have seen that the family

$$\mathcal{F} = \{f_n(z) = nz : z \in U, n = 3, 4, 5, \ldots\}$$

is not normal in U, but $f_n'(z) = n \neq 1, 2$ and $f'(z) - f(z) = n - nz \neq 0$ for all $z \in U$, that is, each $f_n \in \mathcal{F}$ has the property $\mathcal{P}$ in U, thus violating the general Bloch principle.

---

27 L. Rubel, Four counterexamples to Bloch's Principle, *Proc. Amer. Math. Soc.* 98 (1986), 257–260.

There is another useful formalization of the Bloch principle, also based on the framework of Robinson developed by D. Minda,[28] which will not be discussed here.

Normal and Bloch functions

4.41 Definition. A meromorphic function $f(z)$ defined in U is called *normal* if

$$\sup_{z\in U}(1-|z|^2)f^{\#}(z) < \infty.$$

The name in fact comes from the associated normal family

$$F(z) = f\left(e^{i\theta}\frac{z+a}{1+\bar{a}z}\right) \tag{4.11}$$

$0 \le \theta < 2\pi$, $|a| < 1$.

Indeed, for $F(z)$ as in (4.11), and setting $\phi(z) = (e^{i\theta}\frac{z+a}{1+\bar{a}z})$, we have

$$(1-|z|^2)F^{\#}(z) = \frac{(1-|z|^2)|f'(\phi(z))||\phi'(z)|}{1+|f(\phi(z))|^2} = (1-|\phi(z)|^2)f^{\#}(\phi(z))$$

by equation (5.6) in Chapter 5 applied to $\phi(z)$. If $f(z)$ is a normal function, then it follows that $F^{\#}(z)$ is locally uniformly bounded, and hence the family of functions of (4.11) is normal by Marty's theorem.

Conversely, assuming that the family in (4.11) is normal, taking $\theta = 0$ implies that

$$F^{\#}(0) = (1-|\phi(0)|^2)f^{\#}(\phi(0)) = (1-|a|^2)f^{\#}(a),$$

$|a| < 1$, and we conclude that $f(z)$ is a normal function since $F^{\#}(0) \le M < \infty$ for all F, again by Marty's theorem.

Normal functions with such a constraint on the growth of the spherical derivative naturally satisfy the corresponding growth constraint of their characteristic $T_0(r,f)$.

4.42 Theorem. *If $f(z)$ is a normal function in U, then for $|z| = r$,*

$$T_0(r,f) \le C\log\frac{1}{1-r^2}, \quad 0 < r < 1.$$

28 Another approach to Picard's theorem and a unifying principle in geometric function theory, in *Current Topics in Analytic Function Theory*, H. M. Srivastava, S. Owa (Eds.), World Sci. Publ. (1992), 186–200. See also Schiff (1993).

Proof. We have

$$[f^{\#}(z)]^2 \leq \frac{c^2}{(1-r^2)^2},$$

so that

$$S(r) = \frac{1}{\pi}\int_0^{2\pi}\int_0^r [f^{\#}(\rho e^{i\theta})]^2 \rho d\rho d\theta \leq 2c^2 \int_0^r \frac{\rho d\rho}{(1-\rho^2)^2} = c^2 \frac{r^2}{1-r^2}.$$

As a consequence,

$$T_0(r,f) = \int_0^r \frac{S(t)}{t} dt \leq \frac{c^2}{2} \log \frac{1}{1-r^2},$$

as desired. □

4.43 Exercise.
(i) Prove that if $\mathcal{F}$ is a family of meromorphic functions in U that are normal at the origin, then $T_0(r,f) < C$ for $0 < r < R$ and some $R > 0$.
(ii) Show that if $T_0(r,f) < C$ for $0 < r < R$, then for $0 < r_0 < r$, $S(r_0)$ is uniformly bounded for all r_0 sufficiently small.

For an analytic function $f(z)$ in U, replacing the spherical derivative in the definition of a normal function with the ordinary derivative, namely

$$\sup_{z\in U}(1-|z|^2)|f'(z)| < \infty,$$

defines $f(z)$ as a *Bloch function*. Clearly, Bloch functions are normal functions.

5 Hyperbolic geometry

Playfair Axiom

Historically, there was much uneasiness regarding Euclid's 2,000-year-old fifth postulate, which was rephrased by Scottish mathematician John Playfair (1748–1819) into the familiar:

> *Only one line can be drawn through a point that will be parallel to a given line.*

However, there is no indication how this line can be drawn. This led some researchers to scrap the fifth postulate altogether and replace it with something else (maintaining the other four Euclidean postulates), the result being a new type of geometry:

> *Given a line L and a point not on the line, there is* ***at least one*** *straight line that passes through the point that does not intersect L.*

Bolyai/Lobachevsky

This was the approach initiated nearly simultaneously by two mathematicians, János Bolyai[1] (1802–1860) and Nikolai Lobachevsky (1792–1856). One of these new non-Euclidean geometries became known as hyperbolic geometry and is relevant to many fields including complex networks, quantum chaos, biological materials, Riemannian manifolds, and the theory of relativity, among others. In the words of the eminent David Hilbert:

> *The most suggestive and notable achievement of the last century is the discovery of non-Euclidean geometry.*

It must be mentioned in this context that Carl Friedrich Gauss (1777–1855) preceded both Bolyai and Lobachevsky in the discovery of this new geometry but did not publish his results for fear of causing any controversy, and furthermore, for centuries, sailors had been studying spherical geometry.

One of the beautiful features of two-dimensional hyperbolic geometry is that in at least one model, we can do the entire geometry in the open unit disk $U : |z| < 1$. Together with a particular metric, it is known as the "Poincaré disk model" named for Henri Poincaré (1854–1912), another giant of modern science. The connection with a 'hyperboloid' in 3-dimensional space is discussed at the end of this chapter.

First, let us review some preliminaries.

---

1 The father of János Bolyai, Fárkás, labored over many fruitless years trying to prove the parallel postulate and also unsuccessfully tried to dissuade his son from considering it.

https://doi.org/10.1515/9783110757828-005

5.1 Schwarz–Pick lemma. *Let $f(z)$ be analytic in U with $|f(z)| \leq 1$. Then for $|\alpha| < 1$,*

$$\left|\frac{f(z) - f(\alpha)}{1 - \overline{f(\alpha)}f(z)}\right| \leq \left|\frac{z - \alpha}{1 - \bar{\alpha}z}\right| \tag{5.1}$$

and

$$\frac{|f'(\alpha)|}{1 - |f(\alpha)|^2} \leq \frac{1}{1 - |\alpha|^2} \tag{5.2}$$

with equality in both inequalities if and only if $f(z)$ is a Möbius transformation from U to U.

To establish the result recall that

$$S(z) = \frac{z - \alpha}{1 - \bar{\alpha}z} : U \to U$$

is a conformal mapping with $S(\alpha) = 0$. Thus we consider the Möbius transformation $T : U \to U$ defined by

$$T(w) = \frac{w - f(\alpha)}{1 - \overline{f(\alpha)}w},$$

so that $T(f(\alpha)) = 0$. In addition, let $\zeta = S(z)$. Then the composition

$$F(\zeta) = (T \circ f \circ S^{-1})(\zeta) : U \to U$$

is analytic in U with $F(0) = 0$. The Schwarz lemma of Chapter 1 is now applicable, so that

$$|F(\zeta)| \leq |\zeta| = \left|\frac{z - \alpha}{1 - \bar{\alpha}z}\right|. \tag{5.3}$$

Moreover,

$$|F(\zeta)| = |T(f(z))| = \left|\frac{f(z) - f(\alpha)}{1 - \overline{f(\alpha)}f(z)}\right|,$$

completing the proof of inequality (5.1).

For the second inequality (5.2), we use the fact that $|F'(0)| \leq 1$. Then

$$\begin{aligned} F'(0) &= \lim_{h\to 0} \frac{F(h)}{h} = \lim_{h\to 0} \frac{T(f(S^{-1}(h)))}{h} \\ &= \lim_{h\to 0} \frac{f(S^{-1}(h)) - f(\alpha)}{1 - \overline{f(\alpha)}f(S^{-1}(h))}/h, \end{aligned}$$

where $h = S(z) = \frac{z-\alpha}{1-\bar{\alpha}z}$. It follows that

$$\begin{aligned}F'(0) &= \lim_{z\to\alpha} \frac{f(z)-f(\alpha)}{1-\overline{f(\alpha)}f(z)} \Big/ \frac{z-\alpha}{1-\overline{\alpha}z} \\ &= \lim_{z\to\alpha} \frac{f(z)-f(\alpha)}{z-\alpha} \cdot \frac{1-\overline{\alpha}z}{1-\overline{f(\alpha)}f(z)} \\ &= f'(\alpha)\cdot \frac{1-|\alpha|^2}{1-|f(\alpha)|^2},\end{aligned}$$

and the result follows by taking absolute values. Note that equality holds in (5.3) if and only if $F(\zeta) = c\zeta$ with $|c| = 1$. As a consequence,

$$f(z) = T^{-1}\left(c\frac{z-\alpha}{1-\overline{\alpha}z}\right)$$

is a Möbius transformation from U to U, establishing the remainder of the theorem.

An elementary consequence of the theorem leads to a growth estimate for the preceding Möbius function $S(z) : U \to U$.

5.2 Corollary.[2] *If $f(z)$ is analytic in U with $|f(z)| \le 1$, then*

$$|f(z)| \le \frac{|f(0)| + |z|}{1 + |f(0)||z|}.$$

Proof. First, note that for $a, b \in U$,

$$\begin{aligned}\left|\frac{a-b}{1-\overline{a}b}\right|^2 &= 1 - \frac{(1-|a|^2)(1-|b|^2)}{|1-\overline{a}b|^2} \\ &\ge 1 - \frac{(1-|a|^2)(1-|b|^2)}{(1-|a||b|)^2} = \frac{(|a|-|b|)^2}{(1-|a||b|)^2},\end{aligned}$$

so that

$$\left|\frac{a-b}{1-\overline{a}b}\right| \ge \frac{|a|-|b|}{1-|a||b|}. \tag{5.4}$$

In view of inequality (5.1) for $\alpha = 0$ coupled with inequality (5.4),

$$\frac{|f(z)| - |f(0)|}{1-|f(0)||f(z)|} \le \left|\frac{f(z)-f(0)}{1-\overline{f(0)}f(z)}\right| \le |z|.$$

Solving this inequality for $|f(z)|$ gives the desired inequality. □

2 E. Lindelöf, Mémoire sur certaines inéqualités dans la théorie des fonctions monogénes et sur quelques propriétés Nouvelles de ces fonctions dans le voisinage d'un point singulier essential, *Acta Sci. Fenn.* 35 (7) (1909).

5.3 Corollary. *For* $\alpha, z \in U$,

$$\left|\frac{z-\alpha}{1-\bar{\alpha}z}\right| \le e^{\frac{(1-|\alpha|)(|z|-1)}{(|z|+1)}}.$$

Proof. From the preceding corollary with $f(z) = \frac{z-\alpha}{1-\bar{\alpha}z}$, a brief calculation shows that

$$\left|\frac{z-\alpha}{1-\bar{\alpha}z}\right| \le \frac{|z|+|\alpha|}{1+|\alpha||z|} = 1 - \frac{(1-|\alpha|)(1-|z|)}{1+|\alpha||z|} < 1 - \frac{(1-|\alpha|)(1-|z|)}{1+|z|}.$$

Then the inequality $1 - x < e^{-x}$ for $x > 0$ establishes the result. □

An application of the preceding result can be found in Chapter 8 (Theorem 8.18).

Poincaré disk model

The mathematics of the Poincaré disk model $\mathbb{D}$ quite naturally arises from the following considerations stemming from the Schwarz–Pick lemma 5.1.[3]

Let $w = f(z)$ be an analytic function in U with $|f(z)| \le 1$. Then inequality (5.2) for $|z| < 1$ can expressed as

$$\frac{|dw/dz|}{1-|w|^2} \le \frac{1}{1-|z|^2},$$

that is,

$$\frac{|dw|}{1-|w|^2} \le \frac{|dz|}{1-|z|^2}.$$

Now for a curve C given by: $z = z(t)$ in $U, \alpha \le t \le \beta$, we set

$$ds = \frac{2|dz|}{1-|z|^2}$$

as an element of hyperbolic arc length,[4] so that the *hyperbolic length* of C, denoted by $\ell_{\mathbb{D}}(C)$, is given by

$$\ell_{\mathbb{D}}(C) = \int_C ds = \int_C \frac{2|dz|}{1-|z|^2} = \int_\alpha^\beta \frac{2|z'(t)|}{1-|z(t)|^2}dt. \tag{5.5}$$

3 There is also the Beltrami–Klein model of hyperbolic geometry defined in the open unit disk but this topic will not be pursued here.

4 Some authors do not have 2 in the numerator, but there are good reasons for having it there. The two geometries are congruent.

Suppose that $w = f(z)$ is an analytic one-to-one and onto mapping of the disk onto itself (i. e., f is a *biholomorphism*). Then for any curve C in U, by (5.2) we again have

$$\ell_{\mathbb{D}}(f(C)) = \int_{f(C)} \frac{2|dw|}{1-|w|^2} \leq \int_C \frac{2|dz|}{1-|z|^2} = \ell_{\mathbb{D}}(C).$$

Considering the inverse function $g = f^{-1}$ and another application of the Schwarz–Pick lemma implies that $\ell_{\mathbb{D}}(C) \leq \ell_{\mathbb{D}}(f(C))$, and so $\ell_{\mathbb{D}}(C) = \ell_{\mathbb{D}}(f(C))$, and the hyperbolic length is preserved (invariant). Moreover, we can write

$$\frac{|f'(z)|}{1-|f(z)|^2} = \frac{1}{1-|z|^2}, \tag{5.6}$$

that is,

$$\frac{2|dw|}{1-|w|^2} = \frac{2|dz|}{1-|z|^2}.$$

Let us take a point α on the real axis, $0 < \alpha < 1$, so that the hyperbolic length of the line segment L from 0 to α is

$$\ell_{\mathbb{D}}(L) = \int_0^\alpha \frac{2dt}{1-t^2} = \log\frac{1+\alpha}{1-\alpha}.$$

Note that $\ell_{\mathbb{D}}(L) \to \infty$ as $\alpha \to 1$, so that $|z| = 1$ represents the points “at infinity” or “ideal points” for the Poincaré disk.

Poincaré metric

In the preceding discussion, ds is the *Poincaré metric*, which induces a proper metric between two points z_1, z_2 of U and is given by

$$\rho_{\mathbb{D}}(z_1, z_2) = \inf\{\ell_{\mathbb{D}}(C) : C \text{ joins } z_1 \text{ and } z_2 \text{ in } U\}.$$

5.4 Exercise. Show that $\rho_{\mathbb{D}}$ satisfies the triangle inequality

$$\rho_{\mathbb{D}}(z_1, z_2) \leq \rho_{\mathbb{D}}(z_1, z_3) + \rho_{\mathbb{D}}(z_3, z_2).$$

Now for any other curve $C : z = z(t)$ in $U, 0 \leq t \leq \alpha$, connecting 0 and α, we write $z(t) = r(t)e^{i\theta(t)}$, so that $|z(t)|^2 = [r(t)]^2$, and

$$|z'(t)| = \left([r'(t)]^2 + [r(t)\theta'(t)]^2\right)^{1/2} \geq |r'(t)|.$$

Therefore by (5.5)

$$\ell_{\mathbb{D}}(C) \geq \int_0^{\alpha} \frac{2|r'(t)|}{1-|r(t)|^2} dt \geq \int_0^{\alpha} \frac{2d|z|}{1-|z|^2} = \log \frac{1+\alpha}{1-\alpha} = \ell_{\mathbb{D}}(L),$$

and hence the straight-line segment L is a *geodesic*. The uniqueness of a geodesic will be established in the half-plane model, which will imply the same for the Poincaré disk model.

In the general case, for two arbitrary points z_1 and z_2 in U, we take the Möbius transformation

$$w = T(z) = e^{i\delta} \frac{z - z_1}{1 - \overline{z_1} z}, \tag{5.7}$$

where the angle δ is a suitable rotation such that $T(z_2) = p$ lies on the positive real axis in U. As $T(z)$ is an analytic bijective mapping of U to itself, it preserves the hyperbolic distance and hence preserves the geodesic lengths. Therefore the hyperbolic distance between z_1 and z_2 is given explicitly by

$$\rho_{\mathbb{D}}(z_1, z_2) = \rho_{\mathbb{D}}(0, T(z_2)) = \int_0^{T(z_2)} \frac{2d|w|}{1-|w|^2} = \log \frac{1 + \left|\frac{z_1 - z_2}{1 - \overline{z_1} z_2}\right|}{1 - \left|\frac{z_1 - z_2}{1 - \overline{z_1} z_2}\right|} = 2 \tanh^{-1} \left| \frac{z_1 - z_2}{1 - \overline{z_1} z_2} \right|. \tag{5.8}$$

[5]

Furthermore, the inverse Möbius transformation $z = T^{-1}(w)$ maps circles to circles and preserves angles. Since the extended straight-line segment from 0 to $f(z_2)$ is orthogonal to $|z| = 1$, it follows that the extended geodesic connecting the points z_1 and z_2 is also orthogonal to $|z| = 1$ and is an arc of the circle (treating a straight line as a particular case of a circle). See Figure 5.1.

Since the points at $|z| = 1$ are actually at "infinity," note that any geodesics in U do not actually "reach" $|z| = 1$. The geodesics in U play the role of straight lines, like ordinary straight lines in the Euclidean plane.

Hyperbolic length/area

For a hyperbolic circle C_R representing all the points that are at a fixed (Euclidean) radius R from the origin (see Figure 5.2), its circumference is given by its hyperbolic length

$$C_{\mathbb{D}} = \ell_{\mathbb{D}}(C_R) = \int_{C_R} \frac{2|dz|}{1-|z|^2} = \int_0^{2\pi} \frac{2R d\theta}{1-R^2} = \frac{4\pi R}{1-R^2},$$

5 $\tanh^{-1} x = \frac{1}{2} \log\left(\frac{1+x}{1-x}\right)$.

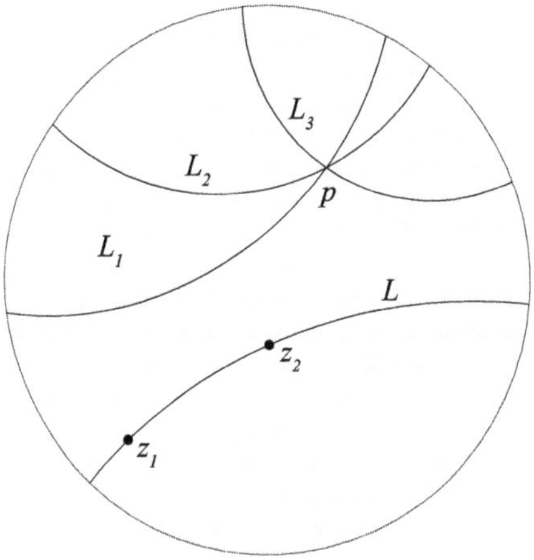

Figure 5.1: In the Poincaré disk model for hyperbolic geometry "straight lines" are arcs of circles that are perpendicular to the boundary at $|z| = 1$. Note that Euclid's fifth postulate is violated as there are infinitely many such "straight lines" passing through a given point p and not intersecting a given line L. Courtesy Katy Metcalf.

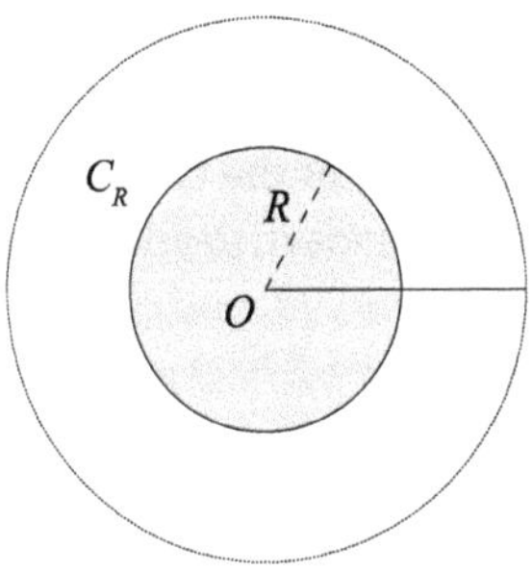

Figure 5.2: The Poincaré disk and interior circle C_R. Courtesy Katy Metcalf.

with $C_{\mathbb{D}} > 2\pi R$. As we have seen above, the hyperbolic radius of the circle is given by

$$R_{\mathbb{D}} = \log \frac{1+R}{1-R},$$

and $R_{\mathbb{D}} > R$. Moreover, $R = \tanh(R_{\mathbb{D}}/2)$.

Note that for fixed r,

$$ds = \frac{2rd\theta}{1-r^2},$$

or for fixed θ,

$$ds = \frac{2dr}{1-r^2},$$

In view of the hyperbolic invariance of $S(z)$ (eq. (5.10)),

$$|z_1 - z_2|^2\left(\frac{|S'(z_1)|}{1-|S(z_1)|^2}\right)\left(\frac{|S'(z_2)|}{1-|S(z_2)|^2}\right) = \frac{|z_1 - z_2|^2}{(1-|z_1|^2)(1-|z_2|^2)},$$

and applying equality (5.11) to the above, we obtain

$$\frac{|S(z_1) - S(z_2)|^2}{(1-|S(z_1)|^2)(1-|S(z_2)|^2)} = \frac{|z_1 - z_2|^2}{(1-|z_1|^2)(1-|z_2|^2)},$$

demonstrating the invariance of the quantity on the right-hand side under the transformation $S(z)$.

Now for any points $z_1, z_2 \in U$, take the Möbius transformation $S(z)$ as in (5.9) with $\beta = -\alpha z_1$, so that $S(z_1) = 0$ and

$$S(z) = \frac{(\alpha/\overline{\alpha})(z - z_1)}{1 - \overline{z}_1 z} = \frac{\gamma(z - z_1)}{1 - \overline{z}_1 z},$$

where $|\gamma| = 1$. Finally, we take a rotation by $\gamma = e^{i\delta}$ so that $S(z_2) = p$ is a point on the positive real axis in U and note that this is just the Möbius transformation (5.7).

Since we have already determined by (5.8) that $p = \tanh \frac{1}{2}\rho_{\mathbb{D}}(0, p)$, we obtain another relation between Euclidean and hyperbolic distances:

$$\frac{|z_1 - z_2|^2}{(1-|z_1|^2)(1-|z_2|^2)} = \frac{p^2}{1-p^2} = \sinh^2\left(\frac{1}{2}\rho_{\mathbb{D}}(z_1, z_2)\right).$$

Again, since $\sinh^2(\frac{x}{2}) = \frac{1}{2}(\cosh x - 1)$, we have another formulation for the hyperbolic distance between two points $z_1, z_2 \in U$:

$$\rho_{\mathbb{D}}(z_1, z_2) = \cosh^{-1}\left(1 + \frac{2|z_1 - z_2|^2}{(1-|z_1|^2)(1-|z_2|^2)}\right). \tag{5.12}$$

Needless to say, there is a comprehensive hyperbolic trigonometry as in the Euclidean case.[8]

Given the notation in Figure 5.3, we have

Hyperbolic law of sines

$$\frac{\sin\alpha}{\sinh a} = \frac{\sin\beta}{\sinh b} = \frac{\sin\gamma}{\sinh c};$$

First hyperbolic law of cosines[9]

$$\cosh c = \cosh a \cosh b - \sinh a \sinh b \cos\gamma;$$

8 See for example, Anderson (2005).

9 There is no version of this in Euclidean geometry.

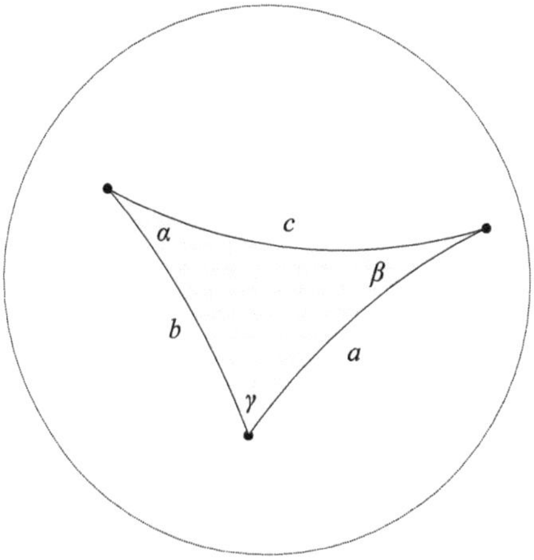

Figure 5.3: A hyperbolic triangle in the Poincaré disk whose sides have hyperbolic lengths a, b, c and opposing angles α, β, γ, respectively. Courtesy Katy Metcalf.

Second hyperbolic law of cosines

$$\cosh c = \frac{\cos\alpha\cos\beta + \cos\gamma}{\sin\alpha\sin\beta}.$$

Taking $\gamma = \frac{\pi}{2}$, we have by the first hyperbolic law of cosines

$$\cosh c = \cosh a \cosh b,$$

which is the *hyperbolic Pythagorean theorem*: for if our triangle is very small, then we have the approximation

$$\cosh x = \frac{e^x + e^{-x}}{2} \approx 1 + \frac{x^2}{2},$$

so that the hyperbolic Pythagorean theorem expression now reads

$$1 + \frac{c^2}{2} \approx \left(1 + \frac{a^2}{2}\right)\left(1 + \frac{b^2}{2}\right),$$

that is, $c^2 \approx a^2 + b^2$, as we expect.

Half-plane model

Another model of hyperbolic geometry can be constructed in the upper half-plane:

$$\mathbb{H} = \{z : \operatorname{Im}(z) = y > 0\}$$

with boundary $\partial\mathbb{H} = \{z = x + iy : y = 0\} \cup \{\infty\}$.

Taking the Möbius transformation $\phi : \mathbb{H} \to U$ given by

$$w = \phi(z) = \frac{iz + 1}{z + i},$$

we have $z = \phi^{-1}(w) = \frac{iw-1}{-w+i}$. Observe that under $\phi(z)$, the real axis is mapped onto $|w| = 1$, and the upper half-plane is mapped conformally to the interior of the unit disk with the positive imaginary axis mapped to the diameter from $-i$ to i. The distance in $\mathbb{H}$ is defined by

$$\rho_{\mathbb{H}}(z_1, z_2) = \rho_{\mathbb{D}}(\phi(z_1), \phi(z_2)).$$

The arc length element in $\mathbb{H}$ obtained from that in the Poincaré disk is

$$ds = \frac{|dw|}{\mathrm{Im}(w)}. \tag{5.13}$$

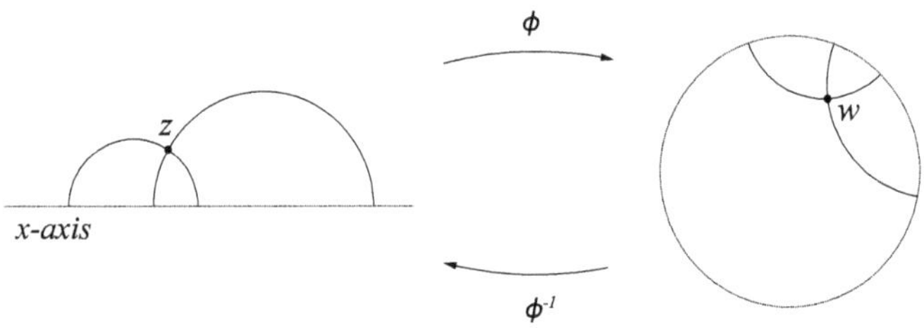

Figure 5.4: The action of the conformal mapping $\phi : \mathbb{H} \to U$ given in the text. The real axis is mapped to the boundary $|w| = 1$. Courtesy Katy Metcalf.

5.5 Exercise. From $ds = \frac{2|dz|}{1-|z|^2}$ and the conformal mapping $\phi : \mathbb{H} \to U$ as in the text, derive equation (5.13).

Then the hyperbolic length of a curve $\gamma(t) : a \le t \le b$ in $\mathbb{H}$ is given by

$$\ell_{\mathbb{H}}(\gamma) = \int_a^b \frac{|\gamma'(t)|}{\mathrm{Im}(\gamma(t))} dt = \int_a^b \frac{2|\phi'(\gamma(t))\gamma'(t)|}{1 - |\phi(\gamma(t))|^2} dt = \ell_{\mathbb{D}}(\phi(\gamma)),$$

and vice-versa, for a curve γ in the Poincaré disk, $\ell_{\mathbb{D}}(\gamma) = \ell_{\mathbb{H}}(\phi^{-1}(\gamma))$, so that geodesics are mapped to geodesics.

5.6 Example. For the line $(t) = k + it$, $a \le t \le b$, we have

$$\ell_{\mathbb{H}}(\lambda) = \int_a^b \frac{dt}{t} = \log \frac{b}{a}.$$

Note that if $\gamma(t) = x(t) + iy(t) : 0 \le t \le 1$ is any curve joining the points $k + ia$ and $k + ib$, then

$$\ell_{\mathbb{H}}(\gamma) = \int_0^1 \frac{|\gamma'(t)|}{\operatorname{Im}(\gamma(t))} dt \ge \int_0^1 \frac{y'(t)}{y(t)} dt = \log\frac{b}{a} = \ell_{\mathbb{H}}(\lambda),$$

showing that the straight-line segment $\lambda(t)$ is a *geodesic*. Observe that we only have equality in the above calculation when $x'(t) = 0$, that is, when $x(t)$ is constant ($= k$) and $|y'(t)| = y'(t) > 0$. In this instance the curve $\gamma(t)$ exactly coincides with $\lambda(t)$, proving that the straight-line segment is the *unique* geodesic from the point $k + ia$ to the point $k + ib$.

Other geodesics are arcs of semicircles meeting the x-axis at right angles since the conformal mapping $\phi^{-1} : U \to \mathbb{H}$ above maps the geodesics of U to geodesics of $\mathbb{H}$ (see the corollary below) as in Figure 5.4; see Figure 5.5 for ideal triangles in both models.

We know by Proposition 1.40 that any Möbius transformation *from the upper half-plane to itself can be written as* $T(z) = \frac{az+b}{cz+d}$ *with* $a, b, c, d \in \mathbb{R}$, $ad - bc = 1$. We will denote all such transformations by $\text{Möb}(\mathbb{H})$.

5.7 Proposition. *For any curve γ in $\mathbb{H}$, its length $\ell_{\mathbb{H}}(\gamma)$ is invariant under any transformation $T \in \text{Möb}(\mathbb{H})$, that is, $\ell_{\mathbb{H}}(\gamma) = \ell_{\mathbb{H}}(T \circ \gamma)$.*

Proof. Letting $\gamma(t) : a \le t \le b$ and $T \in \text{Möb}(\mathbb{H})$, by the chain rule we have

$$\begin{aligned}\ell_{\mathbb{H}}(T \circ \gamma) &= \int_a^b \frac{|(T \circ \gamma)'(t)|}{\operatorname{Im}(T \circ \gamma)(t)} dt \\ &= \int_a^b \frac{|T'(\gamma(t))||\gamma'(t)|}{\operatorname{Im}(T \circ \gamma)(t)} dt\end{aligned}$$

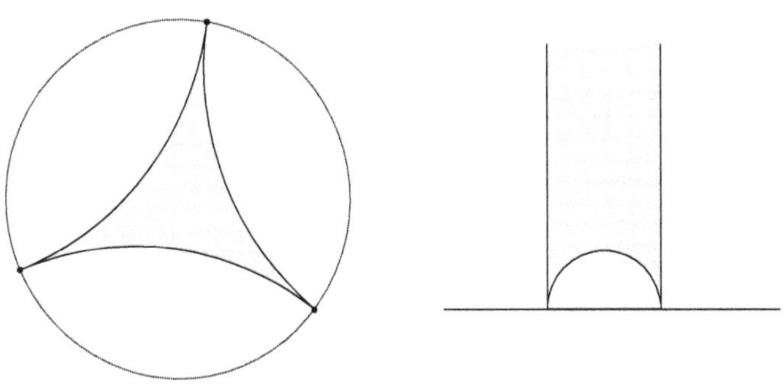

Figure 5.5: (*L*) An ideal triangle in the Poincaré disk (all the vertices meet on the boundary) *and* (*R*) one in the hyperbolic half-plane with one vertex at ∞. Courtesy Katy Metcalf.

$$= \int_a^b \frac{|\gamma'(t)|}{|c\gamma(t)+d|^2} \frac{|c\gamma(t)+d|^2}{\operatorname{Im}(\gamma(t))} dt$$

$$= \int_a^b \frac{|\gamma'(t)|}{\operatorname{Im}(\gamma(t))} = \ell_{\mathbb{H}}(\gamma),$$

establishing the result. □

5.8 Corollary. *Any $T \in$ Möb$(\mathbb{H})$ is an isometry.*

In fact, if γ is a curve joining z_1 to z_2, then $T \circ \gamma$ is a curve from $T(z_1)$ to $T(z_2)$. Since $\ell_{\mathbb{H}}(T \circ \gamma) = \ell_{\mathbb{H}}(\gamma)$ by the proposition, taking the infimum over all curves joining z_1 to z_2 shows that $\rho_{\mathbb{H}}(z_1, z_2) = \rho_{\mathbb{H}}(T(z_1), T(z_2))$.

As a consequence, any geodesic in $\mathbb{H}$ must be unique. For if γ is any geodesic in $\mathbb{H}$ joining the points z_1 and z_2, we can find an isometry $T \in$ Möb$(\mathbb{H})$ that maps γ to the straight-line geodesic λ from ia to ib. If there were some other geodesic γ' in $\mathbb{H}$ joining the points z_1 and z_2 with the same hyperbolic length as γ, then T would map γ' to some curve λ' also connecting the points ia to ib. However,

$$\ell_{\mathbb{H}}(\lambda) = \ell_{\mathbb{H}}(\gamma) = \ell_{\mathbb{H}}(\gamma') = \ell_{\mathbb{H}}(\lambda'),$$

and the uniqueness of the geodesic λ means that $\lambda' \equiv \lambda$, implying that $\gamma' \equiv \gamma$, and this contradiction means that the geodesic γ is unique.

Since geodesics in the half-plane model correspond to geodesics in the Poincaré disk model, it follows that geodesics in the latter are also unique.

The analogue to expression (5.12) in the half-plane model is the following:

5.9 Theorem. *For $z_1, z_2 \in \mathbb{H}$,*

$$\rho_{\mathbb{H}}(z_1, z_2) = \cosh^{-1}\left(1 + \frac{|z_1 - z_2|^2}{2\operatorname{Im}(z_1)\operatorname{Im}(z_2)}\right). \tag{5.14}$$

Proof. Taking $z_1 = ia$, $z_2 = ib$ $(a < b)$, in view of Example 5.6, we have

$$\cosh(\rho_{\mathbb{H}}(z_1, z_2)) = \frac{a^2 + b^2}{2ab} = \frac{a^2 - 2ab + b^2}{2ab} + \frac{2ab}{2ab} = 1 + \frac{|ia - ib|^2}{2\operatorname{Im}(ia)\operatorname{Im}(ib)},$$

which is the desired formulation for the geodesic joining the points ia and ib.

By the preceding corollary any $T \in$ Möb$(\mathbb{H})$ is an isometry, and we claim that the expression $1 + \frac{|z_1 - z_2|^2}{2\operatorname{Im}(z_1)\operatorname{Im}(z_2)}$ is also invariant under the transformation T. To this end, note that

$$|T(z_1) - T(z_2)|^2 = \left|\frac{az_1 + b}{cz_1 + d} - \frac{az_2 + b}{cz_2 + d}\right|^2 = \frac{|z_1 - z_2|^2}{|cz_1 + d|^2 |cz_2 + d|^2}$$

and $\mathrm{Im}(T(z_j)) = \frac{\mathrm{Im}(z_j)}{|cz_j+d|^2}$, $j = 1, 2$. As a consequence,

$$\frac{|T(z_1) - T(z_2)|^2}{\mathrm{Im}(T(z_1))\,\mathrm{Im}(T(z_2))} = \frac{|z_1 - z_2|^2}{\mathrm{Im}(z_1)\,\mathrm{Im}(z_2)},$$

providing the desired invariance. To complete the proof, we choose $T \in \mathrm{M\ddot{o}b}(\mathbb{H})$ that maps the points z_1, z_2 to the points ia, ib, respectively, establishing formula (5.14). □

Regarding the area of a domain D in $\mathbb{H}$ ($z = x + iy$), an area element is $ds^2 = \frac{dxdy}{y^2}$, so that

$$A_{\mathbb{H}}(D) = \iint_D \frac{dxdy}{\mathrm{Im}(z)^2}$$

whenever the integral exists.

Like the hyperbolic length, the hyperbolic area $A_{\mathbb{H}}(D)$ of a domain D in $\mathbb{H}$ is invariant under all $T \in \mathrm{M\ddot{o}b}(\mathbb{H})$.

5.10 Theorem. *For all $T \in \mathrm{M\ddot{o}b}(\mathbb{H})$,*

$$A_{\mathbb{H}}(T(D)) = A_{\mathbb{H}}(D).$$

Proof. Let

$$w = u(x,y) + iv(x,y) = T(z) = \frac{az + b}{cz + d}.$$

Then by the Cauchy–Riemann equations

$$\begin{aligned} dudv &= \left(\frac{\partial u}{\partial x}\frac{\partial v}{\partial y} - \frac{\partial u}{\partial y}\frac{\partial v}{\partial x}\right)dxdy = \left(\left(\frac{\partial u}{\partial x}\right)^2 + \left(\frac{\partial v}{\partial x}\right)^2\right)dxdy \\ &= \left|\frac{\partial}{\partial x}(u + iv)\right|^2 dxdy = \left|\frac{\partial}{\partial x}T(z)\right|^2 dxdy = \left|\frac{d}{dz}T(z)\right|^2 dxdy \\ &= \frac{1}{|cz + d|^4}dxdy. \end{aligned}$$

Now $v = \mathrm{Im}(T(z)) = \frac{y}{|cz+d|^2}$, which means that we now have

$$\begin{aligned} A_{\mathbb{H}}(T(D)) &= \iint_{T(D)} \frac{dudv}{v^2} = \iint_D \frac{|cz + d|^4}{y^2}\frac{1}{|cz + d|^4}dxdy \\ &= \iint_D \frac{dxdy}{y^2} = A_{\mathbb{H}}(D), \end{aligned}$$

establishing the desired invariance. □

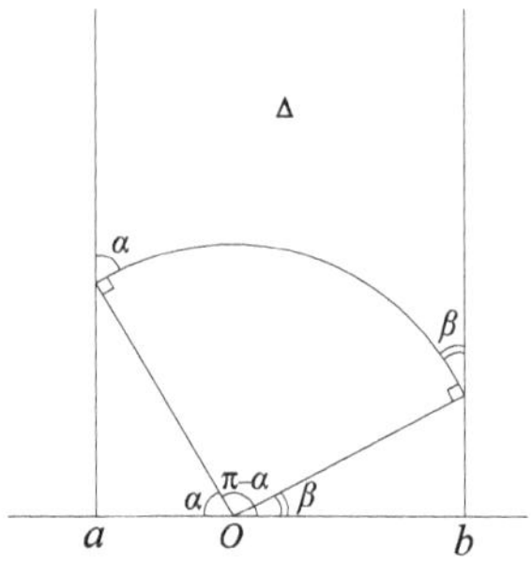

Figure 5.6: The hyperbolic triangle Δ after the Möbius transformation having one vertex at ∞ and one side part of the unit circle. Courtesy Katy Metcalf.

Gauss–Bonnet formula

We are now in a position to prove one of the cornerstones of hyperbolic geometry, the celebrated Gauss–Bonnet formula.

5.11 Gauss–Bonnet formula. *If* Δ *is a hyperbolic triangle in* $\mathbb{H}$ *with angles* α, β, γ, *then its hyperbolic area is given by*

$$A_{\mathbb{H}}(\Delta) = \pi - (\alpha + \beta + \gamma).$$

Proof. We first consider the case where Δ has at least one vertex lying on $\partial\mathbb{H}$. Then there is a Möbius transformation that maps this vertex to the point ∞, with one side a circle and the other two sides parallel lines, preserving both the area and angles of the initial hyperbolic triangle. Then by a translation $T_1(z) = z + b$ and a magnification $T_2(z) = kz$ we can assume that one side of Δ is an arc of the unit circle as in Figure 5.6.

It follows that

$$A_{\mathbb{H}}(\Delta) = \int_a^b \int_{\sqrt{1-x^2}}^{\infty} \frac{1}{y^2} dydx,$$

so that by the substitution $x = \cos\theta$

$$A_{\mathbb{H}}(\Delta) = \int_a^b \frac{1}{\sqrt{1-x^2}} dx = \pi - (\alpha + \beta).$$

In the case where Δ has no vertex on the real axis, we take another Möbius transformation to make one side AC a segment of the infinite vertical line as in Figure 5.7.

Therefore from the preceding case we have

$$\begin{aligned} A_{\mathbb{H}}(\Delta) = A_{\mathbb{H}}(AB\infty) - A_{\mathbb{H}}(CB\infty) &= \pi - (\alpha + \beta + \beta') - (\pi - [(\pi - \gamma) + \beta']) \\ &= \pi - (\alpha + \beta + \gamma), \end{aligned}$$

establishing the formula. □

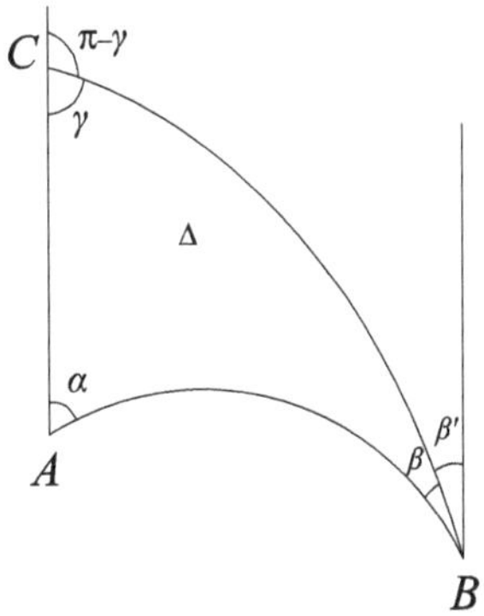

Figure 5.7: The area of the hyperbolic triangle Δ is the difference of the areas of $AB\infty$ and $CB\infty$. Courtesy Katy Metcalf.

Therefore all hyperbolic triangles having the same interior angles have the same hyperbolic area. For example, in the tessellation of the Poincaré disk of Figure 5.8, each hyperbolic triangle has angles $\frac{\pi}{3}$, $\frac{\pi}{3}$, and $\frac{\pi}{4}$, and therefore all have the same hyperbolic area although their Euclidean areas are shrinking to zero as they approach $|z| = 1$. The maximum hyperbolic triangle area is obviously π when $\alpha = \beta = \gamma = 0$ radians, which occurs in the case of an ideal triangle as in Figure 5.5.

Figure 5.8: A tessellation of the Poincaré disk by hyperbolic triangles that each meet at vertices of 6, 6, and 8 triangles. The corresponding angles of each hyperbolic triangle are therefore: 60°, 60°, and 45°, respectively, and as a consequence, all have the same hyperbolic area by the Gauss–Bonnet formula. Moreover, the sum of the interior angles of each hyperbolic triangle is 165° compared to 180° for a Euclidean triangle. The well-known Dutch artist M. C. Escher was duly inspired to capture the essence of infinity incorporating the ideas of hyperbolic geometry. Image public domain.

Hyperbolic geometry plays a significant role in Minkowski spacetime as part of the theory of relativity, and below we provide a brief general discussion as to how it all relates to actual hyperbolas.

Minkowski spacetime

Let us consider the hyperboloid of two sheets (Figure 5.9) but considering only the top surface whose points (x, y, z) satisfy the equation[10]

$$x^2 + y^2 - z^2 = -1 \quad \text{for } z \geq 1.$$

We take the unit disk in the xy-plane with the hyperboloid surface directly above it as in Figure 5.10. Each point on the surface of the hyperboloid can be projected downward into this disk by taking a Euclidean straight line from the point on the hyperbola and connecting it to the point $(0, 0, -1)$. This line intersects the xy-plane at a unique point in the disk. Thus we have a representation of the surface points of the hyperboloid inside the unit disk, which allows the geometry on the hyperboloid to be transferred into the disk.

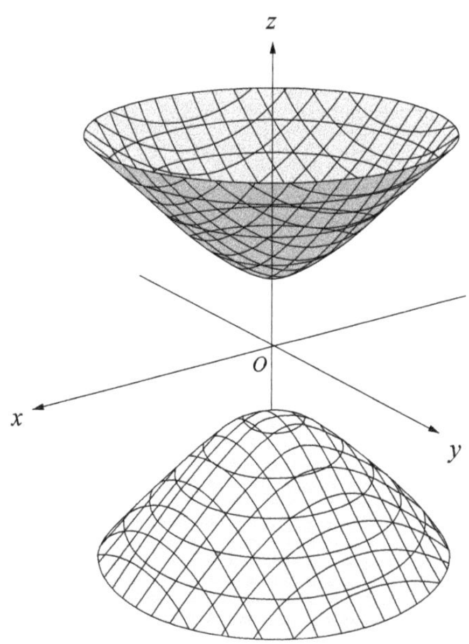

Figure 5.9: The hyperboloid of two sheets with the top surface the one under consideration. Courtesy Katy Metcalf.

10 The two-sheeted hyperboloid can also be defined by $-x^2 - y^2 + z^2 = 1$, resulting in the same two surfaces.

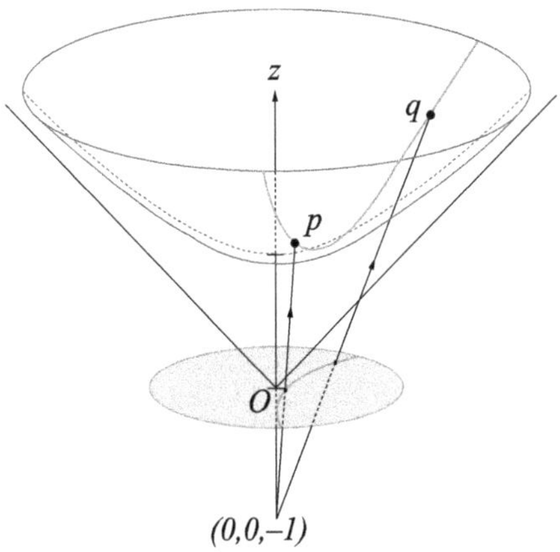

Figure 5.10: A plane passing through two points p and q on the hyperboloid and the origin determining a hyperbolic straight line (geodesic, gray curve) on the hyperboloid. Its counterpart in the Poincaré disk is the circular gray geodesic obtained by the projection discussed in the text. Courtesy Katy Metcalf.

Given two points p, q on the hyperboloid of Figure 5.10, there is a *geodesic* (gray) connecting the two points on the hyperboloid, represented by the points of intersection of a plane passing through those two points and the origin at $(0,0,0)$. Under the preceding correspondence, a geodesic on the hyperbola corresponds to a geodesic in the Poincaré disk, also in gray. It is clear that as a point on the hyperboloid moves further away from the point $(0,0,1)$, the corresponding point in the Poincare disk moves out to the boundary points representing infinity.

If we adjoin a time coordinate to our three spatial coordinates, $\boldsymbol{x} = (x,y,z,t)$, then we now have a point (vector) in a four-dimensional (Euclidean) space. We wish to couple this space with a so-called *indefinite metric* in a somewhat analogous manner to that of Euclidean distance. For an increment (dx, dy, dz, dt), we define an “arc length” by

$$ds^2 = dx^2 + dy^2 + dz^2 - dt^2. \tag{5.15}$$

The geometry is called *pseudo-Euclidean* in view of the minus sign in this expression. The four-dimensional space with this arc length is the *Minkowski spacetime* $\mathcal{M}$, which is a suitable four-dimensional framework for Einstein's theory of special relativity.[11]

Next, let $\mathcal{H}$ be the set of all vectors $\boldsymbol{x} = (x,y,z,t)$ that satisfy the equation

$$x^2 + y^2 + z^2 - t^2 = -1. \tag{5.16}$$

11 Equations such as (5.15) and (5.16) normally involve the velocity of light c in the context of special relativity, but we can set $c = 1$ for convenience.

The hyperboloid $\mathcal{H}$ is a three-dimensional "surface" (that is, a three-dimensional submanifold) inside $\mathcal{M}$ that inherits an arc length from (5.15) and is the three-dimensional analogue of the Poincaré metric described in this chapter.

This framework is indeed a "hyperbolic space" in the sense that by dropping one of the space coordinates, say z, the surface defined by equation (5.16) satisfies

$$x^2 + y^2 - t^2 = -1,$$

which is just a hyperboloid of two sheets as defined above. The direct connection with the Poincaré metric on the unit disk is that for the hyperboloid ($t \geq 0$) with Minkowski metric given by

$$ds^2 = dx^2 + dy^2 - dt^2,$$

there is an isometry with the unit disk possessing the Poincaré metric

$$ds^2 = 4\frac{dx^2 + dy^2}{(1 - x^2 - y^2)^2}.$$

6 Univalent functions

As mentioned in Chapter 1, one-to-one analytic functions are also known as univalent. The terminology slightly varies, and often the term “analytic univalent” is used.

Area Theorem

For a univalent function $f(z)$ on the unit disk $U : |z| < 1$, the following result gives the area of the image.

6.1 Theorem. *If $f(z)$ is an analytic univalent function in $U : |z| < 1$ with Taylor series expansion $f(z) = \sum_{n=0}^{\infty} c_n z^n$, then the area of the image $f(U) = \Omega$ is given by*

$$A(\Omega) = \pi \sum_{n=1}^{\infty} n|c_n|^2.$$

Proof. We have $f = u + iv$ and note that the Jacobian of the transformation is given by $J(x, y) = u_x^2 + v_x^2$, so that by calculus we have

$$A(\Omega) = \iint_U (u_x^2 + v_x^2)dxdy = \int_0^1 \int_0^{2\pi} |f'(re^{i\theta})|^2 r dr d\theta.^1$$

Therefore

$$A(\Omega) = \int_0^1 \int_0^{2\pi} \left(\sum_{n=1}^{\infty} n c_n r^{n-1} e^{i(n-1)\theta} \right) \left(\sum_{m=1}^{\infty} m \overline{c_m} r^{m-1} e^{-i(m-1)\theta} \right) r dr d\theta.$$

Now we need a simple fact that

$$\int_0^{2\pi} e^{ik\theta} d\theta = \begin{cases} 0 & k \neq 0 \\ 2\pi & k = 0. \end{cases}$$

Then, by the uniform convergence of the power series in question,

$$A(\Omega) = \int_0^1 2\pi \sum_{n=1}^{\infty} n^2 |c_n|^2 r^{2n-1} dr = \pi \sum_{n=1}^{\infty} n|c_n|^2,$$

as desired. □

1 Note that $A(\Omega) = D_U(f)$, the Dirichlet integral of f over U discussed in the next chapter.

https://doi.org/10.1515/9783110757828-006

Denoting $U_r : |z| < r < 1$, from the preceding proof we can glean that

$$\text{Area}(f(U_r)) = \pi \sum_{n=1}^{\infty} n|c_n|^2 r^{2n} \geq \pi r^2 |f'(0)|^2.$$

Length–area relations

Considering the closed disk $\overline{U_r} : |z| \leq r < 1$, the length of the image of the boundary $f(|z| = r)$ is given by

$$L(r) = \int_{|z|=r} |f'(z)||dz| = \int_0^{2\pi} |f'(re^{i\theta})| r d\theta,$$

and the Area$(f(U_r)) = A(r)$ is

$$A(r) = \int_0^{2\pi} \int_0^r |f'(\rho e^{i\theta})|^2 \rho d\rho d\theta < \infty.$$

An application of the Cauchy integral formula gives

$$|f'(0)| \leq \frac{1}{2\pi r} \int_{|z|=r} |f'(re^{i\theta})| r d\theta,$$

that is, $L(r) \geq 2\pi r|f'(0)|$. Observe that $L^2(r) \geq 4\pi^2 r^2 |f'(0)|^2$ and also $4\pi A(r) \geq 4\pi^2 r^2 |f'(0)|^2$ by the area theorem. This seems to suggest some close connection between the length $L(r)$ and area $A(r)$ of the enclosed figure. What we are hovering around, here is of course the *isoperimetric inequality*: for any given fixed length L of a curve, the enclosed planar area A satisfies

$$L^2 \geq 4\pi A$$

with equality only for a circle. This inequality has been known since ancient times and is fundamental to the way the world works. It means that the largest enclosed planar area for a given circumference is that of a circle. Various proofs have been given over the years, and it is very important in mathematics and physics.[2]

2 T. Carleman showed that if $f(z)$ is analytic in $|z| \leq 1$ but $\{f(z) : |z| \leq 1\}$ is not necessarily a simply covered image, then

$$4\pi \int_0^{2\pi} \int_0^1 |f'(\rho e^{i\theta})|^2 \rho d\rho d\theta \leq \left(\int_0^{2\pi} |f'(e^{i\theta})| d\theta \right)^2.$$

Using this result, let us consider an annulus $\mathcal{A} : r_0 < r < r_1$, and let Ω be its image under a univalent analytic function $f(z)$. Taking $L(r)$ and $A(r)$ as above and differentiating the latter with respect to r we have

$$A'(r) = \int_0^{2\pi} |f'(re^{i\theta})|^2 r d\theta,$$

and an application of the Cauchy–Schwarz inequality to $L(r)$ yields

$$L^2(r) \leq 2\pi r \int_0^{2\pi} |f'(re^{i\theta})|^2 r d\theta,$$

which gives us

$$4\pi A(r) \leq L^2(r) \leq 2\pi r A'(r).$$

It follows that for $r_0 < r < r_1$,

$$\frac{2}{r} \leq \frac{A'(r)}{A(r)},$$

and integrating from r_0 to r_1,

$$\log \frac{r_1^2}{r_0^2} \leq \log \frac{A(r_1)}{A(r_0)};$$

in other words,

$$\frac{\pi r_1^2}{\pi r_0^2} \leq \frac{A(r_1)}{A(r_0)}.$$

This means that in terms of doubly connected domains Ω bounded by an inner contour C_0 and outer contour C_1, we have the following:

6.2 Theorem. *Among all doubly connected domains conformally equivalent to $\mathcal{A}$, the minimum ratio of the respective areas is attained by the circular annulus.*[3]

Zur Theorie der Minimalflächen, *Math. Zeit.* 9 (1921), 154–160.
Exercise: Verify the inequality for $f(z) = z^n$, $n = 2, 3, 4, \ldots$.

3 This proof is from G. Szegö, Über einige Extremalaufgaben der Potentialtheorie, *Math. Zeit.* 31 (1930), 583–593, although the original result is due to Carleman (1918).

Schlicht class $\mathcal{S}$

To treat univalent functions in a systematic fashion, let us consider the following family of functions on $U : |z| < 1$:

$$\mathcal{S} = \{f(z) \text{ univalent in } U : f(0) = 0, f'(0) = 1\}.$$

This is the family of "schlicht" functions, meaning "simple" in German. Note that if $g(z)$ is univalent in U (hence $g'(z) \neq 0$), then we can define

$$f(z) = \frac{g(z) - g(0)}{g'(0)},$$

so that $f \in \mathcal{S}$. Therefore studying functions of $\mathcal{S}$ is equivalent to studying all univalent functions on any simply connected $D \subsetneq \mathbb{C}$, since any such domain D is conformally equivalent to U by the Riemann mapping theorem (Chapter 4).

Koebe function

The study of schlicht functions was begun by German mathematician Paul Koebe (1882–1945). If $U_{1/4} = \{|z| < 1/4\}$, the Koebe 1/4 theorem (proved in the sequel) states that

$$U_{1/4} \subseteq \bigcap_{f \in \mathcal{S}} f(U),$$

that is, a disk of radius $\frac{1}{4}$ (centered at the origin) is contained in the images of the unit disk by all mappings $f \in \mathcal{S}$. Koebe also showed that the radius $\frac{1}{4}$ is the best possible by noting that the *(Koebe) function*

$$k(z) = \frac{z}{(1-z)^2}$$

belongs to $\mathcal{S}$ and maps U onto the exterior of the line $x \leq -1/4$.[4] Thus in some sense, the Koebe function has the largest possible range.

There are growth constraints on schlicht functions, and again we will find $k(z)$ maximal.

For example, if $z = e^{i\theta}$, then

$$|k(z)| = \frac{r}{1 - 2r\cos\theta + r^2},$$

so that

4 This is easily seen as $k(z) = \frac{1}{4}\left[\left(\frac{1+z}{1-z}\right)^2 - 1\right]$, by unwinding the mappings with $\zeta = \frac{1+z}{1-z}$, which maps the unit disk to the right half-plane.

$$\max_{|z|\le r}|k(z)| = \frac{r}{(1-r)^2},$$

and therefore by Theorem 6.9(a) in the sequel the Koebe function attains the maximum modulus of all functions belonging to $\mathcal{S}$.

Every $f \in \mathcal{S}$ has a Taylor series expansion

$$f(z) = z + a_2 z^2 + a_3 z^3 + \cdots,$$

and computing the Taylor series for the Koebe function, we have

$$k(z) = \sum_{n=1}^{\infty} nz^n = z + 2z^2 + 3z^3 + \cdots.$$

The *rotated Koebe function* is given by

$$k_\alpha(z) = \frac{z}{(1-\alpha z)^2} = \sum_{n=1}^{\infty} n\alpha^{n-1}z^n$$

for $|\alpha| = 1$.

Bieberbach conjecture

The question then naturally arises as to whether the Taylor series coefficients of $k(z)$ are maximal? Ludwig Bieberbach (1886–1982) showed that $|a_2| \le 2$ and conjectured[5] that $|a_n| \le n$, $n = 1, 2, 3, \ldots$, for all $f \in \mathcal{S}$. The general case has a long and colorful history discussed in the sequel. However, there is one case that we can prove without difficulty. See also the discussion regarding a continuous functional in Chapter 7.

Typically real functions

6.3 Definition. An analytic function $f(z)$ in U is called *typically real*[6] if $f(z)$ is real on the real axis and only there.

6.4 Theorem. *If $f \in \mathcal{S}$ is a typically real function, then $|a_n| \le n$ for $= 1, 2, 3, \ldots$.*

Proof. Since f is typically real, its Maclaurin series must have real coefficients. Setting $f(z) = \sum_{n=1}^{\infty} a_n z^n$ and $f = u + iv$, we have

$$v(re^{i\theta}) = \sum_{n=1}^{\infty} a_n r^n \sin n\theta.$$

5 Uber die Koeffizienten derjenigen Potenzreihen, welche eine schlichte Abbildung des Einheitskreises vermitteln, *Sitzung. Preuss. Akad. Wissen.* (1916), 940–955.

6 Typically real functions were introduced by Rogosinski (1931).

L. de Branges had indeed proved the conjecture, and we congratulated him on this great achievement." De Branges had actually proven the Milin conjecture, which in turn implied the Robertson conjecture, which implied the Bieberbach conjecture. The original proof of de Branges was reformulated and simplified by I. M. Milin and others in attendance at the Steklov seminars.[9]

Robertson conjecture

Briefly, the Robertson conjecture is concerned with the class of *odd* analytic functions $\hat{\mathcal{S}}$ of the form

$$\hat{f}(z) = \sqrt{f(z^2)} = \sum_{k=1}^{\infty} b_{2k-1} z^{2k-1} = b_1 z + b_3 z^3 + b_5 z^5 + \cdots \tag{6.1}$$

for $f \in \mathcal{S}$ and $b_1 = 1$, where for $\sqrt{f(z^2)} = z(1 + a_2 z^2 + a_3 z^4 + \cdots)^{1/2}$, we have taken the analytic branch of the latter square root having the value 1 at $z = 0$. Then $\hat{f}(0) = 0$, $\hat{f}'(0) = 1$, and if $\hat{f}(z_1) = \hat{f}(z_2)$, and thus $f(z_1^2) = f(z_2^2)$, then $z_1 = \pm z_2$. If $z_1 = -z_2$, then

$$\hat{f}(z_1) = \hat{f}(-z_2) = -\hat{f}(z_2) = -\hat{f}(z_1),$$

which means that $\hat{f}(z_1) = 0$. Since $\hat{f}(z) = z(1 + b_3 z^2 + b_5 z^4 + \cdots)$, we obtain $z_1 = z_2 = 0$, and hence $\hat{f}$ is univalent and belongs to $\mathcal{S}$.

To relate the coefficients of any $\hat{f} \in \mathcal{S}$ to its corresponding $f \in \mathcal{S}$, a bit of calculation shows that[10]

$$a_n = \sum_{k=1}^{n} b_{2k-1} b_{2(n-k)+1} = b_1 b_{2n-1} + b_3 b_{2(n-2)+1} + \cdots + b_{2n-1} b_1$$

and $b_1 = 1$. Then by the Cauchy–Schwarz inequality we find that[11]

$$|a_n| \le \sum_{k=1}^{n} |b_{2k-1}|^2 = 1 + |b_3|^2 + |b_5|^2 + \cdots + |b_{2n-1}|^2$$

And, specifically, $|a_2| \le 1 + |b_3|^2$ and $|a_3| \le 1 + |b_3|^2 + |b_5|^2$. Since it was known that $|b_3| \le 1$, this immediately gives $|a_2| \le 2$.

9 The proof can be found in de Branges, L., A proof of the Bieberbach conjecture, *Acta Math.* 154 (1985), 137–152.

10 For example,

$$\left(b_1 z + b_3 z^3 + b_5 z^5 + \cdots\right)\left(b_1 z + b_3 z^3 + b_5 z^5 + \cdots\right) = f\left(z^2\right) = z^2 + a_2 z^4 + a_3 z^6 + \cdots$$

gives $a_3 = b_1 b_5 + b_3 b_3 + b_5 b_1$, and so forth as above.

11 Again by Cauchy–Schwarz

$$|a_3|^2 = |b_1 b_5 + b_3 b_3 + b_5 b_1|^2 \le \left(|b_1|^2 + |b_3|^2 + |b_5|^2\right)\left(|b_5|^2 + |b_3|^2 + |b_1|^2\right).$$

A conjecture by M. S. Robertson (1936)[12] claims that for all functions $\hat{f} \in \hat{\mathcal{S}}$,

$$\sum_{k=1}^{n} |b_{2k-1}|^2 \le n, \quad n = 2, 3, 4, \dots$$

($b_1 = 1$), with equality only for rotations of the Koebe function. Robertson proved that $|b_3|^2 + |b_5|^2 \le 2$, that is, $|a_3| \le 3$. Only much later, in 1970, a proof of the Robertson conjecture for $n = 4$ was given by S. Friedland. Clearly, a proof of the Robertson conjecture would imply that of the Bieberbach conjecture.

Milin conjecture

We next consider the Milin conjecture, which concerns the "logarithmic coefficients" of the function

$$\log \frac{f(z)}{z} = 2 \sum_{n=1}^{\infty} c_n z^n \tag{6.2}$$

for $f \in \mathcal{S}$ and $|z| < 1$. Here we take the analytic branch of $f(z)/z$ that vanishes at the origin. The inequality due to Lebedev and Milin showed the relationship between coefficients of (6.1) and (6.2):

$$\sum_{k=1}^{n+1} |b_{2k-1}|^2 \le (n+1) \exp\left\{ \frac{1}{n+1} \sum_{m=1}^{n} \sum_{k=1}^{m} \left(k|c_k|^2 - \frac{1}{k} \right) \right\}, \tag{6.3}$$

$n = 1, 2, 3, \dots$, and subsequently Milin conjectured that for all $f \in \mathcal{S}$,

$$\sum_{m=1}^{n} \sum_{k=1}^{m} \left(k|c_k|^2 - \frac{1}{k} \right) \le 0$$

with equality only for rotations of the Koebe function. From (6.3) this would immediately imply the Robertson conjecture and was proved for $n = 1, 2, 3$ by Grinshpan and by de Branges for all $n = 1, 2, 3, \dots$. The proof by de Branges relies on a deep result from special functions that had been proved earlier by R. Askey and G. Gasper.[13]

In a different vein, the asymptotic Bieberbach conjecture was established by Walter Hayman (1955):[14]

$$\lim_{n\to\infty} \frac{|a_n|}{n} = \alpha(f) \le 1$$

12 A remark on the odd schlicht functions, *Bull. Amer. Math. Soc.* 42 (1936), 366–370.

13 Positive Jacobi polynomial sums. II, *Amer. J. Math.* 98 (1976), 709–737.

14 The asymptotic behavior of p-valent functions, *Proc. London Math. Soc.* 5 (1955), 257–284.

with equality only for a rotation of the Koebe function. Observe that this result is not a consequence of the result of de Branges.

Complementary class Ξ

There is a complementary set to $\mathcal{S}$,

$$\Xi = \left\{ g(z) \text{ univalent in } |z| > 1 : g(z) = z + b_0 + \sum_{n=1}^{\infty} \frac{b_n}{z^n} \right\}.$$

Such functions have a simple pole at ∞. Note that if $f \in \mathcal{S}$, then the function

$$g(z) = \frac{1}{f(\frac{1}{z})} = \frac{1}{\frac{1}{z} + \frac{a_2}{z^2} + \frac{a_3}{z^3} + \cdots} = \frac{z}{1 + \frac{a_2}{z} + \frac{a_3}{z^2} + \cdots} = z + b_0 + \sum_{n=1}^{\infty} \frac{b_n}{z^n}$$

belongs to Ξ, and since $f(w)$ is finite for all $|w| < 1, g(z)$ never equals zero.

Grönwall area theorem

The following result is known as the *area theorem* due to T. H. Grönwall (1877–1932).[15]

6.5 Theorem. *For any* $g \in \Xi$,

(i) *The area of* $E = \mathbb{C} \setminus g(\{|z| > 1\})$ *is given by*

$$A(E) = \pi\left(1 - \sum_{n=1}^{\infty} n|b_n|^2\right),$$

(ii)

$$\sum_{n=1}^{\infty} n|b_n|^2 \leq 1.$$

As a consequence, $|b_n| \leq \frac{1}{\sqrt{n}}$, *and, in particular,* $|b_1| \leq 1$.

Proof. Let C_r be the circle $|z| = r > 1$. For any $g \in \Xi$, let $\Omega_r = \mathbb{C} - g(\{|z| > r\})$, which has a simple closed smooth boundary $\partial\Omega_r = \Gamma_r$. Then Green's theorem for $w = g(z)$ gives[16]

15 Some remarks on conformal representations, *Ann. Math.* 16 (1914–15), 72–76.

16 For a positively oriented piecewise smooth simple closed curve C bounding a domain D in the z-plane and $z = x + iy$, from Green's theorem in the plane, $\int_C Pdx + Qdy = \iint_D \left(\frac{\partial Q}{\partial x} - \frac{\partial P}{\partial y}\right)dxdy$, we have

$$\int_C \bar{z}dz = \int_C (x - iy)dx + (ix + y)dy = 2i \iint_D dxdy.$$

$$A(\Omega_r) = \iint_{\Omega_r} d\Omega_r = \frac{1}{2i}\int_{\Gamma_r} \overline{w}dw.$$

Hence

$$\begin{aligned}
0 \le A(\Omega_r) &= \frac{1}{2}\int_0^{2\pi} \overline{g(re^{i\theta})}g'(re^{i\theta})re^{i\theta}d\theta \\
&= \frac{1}{2}\int_0^{2\pi}\left[\left(re^{-i\theta} + \sum_{m=0}^{\infty}\overline{b_m}r^{-m}e^{im\theta}\right)\left(1 - \sum_{n=1}^{\infty} nb_n r^{-n-1}e^{-i(n+1)\theta}\right)\right]re^{i\theta}d\theta \\
&= \frac{1}{2}\int_0^{2\pi}\left[\left(re^{-i\theta} + \sum_{m=0}^{\infty}\overline{b_m}r^{-m}e^{im\theta}\right)\left(re^{i\theta} - \sum_{n=1}^{\infty} nb_n r^{-n}e^{-in\theta}\right)\right]d\theta \\
&= \pi\left(r^2 - \sum_{n=1}^{\infty} n|b_n|^2 r^{-2n}\right).
\end{aligned}$$

As a consequence,

$$\sum_{n=1}^{\infty} n|b_n|^2 r^{-2n} \le r^2,$$

and (i) and (ii) follow by letting $r \to 1$ (which is permissible as the infinite sum is a decreasing function of r and uniformly bounded above). □

Note that $g(z) = z + e^{i\beta}/z \in \Xi$, so that the bound on $|b_1|$ is strict. It is also evident that equality holds in (ii) if and only if the area of $E = \mathbb{C} - g(\{|z| > 1\})$ is zero.

Armed with the Grönwall area theorem, we are able to give a simple proof of Bieberbach's initial result.

6.6 Corollary. *If $f \in \mathcal{S}$ with $f(z) = z + a_2z^2 + a_3z^3 + \cdots$, then $|a_2| \le 2$.*

Proof. Consider the function

$$f(z^2) = z^2 + a_2z^4 + a_3z^6 + \cdots = z^2(1 + a_2z^2 + a_3z^4 + \cdots).$$

Then we set

$$g(z) = \sqrt{f(z^2)} = z(1 + a_2z^2 + a_3z^4 + \cdots)^{\frac{1}{2}}$$

again taking the single-valued analytic branch of the latter square root function that has the value 1 at the origin. As in our preceding discussion, $g(0) = 0$ and $g'(0) = 1$ and its univalence follows as for the function $\hat{f} \in \mathcal{S}$ implying that $g \in \mathcal{S}$. Consequently, $h(z) = 1/g(1/z) \in \Xi$.

To deal with the latter function of g, we first have[17]

$$g\left(\frac{1}{z}\right)=\frac{1}{z}\left(1+\frac{1}{2}\frac{a_2}{z^2}+\cdots\right),$$

so that we obtain the Laurent series expansion

$$h(z)=z\frac{1}{(1+\frac{1}{2}\frac{a_2}{z^2}+\cdots)}=z\left(1-\frac{1}{2}\frac{a_2}{z^2}+\cdots\right)=z-\frac{1}{2}\frac{a_2}{z}+\cdots.$$

An application of the theorem implies $|a_2|\le 2$. □

6.7 Remark. If $f(z)$ is analytic univalent in $0<|z|<1$ and given by

$$f(z)=\frac{1}{z}+\sum_{n=1}^{\infty}a_nz^n,$$

then

$$\sum_{n=1}^{\infty}n|a_n|^2\le 1,$$

and this fact is also called the area theorem, related to the former by the transformation $w=1/z$.

Following Pommerenke (1975), since $|a_2|\le 2$ for any function $f\in\mathcal{S}$, let us shift this inequality to an arbitrary point $z_0\in U$. To this end, let

$$T(z)=\frac{z+z_0}{1+\overline{z_0}z}$$

be the conformal mapping from the disk to the disk with $T(0)=z_0$ and consider the analytic univalent function $F(z)=f(T(z))$. Although $F(z)$ is not necessarily in $\mathcal{S}$, we have the Taylor series expansion

$$F(z)=F(0)+F'(0)z+\frac{F''(0)z^2}{2}+\cdots.$$

Thus $T'(0)=1-|z_0|^2$ and $T''(0)=-2\overline{z_0}(1-|z_0|^2)$, so that

$$F(z)=f(z_0)+[(1-|z_0|^2)f'(z_0)]z+\frac{1}{2}[(1-|z_0|^2)^2f''(z_0)-2\overline{z_0}(1-|z_0|^2)f'(z_0)]z^2+\cdots.$$

17 Write $(1+a_2z^2+a_3z^4+\cdots)^{\frac{1}{2}}=1+b_1z+b_2z^2+\cdots$, so that

$$(1+b_1z+b_2z^2+\cdots)(1+b_1z+b_2z^2+\cdots)=1+a_2z^2+a_3z^4+\cdots$$

Equating coefficients, we find that $b_1=0$ and $b_2=\frac{1}{2}a_2$ and $g(z)=z(1+\frac{1}{2}a_2z^2+\cdots)$.

To normalize $F(z)$ in order to obtain a schlicht function, as before, we take

$$G(z) = \frac{F(z) - F(0)}{F'(0)} = \frac{F(z) - f(z_0)}{(1 - |z_0|^2)f'(z_0)} = z + \frac{1}{2}\left[(1 - |z_0|^2)\frac{f''(z_0)}{f'(z_0)} - 2\overline{z_0}\right]z^2 + \cdots. \quad (6.4)$$

Since $G(z) \in \mathcal{S}$, we obtain

$$\left|(1 - |z_0|^2)\frac{f''(z_0)}{f'(z_0)} - 2\overline{z_0}\right| \le 4.$$

Dividing each side by $1 - |z_0|^2$, since z_0 was an arbitrary point in U, we arrive at the following:

6.8 Lemma. *For every $f \in \mathcal{S}$,*

$$\left|\frac{f''(z)}{f'(z)} - \frac{2\bar{z}}{1 - |z|^2}\right| \le \frac{4}{1 - |z|^2}.$$

Koebe distortion theorem

The lemma now allows us to establish various growth rate estimates on both f and f'. This is interesting in that univalence turns out to be such a strong determining factor in constraining the growth of functions and their derivatives.

6.9 Theorem. *For $f \in \mathcal{S}$ and $|z| < 1$,*

$$\text{(a)} \quad \frac{|z|}{(1 + |z|)^2} \le |f(z)| \le \frac{|z|}{(1 - |z|)^2},$$

$$\text{(b)} \quad \frac{1 - |z|}{(1 + |z|)^3} \le |f'(z)| \le \frac{1 + |z|}{(1 - |z|)^3},$$

$$\text{(c)} \quad \frac{1 - |z|}{1 + |z|} \le \left|z\frac{f'(z)}{f(z)}\right| \le \frac{1 + |z|}{1 - |z|}.$$

Equality holds in each case for rotations of the Koebe function.

Proof. We will establish the validity of (b) first. For $z = re^{i\theta}$, we can rephrase Lemma 6.8 as

$$\left|\frac{f''(z)}{f'(z)}e^{i\theta} - \frac{2|z|}{1 - |z|^2}\right| \le \frac{4}{1 - |z|^2}.$$

The first term f''/f' means we are dealing with the derivative of $\log f'(z)$, and the second term is again the derivative of $\log u$ for $u = 1 - r^2$. So let us take a branch of $\log f'(z)$ such that $\log f'(0) = 0$, so that differentiating with respect to r we have

$$\left|\frac{\partial}{\partial r}[\log f'(re^{i\theta}) + \log(1 - r^2)]\right| \le \frac{4}{1 - r^2}.$$

Integrating both sides along the line connecting 0 with $|z| = r$, we find that (abusing the notation somewhat)

$$\left| \int_0^{|z|} \frac{\partial}{\partial r} [\log(1 - r^2) f'(re^{i\theta})] dr \right| \le 2 \log\left(\frac{1 + |z|}{1 - |z|} \right),$$

and therefore

$$\left| \log(1 - |z|^2) f'(z) \right| \le 2 \log\left(\frac{1 + |z|}{1 - |z|} \right). \tag{6.5}$$

Rewriting inequality (6.5), we have

$$-2 \log\left(\frac{1 + |z|}{1 - |z|} \right) \le \log[(1 - |z|^2)|f'(z)|] \le 2 \log\left(\frac{1 + |z|}{1 - |z|} \right),$$

and rearranging the terms of this last expression,

$$-2 \log\left(\frac{1 + |z|}{1 - |z|} \right) - \log(1 - |z|^2) \le \log|f'(z)| \le 2 \log\left(\frac{1 + |z|}{1 - |z|} \right) - \log(1 - |z|^2).$$

This simplifies to inequality (b):

$$\frac{1 - |z|}{(1 + |z|)^3} \le |f'(z)| \le \frac{1 + |z|}{(1 - |z|)^3}.$$

For the right-hand inequality of (a), note that we can integrate the derivative, and since $f(0) = 0$,

$$|f(z)| \le \int_0^{|z|} |f'(re^{i\theta})| dr \le \int_0^{|z|} \frac{1 + r}{(1 - r)^3} dr = \frac{|z|}{(1 - |z|)^2}.$$

The proof of the left-hand inequality of (a) is more interesting. Consider any circle $C_r : |z| = r < 1$. Then there is a point $w_0 = f(z_0)$ ($w_0 \neq 0$) on the image curve $f(C_r)$ that has a minimum modulus, that is,

$$|f(z_0)| = \min_{z \in C_r} |f(z)|.$$

Let Γ represent the straight-line path from the origin to the point w_0, and let $\gamma = f^{-1}(\Gamma)$ be the corresponding curve connecting 0 with z_0. Then the length of Γ is just the value $|f(z_0)|$. Hence

$$|f(z_0)| = \int_\Gamma |dw| = \int_\gamma |f'(z)||dz| \ge \int_\gamma |f'(z)| d|z|.$$

The left-hand inequality of (b) is applicable to each value $|f'(z)|, z \in \gamma$, in the preceding integral for $z = \rho e^{i\theta}$ with $0 < \rho < r$:

$$|f(z_0)| \geq \int_0^r \frac{1-\rho}{(1+\rho)^3} d\rho = \frac{r}{(1+r)^2},$$

and likewise for any $z \in C_r$, establishing the inequality. □

Notice that the preceding inequality implies that as $|z| \to 1$, we have $|f(z)| \geq 1/4$ for all $f \in \mathcal{S}$, which is just the Koebe one-quarter theorem.

On the other hand, setting $w = g(z) = \frac{f(z)}{z} = 1 + a_2 z + \cdots$, note that the point $w = 1$ is an interior point of $g(U)$ and $w = 0$ is an exterior point. Thus $\mathrm{dist}(0, \partial(g(U))) < 1$, so that

$$\mathrm{dist}(0, \partial(f(U))) = \varliminf_{|z|\to 1} |f(z)| = \varliminf_{|z|\to 1} |g(z)| \leq 1.$$

To prove (c), note that from (6.4) we have $G(z) \in \mathcal{S}$, and since $F(-z_0) = f(0) = 0$,

$$|G(-z_0)| = \frac{|f(z_0)|}{(1 - |z_0|^2)|f'(z_0)|}.$$

As $G(z) \in \mathcal{S}$, we can apply the inequalities of (a) to $G(z)$, which yield the inequalities of (c) in terms of z_0, and since z_0 is an arbitrary point of U, the proof of the inequality is complete. The reader can check that equality holds in each case for rotations of the Koebe function.

Note that the family $\mathcal{S}$ is locally bounded on compact subsets of U by property (a) of the theorem and hence is a normal family. It is also a compact family. Indeed, if $\{f_n\}$ is a sequence of functions in $\mathcal{S}$ such that $f_n \to f$ uniformly on compact subsets of U, then f is analytic in $U, f(0) = 0$, and $f'(0) = 1$, implying that f is not identically constant. By Corollary 1.29 of the Hurwitz theorem f is univalent. Hence $f \in \mathcal{S}$, and $\mathcal{S}$ is compact.

Furthermore, if $\{f_n\}$ is a sequence of functions in $\mathcal{S}$ such that $f_n \to f$ pointwise in U, then the local boundedness of $\mathcal{S}$ means that we can apply the Vitali–Porter theorem, resulting in the uniform convergence $f_n \to f$ on compact subsets of U and $f \in \mathcal{S}$.

Bloch theorem

We will prove a remarkable theorem dating back to the work of A. Bloch. This version is slightly more restrictive than the more general one as it assumes that our function is analytic in the closed unit disk $|z| \leq 1$.[18]

18 The above theorem can be suitably adapted for the open unit disk U. A more general version and proof can be found in Schiff (1993), p. 112.

Note that for any analytic function $f(z)$ defined in a domain Ω, if $f'(a) \neq 0$ for some point $a \in \Omega$, then although the image of Ω under the mapping $f(z)$ is in general a Riemann surface, there will be a (*schlicht*) disk D_b about the point $b = f(a)$ such that $f(z)$ maps some subdomain of Ω univalently onto D_b. The question naturally arises as to how large the radius of such a disk can be in the special setting that follows.

We first require a simple lemma.

6.10 Lemma. *Suppose that $f(z)$ analytic in the disk Δ_0 with center z_0. If $f(z)$ satisfies*

$$|f'(z) - f'(z_0)| \leq |f'(z_0)|$$

in Δ_0, then $f(z)$ is univalent in Δ_0.

Proof. We will show that for any two points $z_1, z_2 \in \Delta_0, z_1 \neq z_2$ implies that $f(z_1) \neq f(z_2)$. Letting ℓ be a straight-line segment joining z_1 and z_2, observe that

$$|f(z_2) - f(z_1)| = \left| \int_\ell f'(\zeta) d\zeta \right|,$$

and hence it follows that

$$|f(z_2) - f(z_1)| = \left| \int_\ell (f'(\zeta) - f'(z_0) + f'(z_0)) d\zeta \right| \geq |(z_2 - z_1) f'(z_0)| - \int_\ell |f'(\zeta) - f'(z_0)||d\zeta|.$$

By the hypothesis, $\int_\ell |f'(\zeta) - f'(z_0)||d\zeta| < |f'(z_0)||z_2 - z_1|$, and we conclude that $|f(z_2) - f(z_1)| > 0$. □

6.11 Corollary. *If $f(z) = z + a_2 z^2 + a_3 z^3 + \cdots$ is analytic in U with $\sum_{n=2}^{\infty} n|a_n| < 1$, then $f(z)$ is univalent.*

Indeed, with $z_0 = 0$, we have

$$|f'(z) - f'(0)| = \left| \sum_{n=2}^{\infty} n a_n z^{n-1} \right| < 1 = |f'(0)|.$$

Employing the above lemma and the Koebe $\frac{1}{4}$-theorem, we can now establish a particular version of the following theorem. Its validity is somewhat remarkable.

6.12 Theorem (Bloch).[19] *Let $f(z)$ be an analytic function in $\overline{U} : |z| \leq 1$ such that $f'(0) = 1$. Then there exists a subdomain $\Delta \subset U$ on which $f(z)$ is univalent and $f(\Delta)$ contains a disk of radius $b = \frac{1}{24} = 0.041666\ldots$.*

19 Les théorèmes de M. Valiron sur les fonctions entières et la théorie de l'uniformisation, *Ann. Fac. Sci. Univ. Toulouse* (1925), 1–22.

Proof. For $0 \le r \le 1$, define

$$M'(r) = \max_{|z|=r} |f'(z)|,$$

and consider the function $\phi(r) = (1-r)M'(r)$. Clearly, $M'(r)$ is continuous, and so is $\phi(r)$, which also satisfies $\phi(0) = 1$, $\phi(1) = 0$. Then there is $r_0 < 1$ such that $\phi(r_0) = 1$ and $\phi(r) < 1$ whenever $r > r_0$.

Since $\phi(r_0) = 1$, there is a point z_0 with $|z_0| = r_0$, $M'(r_0) = |f'(z_0)|$, and

$$|f'(z_0)| = \frac{1}{1-r_0}.$$

Now we consider the disk $\Delta_{\rho_0} : |z - z_0| < \rho_0 = \frac{1}{2}(1-r_0)$. Note that for any $z \in \Delta_{\rho_0}$ we obtain $|z| < \frac{1}{2}(1+r_0)$, that is, $\Delta_{\rho_0} \subset U_{\frac{(1+r_0)}{2}}$.

Furthermore, since $r_0 < (1+r_0)/2$, by the maximum modulus principle, for $z \in \Delta_{\rho_0}$, we have

$$|f'(z)| \le M'\left(\frac{(1+r_0)}{2}\right) < \frac{\phi(\frac{(1+r_0)}{2})}{1-\frac{(1+r_0)}{2}} < \frac{1}{1-\frac{(1+r_0)}{2}} = \frac{1}{\rho_0}.$$

As a consequence,

$$|f'(z) - f'(z_0)| \le |f'(z)| + |f'(z_0)| < \frac{3}{2\rho_0}.$$

Now we can apply Corollary 1.9 of the Schwarz lemma to the function $g(z) = f'(z) - f'(z_0)$ in Δ_{ρ_0} to obtain

$$|f'(z) - f'(z_0)| < \frac{3|z-z_0|}{2\rho_0^2}.$$

Next, consider the disk $\Delta = \Delta_{\rho_0/3}$ about the point z_0, so that

$$|f'(z) - f'(z_0)| < \frac{1}{2\rho_0} = |f'(z_0)| \quad \text{for } z \in \Delta.$$

Then Lemma 6.10 implies that $f(z)$ is univalent in the disk Δ. It only remains to show that $f(\Delta)$ contains a disk of radius $1/24$.

For $U : |\zeta| < 1$, define the analytic function

$$F(\zeta) = \frac{f(\frac{\rho_0}{3}\zeta + z_0) - f(z_0)}{\frac{\rho_0}{3} f'(z_0)},$$

so that $F(0) = 0$ and $F'(0) = 1$. Then the Koebe $\frac{1}{4}$-theorem implies that $F(U)$ contains a disk of radius $\frac{1}{4}$. Since the denominator in the above expression satisfies

$$\left|\frac{\rho_0}{3}f'(z_0)\right| = \frac{1}{6},$$

we conclude that $f(\Delta)$ contains a disk of radius 1/24 about the point $f(z_0)$. □

6.13 Remarks. The theorem serves only as an indication of what is now generally referred to as Bloch's theorem, which then leads to Bloch's constant. Let

$$\mathcal{F} = \{f \text{ analytic in } U, f'(0) = 1\}.$$

For each admissible $f \in \mathcal{F}$, let $b(f)$ be the supremum of all such values b found in the theorem. Then the *Bloch constant* is defined by

$$\mathcal{B} = \inf_{f\in\mathcal{F}}\{b(f)\},$$

and its value lies between

$$0.4332127\cdots = \frac{\sqrt{3}}{4} + 2\times 10^{-4} < \mathcal{B} \le \frac{\Gamma(\frac{1}{3})\Gamma(\frac{11}{12})}{\sqrt{\sqrt{3}+1}\Gamma(\frac{1}{4})} = 0.4718617\ldots,$$

where Ahlfors and Grunsky (1937) have conjectured that the upper limit is the actual value. The gamma function, which makes a noteworthy appearance above, will be discussed in detail in Chapter 10. Then Bloch's theorem would entail that for each $f \in \mathcal{F}$, there is a subdomain $\Delta \subset U$ such that $f(z)$ is univalent on Δ and $f(\Delta)$ contains a disk of radius $\mathcal{B}$.

Univalent polynomials

Something interesting can also be said about the coefficients of univalent polynomials

$$f(z) = z + a_2 z^2 + a_3 z^3 + \cdots + a_n z^n$$

in the unit disk U, which are actually curious mixes of the identity and multivalent functions. Indeed, the derivative

$$f'(z) = 1 + 2a_2 z + 3a_3 z^2 + \cdots + n a_n z^{n-1} = n a_n\left(z^{n-1} + \cdots + \frac{1}{n a_n}\right)$$

has $n-1$ complex roots $\alpha_1, \alpha_2, \ldots, \alpha_{n-1}$. As a consequence,

$$f'(z) = n a_n (z-\alpha_1)(z-\alpha_2)\ldots(z-\alpha_{n-1}),$$

and since $f'(z) \neq 0$ for all $z \in U$, all roots α_i, $i = 1, 2, \ldots, n-1$, satisfy $|\alpha_i| \ge 1$. Therefore

$$1 = |f'(0)| = n|a_n||\alpha_1||\alpha_2| \cdots |\alpha_{n-1}|,$$

and we obtain the inequality $|a_n| \leq \frac{1}{n}$ for all n.

Normal functions

We conclude this chapter by recalling that a normal function satisfies the condition

$$\sup_{z \in U}(1 - |z|^2)|f^{\#}(z)| < \infty.$$

As it turns out, the univalence is a sufficient condition for an analytic function to be normal. This should not be so surprising as the growth conditions of Theorem 6.9 for schlicht functions indicate how the univalence constrains the growth rate of the function.

6.14 Theorem. *A univalent analytic function $f(z)$ in U is a normal function.*

Proof. Recall that we can write $f(z) = ag(z) + b$ where $g(z)$ is a function belonging to the schlicht class $\mathcal{S}$ with $a = f'(0)$ and $b = f(0)$. Upon applying part (c) of Theorem 6.9 to $g(z)$, namely

$$\left| z\frac{g'(z)}{g(z)} \right| \leq \frac{1 + |z|}{1 - |z|},$$

to the expression

$$\frac{(1 - |z|^2)|f'(z)|}{1 + |f(z)|^2} = \frac{(1 - |z|^2)|ag(z)|}{1 + |ag(z) + b|^2} \cdot \frac{|g'(z)|}{|g(z)|},$$

we find that $(1 - |z|^2)f^{\#}(z)$ remains bounded as $r \to 1$, and thus $f(z)$ is a normal function. □

See the excellent book by Pommerenke (1975) for further details on univalent functions.

7 Harmonic functions

Harmonic functions $u(x, y)$ are C^2-solutions of the partial differential Laplace equation

$$\Delta u = \nabla^2 u = \frac{\partial^2 u}{\partial x^2} + \frac{\partial^2 u}{\partial y^2} = 0$$

or, in polar coordinates for $u(r, \theta)$,

$$\Delta u = \frac{\partial^2 u}{\partial r^2} + \frac{1}{r}\frac{\partial u}{\partial r} + \frac{1}{r^2}\frac{\partial^2 u}{\partial \theta^2} = 0.$$

Such solutions may represent a steady-state[1] heat flow (see Figure 7.1) and electrostatic and gravitational potentials when considering harmonic functions in three dimensions. Harmonic functions can be visualized in one variable as linear functions.

Fundamentals

7.1a Mean value property. *Harmonic functions in a domain Ω satisfy*

$$u(z_0) = \frac{1}{2\pi}\int_0^{2\pi} u(z_0 + re^{i\theta})d\theta$$

whenever the disk $D_r = \{|z - z_0| \leq r\} \subseteq \Omega$.

This follows from the mean-value property for analytic functions with $u = \text{Re}(f)$ for f analytic in Ω. Note that the value at the center of a disk must accommodate all the boundary values of the disk, and since each point of the boundary is equidistant from the center, the central value must naturally be the average of all the boundary values. Multiplying each side by ρ and integrating, we obtain the following:

7.1b Areal mean value property. *As above*

$$u(z_0) = \frac{1}{\pi r^2}\int_0^r\int_0^{2\pi} u(z_0 + \rho e^{i\theta})\rho d\rho d\theta.$$

Moreover, taking the integral mean for $\frac{\partial u}{\partial x} = \text{Re}(f')$, we likewise find that

$$\frac{\partial u}{\partial x}(z_0) = \frac{1}{\pi r^2}\iint_{|z-z_0|\leq r} \frac{\partial u}{\partial x} dxdy$$

and similarly for $\partial u/\partial y$.

1 Does not change with time.

https://doi.org/10.1515/9783110757828-007

7.2 Max–min principle. *A harmonic function on a domain Ω has no maximum or minimum in Ω unless it is identically constant.*

7.3 Proposition. *A continuous function that satisfies the mean value property in a domain also satisfies the max–min principle.*

Proof. Exercise. □

7.4 Uniqueness. *If two functions u_1, u_2 are harmonic in a bounded domain Ω and continuous on $\overline{\Omega}$ with $u_1 = u_2$ on the boundary $\partial\Omega$, then $u_1 \equiv u_2$ in Ω.*

The function $v = u_1 - u_2$ is harmonic in Ω and continuous on $\overline{\Omega}$ and thus takes its max and min on $\overline{\Omega}$. If the max/min of v lies in Ω or on $\partial\Omega$, then $v \equiv 0$ in both cases.

Dirichlet problem. *Given a continuous function f on the boundary $\partial\Omega$ of a domain Ω, find a function u that is harmonic in Ω such that $u = f$ on $\partial\Omega$.*

A domain for which the Dirichlet problem is solvable for every continuous function on its boundary is called a *Dirichlet region*. We will also allow the boundary function f to be piecewise continuous on the boundary so that u should approach the boundary values of f at all its continuity points.

For example, given a continuous (piecewise continuous) temperature distribution on the boundary of a thin ring, we wish to find the steady-state temperature at any point in the interior of the ring.

Poisson formula

Let us return now to the Dirichlet problem for an open disk $|z| < R$ and begin with a given harmonic function in the disk.

7.5 Theorem. *Let u be a harmonic function in the disk $|z| \le R$ (and hence in a slightly larger disk). Then u can be expressed by*

$$u(re^{i\theta}) = \frac{1}{2\pi}\int_0^{2\pi} \frac{R^2 - r^2}{R^2 - 2Rr\cos(\theta - \phi) + r^2} u(Re^{i\phi})\,d\phi.$$

Proof. The expression looks somewhat like the Cauchy integral formula, so we take it as our starting point. Construct an analytic function f in the slightly larger disk containing $|z| \le R$ with $u = \mathrm{Re}(f)$. Let $z = re^{i\theta}$ and $\zeta = Re^{i\phi}$, so that

$$\begin{aligned} f(z) &= \frac{1}{2\pi i}\int_{|\zeta|=R} \frac{f(\zeta)}{\zeta - z}\,d\zeta = \frac{1}{2\pi}\int_0^{2\pi} \frac{\zeta f(\zeta)}{\zeta - z}\,d\phi \\ &= \frac{1}{2\pi}\int_0^{2\pi} \frac{\zeta(\overline{\zeta} - \overline{z}) f(\zeta)}{|\zeta - z|^2}\,d\phi = \frac{1}{2\pi}\int_0^{2\pi} \frac{|\zeta|^2 - \zeta\overline{z}}{|\zeta - z|^2} f(\zeta)\,d\phi. \end{aligned} \tag{7.1}$$

Considering now the reflection point of z, namely $R^2/\overline{z}$, which lies outside the circle $|z| \le R$, we have by Cauchy's theorem

$$0 = \frac{1}{2\pi i} \int_{|\zeta|=R} \frac{f(\zeta)}{\zeta - R^2/\overline{z}} d\zeta = \frac{1}{2\pi} \int_0^{2\pi} \frac{\zeta f(\zeta)}{\zeta - R^2/\overline{z}} d\phi$$

$$= \frac{1}{2\pi} \int_0^{2\pi} \frac{\zeta \overline{z} f(\zeta)}{\zeta \overline{z} - R^2} d\phi = \frac{1}{2\pi} \int_0^{2\pi} \frac{\overline{z} f(\zeta)}{\overline{z} - \overline{\zeta}} d\phi$$

$$= \frac{1}{2\pi} \int_0^{2\pi} \frac{\overline{z}(z - \zeta)}{|\zeta - z|^2} f(\zeta) d\phi = \frac{1}{2\pi} \int_0^{2\pi} \frac{|z|^2 - \overline{z}\zeta}{|\zeta - z|^2} f(\zeta) d\phi. \tag{7.2}$$

Subtracting (7.2) from (7.1) yields

$$f(z) = \frac{1}{2\pi} \int_0^{2\pi} \frac{|\zeta|^2 - |z^2|}{|\zeta - z|^2} f(\zeta) d\phi.$$

Taking real parts of both sides, we get the *Poisson formula* (1823)

$$u(z) = \frac{1}{2\pi} \int_0^{2\pi} \frac{|\zeta|^2 - |z^2|}{|\zeta - z|^2} u(\zeta) d\phi = \frac{1}{2\pi} \int_0^{2\pi} \frac{R^2 - r^2}{R^2 - 2Rr\cos(\theta - \phi) + r^2} u(\zeta) d\phi = P_u(z).$$

The term

$$K(z, \zeta) = \frac{|\zeta|^2 - |z^2|}{|\zeta - z|^2} = \frac{R^2 - r^2}{R^2 - 2Rr\cos(\theta - \phi) + r^2} = P(r, \theta - \phi)$$

is the *Poisson kernel* and gives a weighting to each boundary value $u(\zeta)$ according to its proximity to the interior point z.

Note that the kernel satisfies

$$\frac{R - r}{R + r} \le K(z, \zeta) \le \frac{R + r}{R - r}. \tag{7.3}$$

□

7.6a Example. For $u \equiv 1$,

$$1 = \frac{1}{2\pi} \int_0^{2\pi} K(z, \zeta) d\phi, \tag{7.4}$$

and if $R = 1$ and $\theta = 0$, then

$$1 = \frac{1}{2\pi}\int_0^{2\pi} \frac{1-r^2}{1-2r\cos\phi + r^2}\,d\phi$$

which implies that,

$$\frac{1}{2\pi}\int_0^{2\pi} \frac{r}{1-2r\cos\phi + r^2}\,d\phi = \frac{r}{1-r^2}, \tag{7.5}$$

a result that we have already used in our discussion of the Bieberbach conjecture in Chapter 6.

7.6b Example. We can establish a discrete version of the Poisson formula taking $R = 1$ for simplicity and $u \in H(U) \cap C(\overline{U})$:

$$\begin{aligned} 0 &= \int_0^{2\pi} [P(r,\theta-\phi)u(e^{i\phi}) - u(re^{i\theta})]\,d\phi \\ &= \int_0^{2\pi/m} \left[\sum_{k=0}^{m-1} P\left(r,\theta - \left(\phi + \frac{2k\pi}{m}\right)\right)u\left(e^{i\left(\phi+\frac{2k\pi}{m}\right)}\right) - mu\left(re^{i\theta}\right)\right]d\phi \end{aligned}$$

for each positive integer m. Since the integrand in the last integral is a continuous function of ϕ, there exists at least one value of ϕ, denoted by ϕ_z with $0 < \phi_z < \frac{2\pi}{m}$, for which the integrand is zero. As a consequence,

$$u(re^{i\theta}) = \frac{1}{m}\sum_{k=0}^{m-1} P\left(r,\theta - \left(\phi_z + \frac{2k\pi}{m}\right)\right)u\left(e^{i\left(\phi_z+\frac{2k\pi}{m}\right)}\right),$$

as desired. The *harmonic interpolation points* $\zeta_k = e^{i\left(\phi_z+\frac{2k\pi}{m}\right)}$, $k = 0,1,\ldots,m-1$, are symmetrically arrayed around the unit circle.[2]

However, we really want to start out with only real-valued continuous (piecewise continuous) boundary values $w(\zeta)$ on $|z| = R$. So for such boundary values $w(\zeta)$, let us *define*

$$P_w(z) = u(z) = \frac{1}{2\pi}\int_0^{2\pi} \frac{|\zeta|^2 - |z^2|}{|\zeta - z|^2} w(\zeta)\,d\phi$$

for $\zeta = Re^{i\phi}$.

2 Cf. J. L. Schiff and W. J. Walker, Finite harmonic interpolation, *J. Math. Anal. Appl.*, 86 (2) (1982), 648–658.

First, we claim that $u(z)$ is a harmonic function in $|z| < R$. As noted above, the Poisson kernel can be written as

$$K(z,\zeta) = \operatorname{Re}\left(\frac{\zeta + z}{\zeta - z}\right),$$

so that we have

$$\begin{aligned} u(z) &= \frac{1}{2\pi}\int_0^{2\pi} \operatorname{Re}\left(\frac{\zeta+z}{\zeta-z}\right) w(\zeta)d\phi = \operatorname{Re}\left[\frac{1}{2\pi}\int_0^{2\pi}\left(\frac{\zeta+z}{\zeta-z}\right)w(\zeta)d\phi\right] \\ &= \operatorname{Re}\left[\frac{1}{2\pi i}\int_{|\zeta|=R}\left(\frac{\zeta+z}{\zeta-z}\right)\frac{w(\zeta)}{\zeta}d\zeta\right], \end{aligned}$$

where the quantity inside the brackets is readily seen to be an analytic function of z in $|z| < R$ (exercise). Therefore the function u is harmonic in $|z| < R$.

Next we show that

$$\lim_{z\to\zeta_0} u(z) = w(\zeta_0) = w(\phi_0),$$

where $\zeta_0 = Re^{i\phi_0}$ is a point of continuity of the boundary function $w(z)$, and $|z| < R$. Writing $w(Re^{i\phi}) = w(\phi)$, we consider

$$u(z) - w(\phi_0) = \frac{1}{2\pi}\int_0^{2\pi} K(z,\zeta)[w(\phi) - w(\phi_0)]d\phi$$

in view of (7.4). Now by the continuity of w at the point ϕ_0, for each $\varepsilon > 0$, we can find a suitably small interval $I = (\phi_0 - \delta, \phi_0 + \delta)$ such that

$$|w(\phi) - w(\phi_0)| < \varepsilon$$

for all $\phi \in I$. Next, consider the complement of I on the boundary, $J = \{|\zeta| = R\} - I$. As we are considering the convergence of $z = re^{i\theta} \to Re^{i\phi_0}$, we may restrict the values of z to the wedge formed by $L = (\phi_0 - \frac{\delta}{2}, \phi_0 + \frac{\delta}{2})$ with the origin.

Then with the restriction on the value of θ to L and any $\phi \in J$, we have

$$|Re^{i\phi} - re^{i\theta}| \geq C(R,\delta) > 0.$$

Hence

$$\begin{aligned} |u(z) - w(\zeta_0)| &\leq \frac{1}{2\pi}\int_I K(z,\zeta)|w(\phi) - w(\phi_0)|d\phi + \frac{1}{2\pi}\int_J K(z,\zeta)|w(\phi) - w(\phi_0)|d\phi \\ &\leq \varepsilon + \left\{2\frac{R^2 - r^2}{C^2(R,\delta)}\sup_{\phi\epsilon J}|w(\phi)|\right\}. \end{aligned}$$

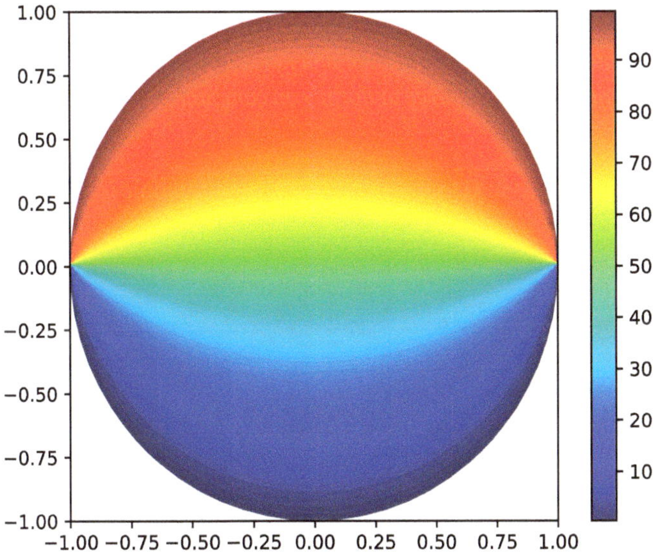

Figure 7.1: Steady-state temperature distribution in the unit disk with boundary values 100° on the top half and 0° on the bottom half. The temperature along the midline is 50° which is also the radial limit value at the two discontinuities ±1. This can be verified explicitly from the solution $u(r,\theta) = 100\big(\big(\frac{1}{2}\big) + \frac{1}{\pi}\tan\big(\frac{2r}{1-r^2}\sin\theta\big)\big)$. Courtesy Anita Kean/Springer.

Now, letting $r \to R$, we have

$$\overline{\lim_{z\to\zeta_0}}\,\big|u(z) - w(\zeta_0)\big| \le \varepsilon,$$

as desired. Thus the Poisson formula solves the Dirichlet problem for a disk.

We obtain the following:

7.7 Corollary. *If $u(z)$ is harmonic in $|z| < R$ and continuous on $|z| = R$, then $u(z) = P_u(z)$ in $|z| < R$.*

The following result shows how the mean value property characterizes harmonic functions.

7.8 Corollary. *Suppose that u is continuous in a domain Ω and satisfies the mean value property on any disk $D_r = \{|z - z_0| \le r\} \subseteq \Omega$. Then u is harmonic in Ω.*

Proof. Choose an arbitrary point $z_0 \in \Omega$ and a disk $D_r \subseteq \Omega$. Since u has continuous boundary values on $|z - z_0| = r$, construct the Poisson integral P_u in $|z - z_0| < r$. Then the function $v = u - P_u$ is continuous and satisfies the mean value property. Therefore v satisfies the max–min principle in $D(r, z_0)$ by Proposition 7.3, and $v = 0$ on $|z - z_0| = r$. As a consequence, $v \equiv 0$ in D_r, that is, $u \equiv P_u$ in D_r and the result follows since z_0 was arbitrary. □

7.9 Corollary. *If $\{u_n\}$ is a sequence of harmonic functions that converge uniformly on compact subsets of a domain Ω to a function u, then u is harmonic in Ω. Furthermore, the partial derivatives $\frac{\partial u_n}{\partial x}$ and $\frac{\partial u_n}{\partial y}$ also converge uniformly on compact subsets to $\frac{\partial u}{\partial x}$ and $\frac{\partial u}{\partial y}$, respectively.*

Proof. The function u is continuous in Ω and by Corollary 7.8 is harmonic in any disk $D = D(r, z_0)$ with $\overline{D}(r, z_0) \subseteq \Omega$, since

$$u_n(z_0) = \frac{1}{2\pi}\int_0^{2\pi} u_n(z_0 + re^{i\theta})d\theta \to \frac{1}{2\pi}\int_0^{2\pi} u(z_0 + re^{i\theta})d\theta$$

by the uniform convergence on compact subsets of Ω.

For the second part of the statement, by the areal mean value for the derivatives we have

$$\frac{\partial(u_n - u)}{\partial x}(z_0) = \frac{1}{\pi r^2}\iint_D \frac{\partial(u_n - u)}{\partial x}dxdy = \frac{1}{\pi r^2}\int_{\partial D}(u_n - u)dy$$

via an application of Green's theorem. Consequently, for any $\varepsilon > 0$,

$$\left|\frac{\partial(u_n - u)}{\partial x}(z_0)\right| \leq \frac{1}{\pi r^2}\int_0^{2\pi}\left|(u_n - u)r\cos\theta\right|d\theta \leq \frac{2\sup_{\partial D}|u_n - u|}{r} < \varepsilon$$

for all n sufficiently large. Taking a compact set $K \subseteq \Omega$, let $\rho = \operatorname{dist}(K, \partial\Omega)$ and $r = \rho/2$. Then since $\overline{D(r, z)} \subseteq \Omega$, for all points $z \in K$, we have

$$\left|\frac{\partial(u_n - u)}{\partial x}(z)\right| < \varepsilon$$

for all n sufficiently large, establishing the uniform convergence $\frac{\partial u_n}{\partial x} \to \frac{\partial u}{\partial x}$ on K and likewise for $\partial u_n/\partial y$. □

It was demonstrated by Stanislaw Zaremba (1911) that there are domains for which there is no solution to the Dirichlet problem.

Indeed, let $\Omega = U - \{0\}$ and define the boundary function

$$f(z) = \begin{cases} 0 & |z| = 1 \\ 1 & z = 0. \end{cases}$$

Any harmonic function u in Ω with boundary values $u = f$ on $\partial\Omega$ satisfies $0 \leq u \leq 1$ and so has a removable singularity[3] at $z = 0$, namely the harmonic function $\hat{u} \equiv 0$

3 In the context of a function u that is harmonic in $\Omega - \{a\}$, the point a is a *removable singularity* of u if there is a harmonic extension $\hat{u}$ of u to Ω. If u is a bounded harmonic function on $\Omega - \{a\}$, then a is a removable singularity. Cf. Axler et al. (2001, Theorem 2.3).

in U. Since the boundary values are taken continuously by both harmonic functions u and $\hat{u}$ in Ω and since the solution is unique with $u(0) \neq \hat{u}(0)$, the original Dirichlet problem has no harmonic solution u.

On the other hand, a solution to the Dirichlet problem is a consequence of the Riemann mapping theorem and the Poisson formula for a suitable domain. First, we require an elementary lemma.

7.10 Lemma. *Let $h(u, v)$ be harmonic in the variables u, v and suppose that*

$$f(z) = u + iv = u(x, y) + iv(x, y)$$

is analytic. Then $H = h \circ f$ is harmonic in the variables x, y.

This is evident since $H(x, y) = h(u(x, y), v(x, y))$ satisfies

$$\Delta H(x, y) = |f'(z)|^2 \Delta h(u, v).$$

The proof is left as an exercise.

7.11 Theorem. *Let Ω be a domain with simple closed boundary. Then the Dirichlet problem is solvable in Ω.*

Proof. Since Ω is simply connected, by the Riemann mapping theorem there is a conformal mapping $F : \Omega \to U$ for $U : \{|w| < 1\}$ such that $F(z)$ has a one-to-one continuous extension to $\partial\Omega$. Given piecewise continuous boundary values f on $\partial\Omega$, the corresponding boundary values of $f \circ F^{-1} = g$ are piecewise continuous on ∂U. With these boundary values, we then solve the Dirichlet problem in the unit disk via the Poisson formula obtaining a harmonic function h. Then by the lemma the function $H = h \circ F$ is a harmonic function in Ω with the desired boundary values f. □

Dirichlet integral/mixed Dirichlet Integral

Let us now consider a general domain Ω of the complex plane and for $u \in C(\overline{\Omega}) \cap C^1(\Omega)$, define

$$D_\Omega(u) = \iint_\Omega |\nabla u|^2 dxdy = \iint_\Omega \left[\left(\frac{\partial u}{\partial x}\right)^2 + \left(\frac{\partial u}{\partial y}\right)^2\right] dxdy.$$

This is an important quantity in potential theory called the *Dirichlet integral.*[4] It is a measure of the smoothness of the gradient field of the function u. If $u \equiv$ constant, then, clearly, $D_\Omega(u) = 0$.

---

4 The term is due to Riemann. The *Dirichlet energy* is sometimes defined as $E(u) = \frac{1}{2}D(u)$ or without the factor of $\frac{1}{2}$ and has found modern applications in image processing and computer vision.

The Dirichlet integral in polar coordinates for $u = u(r,\theta)$ is

$$D_\Omega(u) = \iint_\Omega \left[\left(\frac{\partial u}{\partial r}\right)^2 + \frac{1}{r^2}\left(\frac{\partial u}{\partial \theta}\right)^2\right] r dr d\theta. \tag{7.6}$$

For a continuous function $f(\theta)$ given by a Fourier series on $\partial U : |z| = 1$

$$f(\theta) = \frac{a_0}{2} + \sum_{n=1}^{\infty}(a_n \cos n\theta + b_n \sin n\theta),$$

the solution to the Dirichlet problem in U is given by (exercise)

$$u(r,\theta) = \frac{a_0}{2} + \sum_{n=1}^{\infty} r^n(a_n \cos n\theta + b_n \sin n\theta).$$

Then the Dirichlet integral $D_r(u)$ for $0 < r < 1$ by direct calculation (exercise) is

$$D_r(u) = \pi \sum_{n=1}^{\infty} nr^{2n}(a_n^2 + b_n^2).$$

Therefore, for $0 < r < 1$,

$$\pi \sum_{n=1}^{N} nr^{2n}(a_n^2 + b_n^2) \le D_r(u) \le \pi \sum_{n=1}^{\infty} n(a_n^2 + b_n^2)$$

for each N, and letting $r \nearrow 1$, we have

$$\pi \sum_{n=1}^{N} n(a_n^2 + b_n^2) \le D_U(u) \le \pi \sum_{n=1}^{\infty} n(a_n^2 + b_n^2),$$

that is,

$$D_U(u) = \pi \sum_{n=1}^{\infty} n(a_n^2 + b_n^2).$$

Of course, the infinite series may or may not converge even if $f(\theta)$ is continuous, which is mentioned in footnote 5 in the sequel. Compare this result with Theorem 6.1.

7.12 Exercise. Let $f(\theta)$ be continuous on $|z| = 1$ with Fourier series expansion $f(\theta) = \sum_{n=-\infty}^{\infty} c_n e^{in\theta}$. Show that the solution to the Dirichlet problem in the unit disk is

$$u(re^{i\theta}) = \sum_{n=-\infty}^{\infty} c_n r^{|n|} e^{in\theta}.$$

There is also the *mixed* Dirichlet integral (which is an inner product – exercise) if $D_\Omega(u)$, $D_\Omega(v)$ are both finite:

$$D_\Omega(u, v) = \iint_\Omega (\nabla u \cdot \nabla v) dxdy = \iint_\Omega \left[\frac{\partial u}{\partial x}\frac{\partial v}{\partial x} + \frac{\partial u}{\partial y}\frac{\partial v}{\partial y} \right] dxdy,$$

which satisfies the *parallelogram law*

$$D_\Omega(u + v) = D_\Omega(u) + 2D_\Omega(u, v) + D_\Omega(v).$$

Note that we have the identity

$$4D_\Omega(u, v) = D_\Omega(u + v) - D_\Omega(u - v)$$

and inequality

$$[D_\Omega(u, v)]^2 \le D_\Omega(u) D_\Omega(v)$$

by the Cauchy–Schwarz inequality.

7.13 Proposition. *If $\{u_n\}$ is a sequence of harmonic functions in a bounded domain Ω that converges uniformly on compact subsets of Ω to the harmonic function u, then*

$$D_\Omega(u) \le \varliminf_{n\to\infty} D_\Omega(u_n).$$

Proof. The sequence $\{u_n\}$ of harmonic functions converging uniformly on compact subsets to the harmonic function u implies that their partial derivatives also converge uniformly on compact subsets to the partial derivatives of u by Corollary 7.9. Letting $\{\Omega_\upsilon\}$ be an exhaustion of Ω (cf. footnote 4, Chapter 1), then $D_{\Omega_\upsilon}(u_n) \le D_\Omega(u_n)$ implies

$$D_{\Omega_\upsilon}(u) = \lim_{n\to\infty} D_{\Omega_\upsilon}(u_n) = \varliminf_{n\to\infty} D_{\Omega_\upsilon}(u_n) \le \varliminf_{n\to\infty} D_\Omega(u_n),$$

and the result follows by letting $\Omega_\upsilon \nearrow \Omega$. □

We remark that the proposition holds with the harmonic functions $\{u_n\}$ replaced by analytic functions $\{f_n\}$.

Dirichlet principle

On a bounded domain Ω whose boundary $\partial\Omega$ consists of a finite number of smooth Jordan curves, given a function $f \in C(\overline{\Omega}) \cap C^1(\Omega)$ with $D_\Omega(f) < \infty$, define the family of *admissible* functions:

$$\mathfrak{I} = \{w \in C(\overline{\Omega}) \cap C^1(\Omega) : w = f \text{ on } \partial\Omega \text{ and } D_\Omega(w) < \infty\}.$$

The *Dirichlet principle* is the statement that there is a function $u \in \mathfrak{I}$ such that

$$D_\Omega(u) = \inf_{w \in \mathfrak{I}} D_\Omega(w),$$

and the function u is *harmonic* in Ω, that is, u is the solution to the Dirichlet problem in Ω. Moreover, any other function $w \in \mathfrak{I}$ satisfies $D_\Omega(u) < D_\Omega(w)$.

The original proof by Dirichlet was incorrect as he assumed that the infimum had to be *attained* by some admissible function as all the Dirichlet integrals were nonnegative. Indeed, Riemann used the Dirichlet principle to prove his famous mapping theorem. However, Weierstrass (1870) subsequently showed that such a minimum need not be attained by any admissible function.

It was pointed out by J. Hadamard (1906) that there exist continuously differentiable functions with continuous boundary values for which the Dirichlet problem is solvable and the Dirichlet integral is infinite, thus illustrating that the Dirichlet problem and Dirichlet principle, although related, are not the same.[5]

Therefore considerable effort over the ensuing years went into establishing under what conditions the Dirichlet principle became valid and finding alternative methods for solving the Dirichlet problem. Only in 1899 David Hilbert salvaged the Dirichlet principle by providing suitable conditions on the boundary of the domain Ω for the principle to hold. See Monna (1975) for a fascinating discussion of the whole saga.

In the present section, we will assume as before that Ω is a bounded domain whose boundary $\partial\Omega$ consists of finitely many smooth Jordan curves. Then we can show now in this regard the following:

7.14 Theorem. *If $u \in \mathfrak{I}$ is a harmonic solution to the Dirichlet problem with continuous first partial derivatives on $\partial\Omega$, then $D_\Omega(u) < D_\Omega(w)$ for any other function $w \in \mathfrak{I}$, that is, u uniquely minimizes the Dirichlet integral among all admissible functions.*

Proof. Choose any $w \in \mathfrak{I}$ and let $v = w - u$. From the parallelogram law we have

$$D_\Omega(w) = D_\Omega(u) + 2D_\Omega(u, v) + D_\Omega(v) = D_\Omega(u) + 2\iint_\Omega (\nabla u \cdot \nabla v)\,dxdy + D_\Omega(v).$$

Regarding the mixed Dirichlet integral, by Green's first identity we have

$$\iint_\Omega (v\Delta u + \nabla u \cdot \nabla v)\,dxdy = \int_{\partial\Omega} v\frac{\partial u}{\partial n}\,ds,^{6} \tag{7.7}$$

5 Hadamard's example was $u(r,\theta) = \sum_{n=1}^{\infty} n^{-2} r^{n!} \sin(n!\theta)$, which is harmonic in U (since each term is and hence the partial sums are harmonic) and continuous on $|z| \le 1$, but a direct calculation from (7.6) by the absolute uniform convergence of the series yields (exercise)

$$D_U(f) = \pi \sum_{n=1}^{\infty} \frac{n!}{n^4} = \infty.$$

A similar example was given by Riemann's student Friedrich Prym in 1871.

and since $\Delta u = 0$ and $v|\partial\Omega = 0$, we obtain

$$D_\Omega(w) = D_\Omega(u) + D_\Omega(v) \geq D_\Omega(u).$$

Moreover, if $D_\Omega(w) = D_\Omega(u)$, then $D_\Omega(v) = 0$, so that $v \equiv$ constant. However, $v = 0$ on $\partial\Omega$ implies $v \equiv 0$, so that $w \equiv u$. □

From the proof we have the following:

7.15 Corollary. *If $u, v \in C^1(\overline{\Omega})$ and u is harmonic in Ω, then*

$$D_\Omega(u,v) = \int_{\partial\Omega} v\frac{\partial u}{\partial n}ds, \quad D_\Omega(u) = \int_{\partial\Omega} u\frac{\partial u}{\partial n}ds,$$

and taking $v = 1$,

$$\int_{\partial\Omega} \frac{\partial u}{\partial n}ds = 0.$$

The latter integral is called the *flux* of u. We also have the following converse of the latter equation.

7.16 Theorem. *Let $u \in C^1(\Omega)$ be such that*

$$\int_{\gamma} \frac{\partial u}{\partial n}ds = 0$$

for all circles γ interior to Ω. Then u is harmonic in Ω.

Proof. For an arbitrary point $z_0 \in \Omega$, let c_r be any circle in Ω with center z_0. Then

$$0 = \int_{c_r} \frac{\partial u}{\partial n}ds = \int_0^{2\pi} \frac{\partial u(r,\theta)}{\partial r}r\,d\theta.$$

For fixed r, this implies

$$\int_0^{2\pi} \frac{\partial u(r,\theta)}{\partial r}d\theta = 0,$$

so that integrating from 0 to r and interchanging the order of integration, we obtain

$$\int_0^{2\pi} (u(z_0 + re^{i\theta}) - u(z_0))d\theta = 0$$

---

6 In this form, $\frac{\partial u}{\partial n}$ represents the derivative in the direction of the outer normal.

and arrive at the mean-value property

$$u(z_0) = \frac{1}{2\pi}\int_0^{2\pi} u(z_0 + re^{\imath\theta})d\theta.$$

Since u is continuous in Ω, u is harmonic there by Corollary 7.8. □

7.17 Proposition. *If $u \in C^1(\overline{\Omega})$ is harmonic in Ω and*

$$\frac{\partial u}{\partial n} = 0$$

on $\partial\Omega$, then u is constant in Ω.

Proof. By Corollary 7.15

$$D_\Omega(u) = \int_{\partial\Omega} u\frac{\partial u}{\partial n}ds,$$

implying that

$$\iint_\Omega |\nabla u|^2 dxdy = 0.$$

Therefore $|\nabla u|^2 = 0$, which means that $u_x = u_y = 0$, and hence $u \equiv$ constant. □

7.18 Proposition. *The Dirichlet integral is invariant under conformal mappings.*

Proof. For $u = u(x,y)$ and $v = v(x,y)$ representing a conformal mapping from the domain Ω to the domain Ω', by the Cauchy–Riemann equations for $\phi(u,v)$ defined on Ω' and some direct calculation (exercise) we have

$$(\phi_x^2 + \phi_y^2) = (\phi_u^2 + \phi_v^2)\frac{\partial(u,v)}{\partial(x,y)},^7$$

implying

$$\iint_\Omega (\phi_x^2 + \phi_y^2)dxdy = \iint_\Omega (\phi_u^2 + \phi_v^2)\frac{\partial(u,v)}{\partial(x,y)}dxdy = \iint_{\Omega'} (\phi_u^2 + \phi_v^2)dudv,$$

establishing the result. □

The Dirichlet principle can be established for the unit disk U although the details are rather technical and can be found in the text by Courant (1950). If this is the

7 Recall that $\frac{\partial(u,v)}{\partial(x,y)} = \begin{vmatrix} u_x & u_y \\ v_x & v_y \end{vmatrix}$ and $\text{Area}(\Omega') = \iint_{\Omega'} dudv = \iint_\Omega \frac{\partial(u,v)}{\partial(x,y)}dxdy$.

case, then the preceding proposition means that the Dirichlet principle is valid in any domain conformally equivalent to U whenever the conformal mapping extends continuously to the boundary.

Liouville-type theorem

It is clear that any function u that is harmonic in the entire plane and bounded above (resp., bounded below) must reduce to a constant (by Liouville's theorem) by simply considering the modulus of the analytic function e^f (resp., e^{-f}) with $f = u + iv$ analytic in $\mathbb{C}$. Furthermore, we have a Liouville-type theorem regarding the Dirichlet integral.

7.19 Theorem. *If u is harmonic in $\mathbb{C}$ and $D_{\mathbb{C}}(u) < \infty$, then $u \equiv$ constant in $\mathbb{C}$.*

Indeed, note that regarding the Dirichlet integral, if $u = \mathrm{Re}(f)$ on a domain Ω for f analytic on Ω, then

$$D_\Omega(u) = D_\Omega(f) = \iint_\Omega |f'(z)|^2 dxdy.$$

In the case where $\Omega = \mathbb{C}$ and $D(u) < \infty$ in the plane, the analytic function $f(z)$ in $\mathbb{C}$ satisfies

$$f'(z_0) = \frac{1}{\pi R^2} \iint_{|z-z_0|<R} f'(z)dxdy$$

on any disk $|z - z_0| < R$ (exercise). Therefore by the Cauchy–Schwarz inequality we have

$$|f'(z_0)|^2 \le \left(\frac{1}{\pi^2 R^4} \iint_{|z-z_0|<R} |f'(z)|^2 dxdy\right)\left(\iint_{|z-z_0|<R} dxdy\right) \le \frac{1}{\pi R^2} D_\Omega(f). \tag{7.8}$$

As R is arbitrarily large, $f'(z_0) = 0$, so that $f \equiv$ constant in $\mathbb{C}$, implying $u \equiv$ constant likewise.

We will see other Liouville-type theorems in the sequel.[8]

7.20 Corollary. *If $f(z)$ is an entire function with $D_{\mathbb{C}}(f) < \infty$, then $f \equiv$ constant in $\mathbb{C}$.*

Moreover, setting $\Omega = U$ and $R = 1 - |z_0|$ in (7.8), we can write

$$(1 - |z|)|f'(z)| \le \sqrt{\frac{1}{\pi} D_U(f)}, \quad z \in U, \tag{7.9}$$

8 It is certainly the case that if u is bounded and harmonic in $\mathbb{C}$, then u reduces to a constant (Corollary 7.36). But more strongly, if $\underline{\lim}_{r\to\infty} m(r)/\log r = 0$, where $m(r) = \max_{|z|=r} u(z)$, then u is constant (see Chapter 8 as the same result holds for subharmonic functions).

which implies

$$(1-|z|^2)|f'(z)| \le 2(1-|z|)|f'(z)| \le 2\sqrt{\frac{1}{\pi}D_U(f)}.$$

From this we can conclude the following:

7.21 Corollary. *Dirichlet-finite analytic functions in U are Bloch functions.*

7.22 Corollary. *Let*

$$\mathcal{F} = \{f \text{ analytic in } U : D_U(f) \le M\}.$$

Then $\mathcal{F}$ is a normal and compact family in U.

Proof. From (7.9) we have

$$|f'(z)| \le \frac{1}{1-|z|}\sqrt{M/\pi},$$

so that the corresponding family $\mathcal{F}' = \{f' : f \in \mathcal{F}\}$ is locally uniformly bounded, and hence $\mathcal{F}$ is normal by Marty's theorem 4.27. Moreover, if $f_n \to f$ uniformly on compact subsets of U, then the derivatives do likewise, so that as in Proposition 7.13,

$$D_U(f) \le \varliminf_{n\to\infty} D_U(f_n) \le M,$$

proving that $\mathcal{F}$ is compact. □

7.23 Remark. By taking $u = \mathrm{Re}(f)$ in U we have that

$$\mathcal{F}_u = \{u \text{ harmonic in } U : D_U(u) \le M\}$$

is normal and compact.

The Dirichlet integral is in fact an example of a more general type of functional.

Continuous functionals

7.24 Definition. If $\mathcal{F}$ is a family of functions defined on a domain Ω, then a mapping

$$J : \mathcal{F} \to \mathbb{R}$$

is called a *lower semicontinuous functional on* $\mathcal{F}$ if whenever $f_n \to f$ uniformly on compact subsets of Ω, *then*

$$J(f) \le \varliminf_{n\to\infty} J(f_n).$$

As was shown in Proposition 7.13, the Dirichlet integral is a lower semicontinuous functional on the family of harmonic functions in Ω.

Although Weierstrass pointed out that the infimum of a functional defined on a given family of functions is not necessarily the minimum value attained by any member of the family, an elementary result demonstrating when a minimum is attained is the following:

7.25 Theorem. *If $\mathcal{F}$ is a nonempty normal compact family of functions defined on a domain Ω and J is a lower semicontinuous functional defined on $\mathcal{F}$, then there exists a function $f_0 \in \mathcal{F}$ such that*

$$J(f_0) = \inf_{f \in \mathcal{F}} J(f).$$

Proof. Let

$$-\infty \leq d = \inf_{f \in \mathcal{F}} J(f).$$

Then there is a sequence $\{f_n\}$ in $\mathcal{F}$, known as a *minimizing sequence*, such that

$$d = \lim_{n \to \infty} J(f_n).$$

As the family is normal and compact, there is a subsequence $\{f_{n_k}\}$ in $\mathcal{F}$ such that $f_{n_k} \to f_0 \in \mathcal{F}$ uniformly on compact subsets of Ω. Since J is a lower semi continuous functional,

$$d \leq J(f_0) \leq \varliminf_{k \to \infty} J(f_{n_k}) = \lim_{k \to \infty} J(f_{n_k}) = d,$$

which shows that $J(f_0) = d > -\infty$, establishing the result.

Likewise, for a *continuous functional*[9] $J : \mathcal{F} \to \mathbb{C}$ where $\mathcal{F}$ is normal compact, there exists a function $f_0 \in \mathcal{F}$ such that

$$|J(f_0)| = \sup_{f \in \mathcal{F}} |f(z)|,$$

and similarly for the *infimum*. □

For the notion of a continuous functional on a normal compact family, there is an analogue with the extreme value theorem: *A continuous function on a compact set in a metric space attains it maximum and minimum in the set.* This notion got Riemann into trouble in the proof of his mapping theorem. In the proof given in Chapter 4,

9 $J : \mathcal{F} \to \mathbb{C}$ is a *continuous functional* if whenever $f_n \to f$ in $\mathcal{F}$ uniformly on compact subsets of Ω, then $|J(f)| = \lim_{n \to \infty} |J(f_n)|$.

the family of functions $\mathcal{F}$ is shown to be a normal compact family, and the desired mapping function is precisely the extremal solution for a continuous functional as above.

7.26 Exercise. Let $\mathcal{F}$ be a family of analytic functions in U such that $|f(z)| < 1$. Show that for any $z_0 \in U$,

$$J(f) = \frac{f'(z_0)}{1 - |f(z_0)|^2}$$

is a continuous functional on $\mathcal{F}$.

There is also a direct relation to the extremal problem of the Bieberbach conjecture of Chapter 6. For each analytic function in U given by $f(z) = z + \sum_{n=2}^{\infty} a_n z^n$ (these belong to the class $\mathcal{S}$ discussed in Chapter 6), define $A_n(f) = a_n$, $n = 2, 3, 4, \ldots$, and consider a sequence $\{f_k\}$ in $\mathcal{S}$ such that $f_k \to f \in \mathcal{S}$ uniformly on compact subsets of U. Writing

$$f_k(z) = z + \sum_{n=2}^{\infty} a_n^{(k)} z^n, \quad f(z) = z + \sum_{n=2}^{\infty} a_n z^n,$$

for $c_r : |z| = r < 1$ and fixed n, we have

$$\lim_{k\to\infty} A_n(f_k) = \lim_{k\to\infty} \frac{1}{2\pi i} \int_0^{2\pi} \frac{f_k(\zeta)}{\zeta^{n+1}} d\zeta = \frac{1}{2\pi i} \int_0^{2\pi} \frac{f(\zeta)}{\zeta^{n+1}} d\zeta = A_n(f).$$

Therefore A_n is a continuous functional on the set S, which has been shown to be normal and compact, and hence there is some function $f_0 \in S$ such that $A_n(f_0) = \max |a_n|$. We know the function f_0 is the Koebe function $k(z) = z + 2z^2 + 3z^3 + \cdots$.

To give the flavor of the relationship of Theorem 7.25 to the Dirichlet principle, let

$$0 \le d = \inf_{w \in \mathcal{F}} D_\Omega(w)$$

for a suitably defined set of admissible functions $\mathcal{F}$, and let $\{u_n\}$ be a minimizing sequence $\{u_n\}$ in $\mathcal{F}$ with

$$d = \lim_{n\to\infty} D_\Omega(u_n).$$

Hilbert[10] was able to show that there is a subsequence $\{u_{n_k}\}$ that converges to a function $u \in \mathcal{F}$ such that $D_\Omega(u) = d$. This minimizing function u turns out to be harmonic

10 Über das Dirichletsche Prinzip, *Math. Ann.* 59 (1904), 161–186. See also Tsuji (1959), where a normal family argument is used, and Courant (1950).

and thus the solution to the Dirichlet problem. This revived interest in the Dirichlet principle at the beginning of the 20th century.

Indeed, if we take a small liberty and assume that the minimizing function $u \in C^2(\Omega) \cap C^1(\overline{\Omega})$, then we can again apply Green's first identity (7.7):

$$\iint_{\Omega} (v\Delta u)dxdy + D_{\Omega}(u,v) = \int_{\partial\Omega} v\frac{\partial u}{\partial n}ds. \tag{7.10}$$

Considering an arbitrary function $v \in C(\overline{\Omega}) \cap C^1(\Omega)$ with $v = 0$ on $\partial\Omega$ implies that the right-hand side of (7.10) equals zero. For the mixed Dirichlet integral, given any $t \in \mathbb{R}$, the function $w = u + tv$ belongs to the family $\mathfrak{I}$, which means that the function

$$\phi(t) = D_{\Omega}(u + tv) = D_{\Omega}(u) + 2tD_{\Omega}(u,v) + t^2 D_{\Omega}(v)$$

has a minimum at $t = 0$. Thus $\phi'(0) = 0$, and, consequently, $D_{\Omega}(u,v) = 0$. We are now left with from (7.10)

$$\iint_{\Omega} (v\Delta u)dxdy = 0.$$

If it is not the case that $\Delta u = 0$, suppose that $\Delta u > 0$ as some point $p \in \Omega$ and hence in some disk D about p. We are now free to take a smooth function $v > 0$ in D with $v = 0$ in the complement D', so that

$$0 < \iint_{D} (v\Delta u)dxdy = \iint_{\Omega} (v\Delta u)dxdy,$$

a contradiction, which proves that u is harmonic in Ω.

Thus in this sense the minimizing function is the solution to the Dirichlet problem. This is the converse of Theorem 7.14, where we showed that the harmonic solution to the Dirichlet problem was a minimizing function.

Note also that on the space of functions $\mathcal{D}_{\Omega} = \{u \in C^1(\Omega) \text{ and } D_{\Omega}(u) < \infty\}$, for a bounded open set Ω, the mixed Dirichlet integral induces the *norm*

$$\|u\|_{\Omega} = [D_{\Omega}(u)]^{1/2} = [D_{\Omega}(u,u)]^{1/2} = \left(\iint_{\Omega} \left[\left(\frac{\partial u}{\partial x}\right)^2 + \left(\frac{\partial u}{\partial y}\right)^2\right] dxdy\right)^{1/2}.$$

Identifying all functions in $\mathcal{D}_{\Omega}$ that differ by a constant, we have a normed linear space $\hat{\mathcal{D}}_{\Omega}$.

(Exercise: Prove the triangle inequality.)

Green's function

This is also referred to as the *Green function*. It arises from considerations of a region Ω that is a vacuum bounded by a perfectly conducting surface. Then a positive point charge located at a point $z_0 \in \Omega$ induces a negative charge over $\partial\Omega$, and the electrostatic potential in Ω suitably normalized will have the properties of Green's function.

7.27 Definition. Let Ω be a domain in the complex plane bounded by finitely many contours with fixed point $z_0 \in \Omega$. *Green's function* $G(z, z_0)$ with pole at z_0 is a function satisfying

(i) $G(z, z_0) = \log \frac{1}{|z-z_0|} + h(z)$, where $h(z)$ is harmonic in Ω and continuous on $\overline{\Omega}$;

(ii) $G(\zeta, z_0) = 0$ for each $\zeta \in \partial\Omega$.

Therefore Green's function is a harmonic function of $z \neq z_0$ in Ω possessing a *logarithmic singularity* at the point z_0, which vanishes on the boundary of Ω. It is also unique; indeed, suppose that $G_1(z, z_0)$ and $G_2(z, z_0)$ are two Green's functions for Ω with the same pole. Then the function $H(z) = G_1(z, z_0) - G_2(z, z_0)$ has a removable singularity at z_0 and is thus harmonic in Ω by (i), and by (ii) $H = 0$ on $\partial\Omega$. By the max–min principle for harmonic functions, $H(z) \equiv 0$ and $G_1(z, z_0) \equiv G_2(z, z_0)$.

7.28 Example. For the disk $|z| < R$, $G(z, 0) = \log \frac{R}{|z|}$ with a pole at the origin. Note that as $R \to \infty$, the obvious implication is that there is no Green's function for the complex plane.

It is also worth noting that since $G(z, z_0) \to \infty$ as $z \to z_0$, there is an arbitrarily small disk $D = D(z_0, r) \subseteq \Omega$ with $G(z, z_0) > 0$ on ∂D, so that the max–min principle applied to $\Omega - D$ allows us to conclude that $G(z, z_0) > 0$ in Ω.

If our domain Ω is bounded by a simple closed contour, then the Riemann mapping theorem provides a conformal mapping $F(z) : \Omega \to U$ such that $F(z_0) = 0$. Moreover, in this instance, $F(z)$ has a one-to-one continuous extension to the boundary as mentioned previously, and $|F(\zeta)| = 1$ for $\zeta \in \partial\Omega$. Then Green's function for Ω is given by

$$G(z, z_0) = -\log\left|F(z)\right|.$$

Indeed, $G(z, z_0) = \mathrm{Re}(-\log F(z))$ is harmonic in Ω except at the point z_0 and takes the zero values on $\partial\Omega$. Moreover, in a neighborhood $z = z_0$, we have

$$F(z) = c_1(z - z_0) + c_2(z - z_0)^2 + \cdots$$

with $c_1 \neq 0$. Let

$$g(z, z_0) = -\log F(z) = -\log\left[c_1(z - z_0)\left(1 + \frac{c_2}{c_1}(z - z_0) + \cdots\right)\right]$$

$$= -\log(z - z_0) - \log c_1 - \log\left(1 + \frac{c_2}{c_1}(z - z_0) + \cdots\right)$$
$$= -\log(z - z_0) + g_1(z, z_0),$$

where $g_1(z, z_0)$ is analytic in a neighborhood of $z = z_0$. Then

$$G(z, z_0) = \operatorname{Re}(g(z, z_0)) = -\log|z - z_0| + \operatorname{Re}(g_1(z, z_0)) = -\log|z - z_0| + g_2(z, z_0)$$

shows that $G(z, z_0)$ possesses the required logarithmic singularity and $g_2(z, z_0)$ is harmonic in Ω. Since Green's function is unique, the statement is proved.

Thus, for example, Green's function for the unit disk with singularity at $z_0 \in U$ is given by

$$G(z, z_0) = -\log\left|\frac{z - z_0}{1 - \overline{z_0}z}\right|,$$

and

$$G(z, i) = -\log\left|\frac{z - i}{z + i}\right|$$

is Green's function for the upper half-plane with singularity at $z_0 = i$.

Solution to Dirichlet problem via Green's function

The Green's function is actually intimately connected to the Dirichlet problem.

7.29 Theorem. *Green's function exists for any bounded domain for which the Dirichlet problem is solvable.*

In fact, for a bounded domain Ω in $\mathbb{C}$ and $z_0 \in \Omega$, let $h(z)$ be a harmonic solution to the Dirichlet problem in Ω such that for $\zeta \in \partial\Omega$,

$$h(\zeta) = \log|\zeta - z_0|.$$

Then the function given by

$$G(z, z_0) = h(z) - \log|z - z_0|$$

is Green's function for Ω.

A more interesting result is the following converse formulation for the solution to the Dirichlet problem.

7.30 Theorem. *Let $G(z, z_0)$ be Green's function for a bounded domain Ω in $\mathbb{C}$ whose boundary C consists of finitely many smooth Jordan curves. If $u(z)$ is the solution to the Dirichlet problem with continuous boundary values $u(\zeta)$ on C, then*

$$u(z_0) = \frac{1}{2\pi} \int_C u(\zeta) \frac{\partial}{\partial n} G(\zeta, z_0) ds,$$

where $\frac{\partial}{\partial n}$ represents the directional derivative in the direction of the inner normal, and the integral is with respect to arc length.

Proof. We will prove the result under the assumption that both functions $u(z)$ and $G(z, z_0)$ are continuously differentiable on the boundary C. This would be the case if the boundary curves are piecewise analytic, for then we could achieve the desired differentiability by the Schwarz reflection principle. The extension from the present restricted version to the general case of the theorem is carried out by Hille (1962).[11]

The idea is to excise a small circle around the pole: $D_r = \{0 < |z - z_0| \le r\} \subset \Omega$ and let $\Omega' = \Omega - D_r$. We then apply Green's second identity to the domain Ω' in the form

$$\iint_{\Omega'} (u\Delta v - v\Delta u) dxdy + \int_{\partial\Omega'} \left(u\frac{\partial v}{\partial n} - v\frac{\partial u}{\partial n} \right) ds = 0. \tag{7.11}$$

Here we put $v = G(z, z_0)$ and note that $\partial\Omega' = C + \partial D_r$. Since both $u(z)$ and $G(z, z_0)$ are harmonic in Ω',

$$0 = \int_{\partial\Omega'} \left(u\frac{\partial G}{\partial n} - G\frac{\partial u}{\partial n} \right) ds,$$

and since $G(\zeta, z_0) = 0$ for $\zeta \in C$, (7.11) reduces to

$$-\int_C u\frac{\partial G}{\partial n} ds = \int_{\partial D_r} u\frac{\partial G}{\partial n} ds - \int_{\partial D_r} G\frac{\partial u}{\partial n} ds = 0. \tag{7.12}$$

We have therefore transferred the integration to the boundary of the disk D_r, where it is more amenable to calculation. See Figure 7.2.

Note that on $\partial D_r : |z - z_0| = r$ the inner normal derivative with respect to the domain Ω' is given by

$$\frac{\partial}{\partial n} = \frac{\partial}{\partial r}.$$

Thus on $|z - z_0| = r$ Green's function

$$G(z, z_0) = -\log r + h(z)$$

---

11 Cf. pp. 399–405.

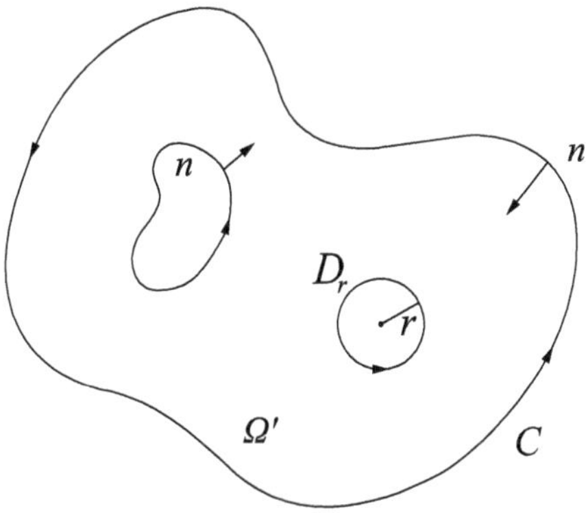

Figure 7.2: A small disk is removed around the singularity z_0 of Green's function whose boundary is in the negative orientation with respect to the region Ω'. Courtesy Katy Metcalf.

satisfies

$$\frac{\partial G}{\partial n} = -\frac{1}{r} + \frac{\partial h}{\partial r}.$$

Regarding the second integral of (7.12), as $u(z)$ and $h(z)$ are harmonic in D, we have

$$\int_{\partial D_r} u\frac{\partial G}{\partial n}ds = \int_0^{2\pi} u\left(-\frac{1}{r} + \frac{\partial h}{\partial r}\right)rd\theta = -2\pi u(z_0) + O(r)$$

as $r \to 0$. Considering the last integral of (7.12),

$$\int_{\partial D_r} G\frac{\partial u}{\partial n}ds = \int_0^{2\pi} (-\log r + h)\frac{\partial u}{\partial n}rd\theta = O(-r\log r) + O(r) \tag{7.13}$$

as $r \to 0$. Combining (7.12) and (7.13) with the preceding equality and letting $r \to 0$ establish our desired equality of the theorem. □

Note that if the domain is the disk $D : |z| < R$, then the integration factor $\partial G/\partial n$ in the theorem becomes the Poisson kernel.[12] Taking an outer normal in the theorem results in the sign of the integral in the equation becoming minus.

The above representation as well as the Poisson and Herglotz formulas (the latter is Theorem7.38) for the harmonic solution to the Dirichlet problem in a bounded domain Ω are particular cases of the general result of the Riesz representation theorem. Indeed, let $\Gamma = \partial\Omega$ and denote by $C(\Gamma)$ the space of continuous functions on Γ. Then for any $z \in \Omega$ and for $f \in C(\Gamma)$, define the functional

$$F_z : f \to H_f^{\Omega}(z),$$

12 See Garabedian (1964, pp. 247–248).

where H_f^Ω is the solution to the Dirichlet Problem in Ω with boundary values f. Clearly, if $f \geq 0$ on Γ, then $H_f^\Omega \geq 0$ on Ω, and F_z is a nonnegative linear functional. By the RRT there is a nonnegative measure μ_z^Ω on Γ such that the solution to the Dirichlet problem has the form

$$H_f^\Omega(z) = \int_\Gamma f(\zeta)d\mu_z^\Omega(\zeta). \tag{7.14}$$

This measure is called the *harmonic measure*, which will be discussed in a different context presently. There is another method for solving the Dirichlet problem introduced by Oskar Perron and based on subharmonic functions which we take up in the next chapter.

Harmonic measure

Two closely related notions are a "random walk" solution of a Dirichlet problem and a harmonic measure mentioned above. This latter concept was introduced by R. Nevanlinna in a series of lectures at the ETH, Zurich, in 1928–1929.[13] The function u_f in the solution to the Dirichlet problem via harmonic measure with respect to an arc $E \subseteq \{|z| = 1\}$ turns out to be the same as that given by the "hitting probability" random walk, which we discuss subsequently; see Figure 7.3. Note that if the arc lies between two angles, $E = (\theta_1, \theta_2)$, then by the Poisson formula, for $z = re^{i\phi}$ and taking the characteristic function χ_E of E,

$$u_{\chi_E}(z) = \frac{1}{2\pi}\int_{\theta_1}^{\theta_2} \frac{1 - r^2}{1 - 2r\cos(\theta - \phi) + r^2} d\theta = \omega(z, \theta_1, \theta_2) \tag{7.15}$$

defines the harmonic measure of E, denoted by $\omega(z, \theta_1, \theta_2)$, and thus $\omega(0, \theta_1, \theta_2) = \frac{\theta_2 - \theta_1}{2\pi}$. In the latter instance the harmonic measure is a measure of the length of the arc relative to the point at the origin. From the probabilistic interpretation the harmonic measure of an arc is the size of the arc as "seen" from the point z, and each point of the domain gives rise to a different measure.

In view of (7.14), we can rewrite the Poisson formula in terms of harmonic measure for a piecewise continuous boundary function $f(e^{i\theta})$ as

$$u_f(z) = \int_0^{2\pi} f(e^{i\theta})d\omega(z, 0, \theta).$$

13 First published in, Das harmonische Mass von Punktmengen und seine Anwendung in der Funktionentheorie, 8. *Cong. Math. Scand.*, Stockholm, 1934.

It will be instructive to show that if $f(\zeta)$ is continuous at the point $\zeta_0 = e^{i\theta_0}$, then $u_f(z) \to f(\zeta_0)$ uniformly as $z \to \zeta_0$. To this end, given $\varepsilon > 0$, and suppose that $|f(\zeta) - f(\zeta_0)| < \varepsilon$ whenever ζ_0 belongs to some interval $\theta_1 \le \theta \le \theta_2$. Since $|f(\zeta)| \le M < \infty$ on the unit circle,

$$\begin{aligned} |u_f(z) - f(\zeta_0)| &= \left| \int_{\theta=0}^{\theta=2\pi} (f(e^{i\theta}) - f(e^{i\theta_0}))d\omega \right| \\ &\le \int_{\theta_1}^{\theta_2} |f(\zeta) - f(\zeta_0)|d\omega + \int_{\theta_2}^{\theta_1+2\pi} |f(e^{i\theta}) - f(e^{i\theta_0})|d\omega \\ &\le \varepsilon + 2M\omega(z, \theta_2, \theta_1 + 2\pi). \end{aligned}$$

Observe that the harmonic measure $\omega(z, \theta_2, \theta_1 + 2\pi)$ vanishes uniformly on the complementary arc defined by $\theta_1 < \theta < \theta_2$, so that for some $r_\varepsilon > 0$ and $|z - \zeta_0| < r_\varepsilon$, we have $\omega(z, \theta_2, \theta_1 + 2\pi) < \varepsilon$. It follows that for $|z - \zeta_0| < r_\varepsilon$,

$$|u_f(z) - f(\zeta_0)| < \varepsilon(2M + 1),$$

that is, $u_f(z) \to f(\zeta_0)$ uniformly as $z \to \zeta_0$, concluding the proof.

We can analogously show (exercise) that at a point of discontinuity $\zeta_0 = e^{i\theta_0}$, if $f(e^{i(\theta_0+0)})$ and $f(e^{i(\theta_0-0)})$ are the right- and left-hand limits, respectively, then the radial limit has the value

$$\lim_{r\to 1^-} u_f(re^{i\theta_0}) = \frac{1}{2}(f(e^{i(\theta_0+0)}) + f(e^{i(\theta_0-0)})).$$

More generally, let Ω be a domain whose boundary Γ consists of finitely many smooth disjoint Jordan curves. If α is an arc of Γ, then the *harmonic measure* of the arc α with respect to the domain Ω at the point z, denoted by $\omega(z, \alpha, \Omega)$,[14] is the harmonic function that takes the boundary values 1 on α and 0 on $\Gamma - \alpha$. The existence follows from the solution to the Dirichlet problem[15] for a piecewise continuous function, and the harmonic measure is unique at each point of Ω. Moreover, $0 \le \omega(z, \alpha, \Omega) \le 1$, and if $\alpha = \partial\Omega$, then $\omega(z, \alpha, \Omega) = 1$.

The notion of a harmonic measure will be extended to the more general setting of Borel sets on the boundary in Chapter 8.

14 Or simply $\omega(z, \alpha)$ when the domain is apparent.

15 A very general method for finding the solution to the Dirichlet problem is discussed at length in the next chapter.

7.31 Example. If the domain Ω is as in Theorem 7.30 and α is an arc on $\partial\Omega$, then

$$\omega(z,\alpha,\Omega) = \frac{1}{2\pi}\int_{\alpha} \frac{\partial}{\partial n} G(\zeta, z_0)ds$$

in view of the result of that theorem.

7.32 Example. If Ω is the annulus $0 < r_1 < |z| < r_2$ and $E = |z| = r_1$, then

$$\omega(z,E,\Omega) = \log\left(\frac{|z|}{r_2}\right) / \log\left(\frac{r_1}{r_2}\right).$$

Another immediate property is that if α and β are two disjoint arcs of $\partial\Omega$, then $\omega(z,\alpha,\Omega) + \omega(z,\beta,\Omega) = \omega(z,\alpha+\beta,\Omega)$, that is, the harmonic measure is *additive*. Furthermore, the harmonic measure of any finite set of points is zero, and the harmonic measure of any continuum in $\partial\Omega$ is always positive. The harmonic measure can be related to various other measures.

For example, a point set A has *logarithmic measure zero* if for each $\varepsilon > 0$, the set A can be covered by a family of disks D_n with radii r_n satisfying the condition

$$\sum \frac{1}{\log^{+}\frac{1}{r_n}} < \varepsilon.$$

Then any closed point set with logarithmic measure zero also has harmonic measure zero.[16]

One useful property is the *invariance* of the harmonic measure under conformal mapping of the domain. Indeed, suppose that the domain Ω is mapped analytically one-to-one onto the domain Ω' with a continuous extension to the boundary $\partial\Omega = \alpha+\beta$. Then the arcs α,β are mapped to arcs α',β' of $\partial\Omega'$, respectively. Since harmonicity is preserved under conformal mappings of the domain (Lemma 7.10), the resulting harmonic measure on Ω' satisfies

$$\omega(z',\alpha',\Omega') = \omega(z,\alpha,\Omega).$$

If we parameterize the boundary $\partial\Omega$ as $\zeta = \zeta(t)$, $0 \le t < 1$, let the arc $\alpha(t)$ represent that part of the boundary from $\zeta(0)$ to $\zeta(t)$. The following result is exactly what we would expect.

7.33 Proposition. *For any fixed $z \in \Omega$, the harmonic measure $\omega(z,\alpha(t),\Omega)$ is a continuous increasing function of the parameter t.*

---

16 See Nevanlinna (1970, p. 147).

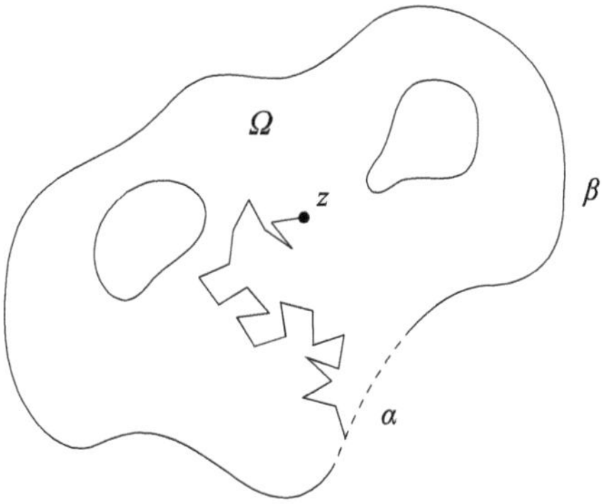

Figure 7.3: A Brownian path from the point z to the boundary. The probability of starting at z and first hitting a point of the arc α is the value of the harmonic measure $\omega(z, \alpha, \Omega)$. Clearly, if $\alpha = \partial\Omega$, then $\omega(z, \alpha, \Omega) \equiv 1$. Courtesy Katy Metcalf.

Proof. To show the monotonicity, letting $\Delta\alpha$ correspond to that part of the boundary corresponding to the interval $(t, t + \Delta t)$, it is clear by additivity property of harmonic measure that $\Delta\omega = \omega(z, \alpha(t + \Delta t)) - \omega(z, \alpha(t)) = \omega(z, \Delta\alpha) > 0$.

For the continuity property, let $\zeta = \zeta(t)$ be an endpoint of the arc $\Delta\alpha$, and let $D(\zeta, r)$ be the smallest disk centered at ζ that contains $\Delta\alpha$. Note that $\Delta t \to 0$ implies that $r \to 0$ by the continuity of $\zeta(t)$. We now take another disk centered at the point ζ, namely one that encompasses the entire domain Ω, and call it $D(\zeta, R)$. Then for $z \in \Omega$, the function defined by

$$u(z) = \log \frac{R}{|z - \zeta|} / \log \frac{R}{r}$$

is a nonnegative harmonic function in Ω with $u(z) > 1$ in the disk $D(\zeta, r)$ and hence larger than 1 on $\Delta\alpha$. As a consequence, $u(z) \geq \omega(z, \Delta\alpha)$. We conclude that for fixed $z \in \Omega$, as $\Delta t \to 0$ and hence as $r \to 0$, we have $u(z) \to 0$ and likewise $\Delta\omega \to 0$, proving that ω is right-continuous, and we can similarly show that ω is left-continuous. □

It follows that as $\zeta(t)$ traces out the boundary $\partial\Omega$ as t varies from 0 to 1, the harmonic measure ω at each point is increasing from 0 to 1 and conversely.

Random walks

As has been suggested, harmonic measure has a deep connection with probability theory. Let $E \subseteq \{|z| = 1\}$ be an arc of the boundary of the open unit disk U and consider the piecewise continuous boundary function

$$f = \begin{cases} 1 & \text{on } E \\ 0 & \text{on } E' = \{|z| = 1\} - E. \end{cases}$$

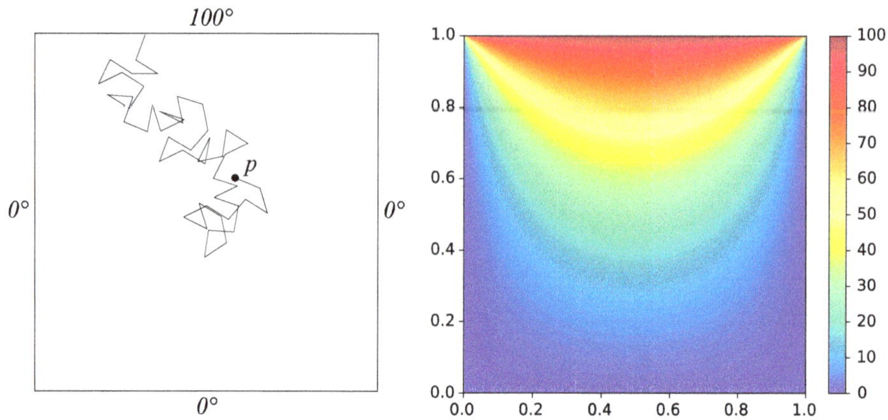

Figure 7.4: (*L*) Determining the temperature at a point p in the interior of a square by means of a random walk with boundary values as indicated. (*R*) The realized steady-state temperature distribution in the square region which was actually done via a cellular automata routine using small squares instead of individual points. Courtesy (*L*) Katy Metcalf, (*R*) Anita Kean.

Suppose that u_f is the solution to the Dirichlet problem, that is, u_f is harmonic in U, and $u_f = f$ on $|z| = 1$. It was shown by Shizuo Kakutani in 1944[17] that

$$u_f(z_0) = \text{Probability that a random walk starting at } z_0 \in U$$
$$\textbf{first} \text{ hits a point of the boundary at a point of } E.$$

This probability is of course the *harmonic measure* $\omega(z, E, U)$.

Changing the setting slightly, let us put a temperature distribution of 100° on the top side of a square and 0° on the other three sides as in Figure 7.4 (L). We can determine the temperature at any point p interior to the square by taking a random walk (short steps in a random direction) until the path of the walk hits one of the sides of the rectangle.

Such a procedure is easily implemented on a computer. Once a random walk has hit a side, a new random walk is started over again from p, and the procedure repeated for say 1,000 runs. The temperature at p is then given by the tally

$$Temp(p) \approx \frac{\#\text{ of hits of } 100^\circ\ \textit{first}}{\text{total }\#\text{ of random walks}} \times 100.$$

7.34 Exercise. By the random walk method write a simple computer program and determine the temperature at the point $(0, 0.5)$ for a temperature distribution of 100° on the top half of the unit circle and 0° on the bottom half of the unit circle.

17 On Brownian motion in n-space, *Proc. Imp. Acad. Tokyo*, 20 (9) (1944), 648–652.

Positive harmonic functions/class $h^1(U)$

In this section, we examine harmonic functions that are positive as they have special characteristics. In general, by positive we mean nonnegative so as not to be overly pedantic.

Harnack inequality

One immediate consequence of the Poisson formula and the bounds given in (7.3) is the inequality due to Axel Harnack (1851–1888). Without taking any credit away from Herr Harnack, it is the sort of result that is rather obvious from simply looking at the Poisson kernel.

7.35 Theorem (Harnack Inequality).[18] *Let u be harmonic in $|z| < R$ and continuous on $|z| \le R$. If $u \ge 0$ in $|z| \le R$, then*

$$\frac{R-|z|}{R+|z|}u(0) \le u(z) \le \frac{R+|z|}{R-|z|}u(0), \quad |z| < R.$$

7.36 Corollary. *A positive harmonic function in the entire complex plane is identically constant.*

The proof follows from Harnack's inequality by letting $R \to \infty$. We could also construct a harmonic conjugate v of u and consider the analytic function $F = e^{-(u+iv)}$.

7.37 Exercise. If $f(z)$ *is analytic in* $|z| \le R$, *then for* $0 < r < R$, *prove the following adjunct to the Hadamard–Borel–Carathéodory Theorem* 1.10:

$$A(R) \le \frac{R-r}{R+r}\operatorname{Re}(f(0)) + \frac{2r}{R+r}A(R).$$

Herglotz theorem

In a similar vein to the Poisson formula, we have the theorem of Gustav Herglotz (1881–1953)[19] in terms of a Poisson–Stieltjes integral.

7.38 Theorem. *If u is a positive harmonic function in U, then there is a nondecreasing function $\mu(t)$ on $[0, 2\pi]$ such that*

---

18 Die Grundlagen der Theorie des logarithmischen Potentiales und der eindeutigen Potentialfunktion in der Ebene, Leipzig, 1887.

19 Über Potenzreihen mit positive reellen Teil im Einheitskreis, *Ber. Verhandl. Kön. Sächs. Gesell. Wiss. Leipzig* 63 (1911), 501–511.

$$u(z) = \frac{1}{2\pi} \int_0^{2\pi} \frac{1 - |z|^2}{|\zeta - z|^2} d\mu(t), \quad \zeta = e^{it}, |z| = r < 1. \tag{7.16}$$

Proof. Define the nondecreasing function for $0 < r < 1$

$$\mu_r(\tau) = \int_0^{\tau} u(re^{i\theta}) d\theta. \tag{7.17}$$

By taking a partition of $[0, 2\pi]$, $0 = \tau_0 < \tau_1 < \cdots < \tau_n = 2\pi$, we then can write

$$\sum_{k=0}^{n-1} |\mu_r(\tau_{k+1}) - \mu_r(\tau_k)| = \int_0^{2\pi} u(re^{i\theta}) d\theta = 2\pi u(0).$$

Thus the family of functions $\{\mu_r\}$ is of bounded total variation and uniformly bounded since $0 \le \mu_r(\tau) \le u(0)$ for $0 < r < 1$. Taking a sequence $\{\mu_{r_n}\}$ with $r_n \to 1$, by the Helly selection theorem[20] there exists a subsequence, which we also denote $\{\mu_{r_n}\}$, converging to a function μ of bounded variation (and in this case, nondecreasing) on $[0, 2\pi]$. From expression (7.17), due to the continuity of the Poisson kernel in the variable t ($\zeta = e^{it}$), it follows that

$$\frac{1}{2\pi} \int_0^{2\pi} \frac{1 - |z^2|}{|\zeta - z|^2} d\mu(t) = \lim_{n\to\infty} \frac{1}{2\pi} \int_0^{2\pi} \frac{1 - |z^2|}{|\zeta - z|^2} d\mu_{r_n}(t) = \lim_{n\to\infty} \frac{1}{2\pi} \int_0^{2\pi} \frac{1 - |z^2|}{|\zeta - z|^2} u(r_n e^{it}) dt. \tag{7.18}$$

The functions $u_n(z) = u(r_n z)$, $z \in U$, are harmonic in U and continuous on ∂U. Therefore the Poisson integral part in (7.18) represents the function value $u_n(z)$, so that the limit in (7.18) is

$$\frac{1}{2\pi} \int_0^{2\pi} \frac{1 - |z^2|}{|\zeta - z|^2} d\mu(t) = \lim_{n\to\infty} u_n(z) = u(z),$$

proving the theorem. □

20 Helly selection theorem: For any sequence of real-valued functions $\{v_n\}$ on the interval $[a, b]$ that is uniformly bounded and of uniformly bounded total variation on $[a, b]$, there exists a subsequence $\{v_{n_k}\}$ that converges on $[a, b]$ to a function v of bounded variation in the sense that for any real-valued continuous function $\lambda(t)$ on $[a, b]$,

$$\int_a^b \lambda(t) dv_{n_k}(t) \to \int_a^b \lambda(t) dv(t)$$

as $k \to \infty$.

An equation of the form (7.16) for a function $\mu(t)$ of bounded variation on $[0, 2\pi]$ is a *Poisson–Stieltjes integral*. In this regard, note that if f is integrable on $[0, 2\pi]$, then

$$\mu(t) = \int_0^t f(\tau)d\tau$$

is a (continuous) function of bounded variation on $[0, 2\pi]$. The proof is left as an exercise. Thus the ordinary Poisson integral can be regarded as a particular case of the Poisson–Stieltjes integral.

In the converse direction, if μ is a function of bounded variation, then μ is the difference of two nondecreasing functions, and their Poisson–Stieltjes integrals represent two positive harmonic functions (exercise).

7.39 Example. Let $f(z)$ be analytic in U such that $|f(z)| > 1$, so that $g(z) = \log|f(z)|$ is a positive harmonic function in U. Applying the Herglotz representation (7.16) with $|z| = r$, we find that

$$\log M(r,f) \le \frac{1+r}{1-r}\log|f(0)| \tag{7.19}$$

where $M(r,f) = \max_{|z|=r}|f(z)|$. From this inequality written in the form $|f(z)| \le |f(0)|^{\frac{1+r}{1-r}}$ we can appreciate just how much the growth of the function $f(z)$ is controlled by its value at the origin. Similarly, if $|f(z)| > \alpha > 0$, we leave to the reader to obtain a growth estimate for $M(r,f)$ similar to (7.19).

Although a positive harmonic function in U can approach the value zero on the boundary, it cannot do so too rapidly without becoming degenerate.

7.40 Proposition. *Suppose that $h(z)$ is a positive harmonic function in U such that*

$$\underline{\lim}_{z\to\zeta_0}\frac{h(z)}{1-|z|} = 0$$

for some point $\zeta_0 \in \partial U$. Then $h \equiv 0$ in U.

Proof. From the Herglotz representation for $h(z)$ by Fatou's lemma we have with $\zeta = e^{it}$

$$0 = \underline{\lim}_{z\to\zeta_0}\frac{h(z)}{1-|z|} \ge \frac{1}{2\pi}\int_0^{2\pi}\underline{\lim}_{z\to\zeta_0}\frac{d\mu(t)}{|\zeta - z|^2} = \frac{1}{2\pi}\int_0^{2\pi}\frac{d\mu(t)}{|\zeta-\zeta_0|^2} \ge 0.$$

A fortiori $d\mu \equiv 0$, and thus $h \equiv 0$. □

A simple thing we can do with positive harmonic functions is subtracting them, but let us first define a new class of harmonic functions.

7.41 Definition. A harmonic function $u \in h^1(U)$ if the integral mean remains bounded in U, that is,

$$\frac{1}{2\pi}\int_0^{2\pi} |u(re^{i\theta})|d\theta < M < \infty$$

for all $< r < 1$.

This class of harmonic functions is indeed related to positive harmonic functions.

7.42 Theorem. *If $u = u_1 - u_2$, where $u_1, u_2 \geq 0$ are harmonic, then $u \in h^1(U)$. Conversely, any $u \in h^1(U)$ is the difference of two nonnegative harmonic functions.*

Proof. If $u = u_1 - u_2$, then

$$\frac{1}{2\pi}\int_0^{2\pi} |u(re^{i\theta})|d\theta \leq \frac{1}{2\pi}\int_0^{2\pi} u_1(re^{i\theta})d\theta + \frac{1}{2\pi}\int_0^{2\pi} u_2(re^{i\theta})d\theta = u_1(0) + u_2(0).$$

One rather interesting way to see the converse for $u \in h^1(U)$ is to let $f = u + iv$ and consider the Nevanlinna characteristic of e^f:

$$T(r, e^f) = m(r, e^f) = \frac{1}{2\pi}\int_0^{2\pi} \frac{u(re^{i\theta}) + |u(re^{i\theta})|}{2}d\theta = \frac{u(0)}{2} + \frac{1}{4\pi}\int_0^{2\pi} |u(re^{i\theta})|d\theta < \infty.$$

Therefore e^f has bounded characteristic, and by Theorem 3.29, $e^f = \phi_1/\phi_2$, where ϕ_1 and ϕ_2 are nonzero bounded analytic functions. Without loss of generality, we can assume that $|\phi_i| < 1$, $i = 1, 2$. Then $f = \log\phi_2^{-1} - \log\phi_1^{-1}$, and taking real parts proves the statement. □

Radial limits

The Poisson–Stieltjes integral of positive harmonic functions allows us to establish the existence of radial limits a. e. of h^1-functions. This is essentially a theorem of Fatou, who proved that the radial limits of a bounded analytic function in U exist for all points on $|z| = 1$ except for a set of linear measure zero.[21] We prove the result for the more general *symmetric derivative* of a function $\mu(t)$ given by

$$D\mu(\theta_0) = \lim_{t\to 0}\frac{\mu(\theta_0 + t) - \mu(\theta_0 - t)}{2t}.$$

21 Séries trigonométriques et series de Taylor, *Acta Math.* 30 (1906), 335–400.

When the usual derivative exists at a point, then the symmetric derivative exists, but the converse is not true.

Let us now represent the Poisson kernel $\frac{1-|z|^2}{|\zeta-z|^2}$ as $P(r,\theta-t)$ for $z=re^{i\theta}$ and $\zeta=e^{it}$.

7.43 Theorem. *In the unit disk U, if $u(z)$ has the representation*

$$u(re^{i\theta})=\frac{1}{2\pi}\int_{-\pi}^{\pi}P(r,\theta-t)d\mu(t),$$

where $\mu(t)$ is a function of bounded variation, then whenever $D\mu(\theta_0)$ exists,[22] the radial limit is given by

$$\lim_{r\to1^-}u(re^{i\theta})=D\mu(\theta_0).$$

Proof. Without loss of generality, take $\theta_0=0$ and let $\gamma=D\mu(0)$. Then integrating by parts we get

$$\begin{aligned}u(r)-\gamma&=\frac{1}{2\pi}\int_{-\pi}^{\pi}P(r,t)(d\mu(t)-\gamma dt)\\&=\frac{1}{2\pi}P(r,t)(\mu(t)-\gamma t)|_{-\pi}^{\pi}-\frac{1}{2\pi}\int_{-\pi}^{\pi}(\mu(t)-\gamma t)\left(\frac{\partial P}{\partial t}\right)dt,\end{aligned}\tag{7.20}$$

so that the first term tends to zero ($t\neq0$) as $r\to1$. Computing the partial derivative

$$-\frac{\partial P}{\partial t}=\frac{2r(1-r^2)\sin t}{(1-2r\cos t+r^2)^2},$$

we see that for $0<t\le\pi$, the function $F(t)=t(-\frac{\partial P}{\partial t})\ge0$, and likewise for $-\pi\le t\le0$ with $F(t)=F(-t)$.

Furthermore, for $0<\delta\le|t|\le\pi$,

$$\left|\frac{\partial P}{\partial t}\right|\le\frac{2r(1-r^2)}{(1-2r\cos\delta+r^2)^2},$$

so that $|\frac{\partial P}{\partial t}|\to0$ as $r\to1$ in the interval $0<\delta\le|t|\le\pi$. Thus we break the integral in (7.20) into $\int_{-\pi}^{\pi}=\int_{\delta\le|t|\le\pi}+\int_{-\delta}^{\delta}$, and hence the first term tends to zero as $r\to1$. Regarding the second term, a bit of calculation shows that

$$\int_{-\delta}^{\delta}=-\frac{1}{2\pi}\int_{-\delta}^{\delta}(\mu(t)-\gamma t)\left(\frac{\partial}{\partial t}P(r,t)\right)dt=\frac{1}{\pi}\int_{0}^{\delta}\left(\frac{\mu(t)-\mu(-t)}{2t}-\gamma\right)\left(-t\frac{\partial}{\partial t}P(r,t)\right)dt.$$

22 A function of bounded variation has derivatives existing almost everywhere.

Then, for $\varepsilon > 0$, we can find $\delta > 0$ sufficiently small such that

$$\left|\frac{\mu(t) - \mu(-t)}{2t} - \gamma\right| < \varepsilon$$

for $0 < t \le \delta$. As a consequence,

$$\left|\int_{-\delta}^{\delta} \cdots\right| \le \frac{\varepsilon}{\pi}\int_0^{\delta}\left|t\frac{\partial P}{\partial t}\right|dt \le \frac{\varepsilon}{\pi}\int_0^{\pi} t\left(-\frac{\partial P}{\partial t}\right)dt = \frac{\varepsilon}{2\pi}\int_{-\pi}^{\pi} t\left(-\frac{\partial P}{\partial t}\right)dt < 2\varepsilon$$

as $r \to 1$, upon applying integration by parts to this last integral. We conclude that $u(r) \to \gamma$ as $r \to 1$, establishing the theorem. □

Since each $u \in h^1(U)$ can be expressed as the difference of two positive harmonic functions, each of which has a Poisson–Stieltjes integral of the form given in the theorem, we maintain the following:

7.44 Corollary. *If $u \in h^1(U)$, then its radial limits exist almost everywhere on $|z| = 1$.*

It is evident now that a bounded analytic function in U has radial limits a. e. Even more is the case. Analytic functions from the class $H^1(U)$ whose integral means are bounded in U, that is,

$$\frac{1}{2\pi}\int_0^{2\pi} |f(re^{i\theta})|d\theta < M < \infty$$

for $0 < r < 1$, have radial limits a. e., since their real and imaginary parts belong to $h^1(U)$.

7.45 Exercise.

(i) Let $u \ge 0$ be a harmonic function on a domain Ω, and let

$$\mathfrak{H}_u = \{v \text{ bounded harmonic in } \Omega : v \le u\}.$$

Prove that the function

$$\mathcal{B}u(z) = \sup_{v \in \mathfrak{H}_u} v(z)$$

is harmonic in Ω and $0 \le \mathcal{B}u \le u$.

If $\mathcal{B}u = u$, then u is called *quasibounded*, and if $\mathcal{B}u = 0$, then u is called *singular*. These classes were introduced by M. Parreau (1951)[23] in the context of Riemann

23 Sur les moyennes des fonctions harmoniques et analytiques et la classification des surfaces de Riemann, *Ann. Inst. Fourier (Grenoble)* 3 (1951), (1952), 103–197.

surfaces. For example, the function $u = -\log|z|$ is singular in $0 < |z| < 1$, since any bounded harmonic function $v \le u$ has a removable singularity at the origin, and being zero on $|z| = 1$ means that $v \equiv 0$.

(ii) Prove that $a\mathcal{B}(u) = \mathcal{B}(au)$ and $\mathcal{B}(u_1 + u_2) = \mathcal{B}(u_1) + \mathcal{B}(u_2)$.

(iii) Prove that $\mathcal{B}^2 = \mathcal{B}(\mathcal{B}) = \mathcal{B}$.

(iv) Prove that if $u \ge 0$ is harmonic in a domain Ω, then it has the (Parreau) decomposition

$$u = q + s,$$

where q is a quasibounded function, and s is singular.

Normal families of harmonic functions

As with analytic functions, we can define the notion of normality with respect to a family of harmonic functions.

7.46 Definition. A family $\mathcal{H}$ of harmonic functions defined in a domain Ω is *normal* if every sequence of functions $\{u_n\}$ belonging to $\mathcal{H}$ contains a subsequence that converges uniformly on compact subsets of Ω either to a harmonic function or to $\pm\infty$.

Normality in a domain is also equivalent to pointwise normality just as it is in the analytic case and can be proved directly by a diagonal method (exercise).

We might expect that for a normal family of analytic functions, the corresponding families of real (or imaginary) parts would also be normal. However, this is *not* the case as the following family of functions shows:

$$f_n(z) = nx + i(ny + n^2) = u_n(x, y) + iv_n(x, y), \quad z \in U, n = 1, 2, 3, \dots.$$

Then $f_n(z)$ is clearly analytic, and $f_n(z) \to \infty$ uniformly in U, and hence the family $\{f_n\}$ is normal in U, but the family of harmonic real parts $\{u_n\}$ is not normal in U.

The problem here was the infinite limit, and if we remove this possibility, we can say the following:

7.47 Proposition. *For normal families of analytic functions $\mathcal{F}$ that do not admit the value ∞ as a limit, the family of real parts $\mathcal{H} = \operatorname{Re}\mathcal{F}$ is a normal family.*

Conversely, if $\mathcal{H} = \{u\}$ is a normal family of harmonic functions defined on a simply connected domain Ω, the corresponding family of analytic functions $\mathcal{F}$ defined by $f = \operatorname{Re}(u)$ is also normal in Ω. This is so because if $\{f_n\} \subseteq \mathcal{F}$ and $\{u_n\}$ are the corresponding harmonic functions in $\mathcal{H}$, then there is a convergent subsequence $\{u_{n_k}\}$ that converges normally either to $\pm\infty$ or to a harmonic function. In the former case, as $|f_{n_k}| \ge |u_{n_k}|$, the same applies to the subsequence $\{f_{n_k}\}$. On the other hand, if $u_{n_k} \to u$ uniformly for u harmonic in Ω, then $\{u_{n_k}\}$ is locally uniformly bounded. Hence the partial derivatives are also locally bounded (why?), and in view of

$$f'_{n_k} = \frac{\partial u_{n_k}}{\partial x} - i\frac{\partial u_{n_k}}{\partial y},$$

we deduce that the family $\{f'_{n_k}\}$ is locally bounded, and hence $\mathcal{F}$ is normal in Ω by Marty's theorem 4.27.

Just as in the case of families of analytic functions, we have the following:

7.48 Theorem. *A locally bounded family of harmonic functions $\mathcal{H}$ defined on a domain Ω is normal.*

Proof. The goal is to construct a corresponding family of analytic functions in an obvious manner. Take any $z_0 \in \Omega$ and a sequence $\{h_n\}$ in $\mathcal{H}$. As the family $\mathcal{H}$ is locally bounded, there is a disk $D(z_0, r) \subseteq \Omega$ on which $|h_n| \leq M$ for all $= 1, 2, 3, \ldots$. In the disk $D(z_0, r)$, we can construct an analytic function f_n for which $h_n = \operatorname{Re} f_n$, and hence for the analytic function $g_n = e^{f_n}$, we have

$$e^{-M} \leq |g_n| = e^{h_n} \leq e^M.$$

It follows by Montel's Theorem 4.4 that there is a subsequence $\{g_{n_k}\}$ such that g_{n_k} converges uniformly on compact subsets of $D(z_0, r)$ to an analytic function g in $D(z_0, r)$ satisfying $e^{-M} \leq g \leq e^M$. In view of the uniform continuity of the log function on the closed interval $[e^{-M}, e^M]$,

$$h_{n_k} = \log|g_{n_k}| \to \log g = h$$

uniformly on compact subsets of $D(z_0, r)$. Thus h is harmonic in $D(z_0, r)$, and since z_0 was arbitrary, h is harmonic in Ω. □

In keeping with the Bloch principle, since a positive harmonic function in the complex plane reduces to a constant, we have the following:

7.49 Theorem. *The family of positive harmonic functions $\mathcal{H}^+$ of a domain Ω is normal.*

Proof. We will show that $\mathcal{H}^+$ is normal at an arbitrary point $z_0 \in \Omega$. For a sequence $\{u_n\}$ in $\mathcal{H}^+$, the sequence of values $\{u_n(z_0)\}$ has a subsequence $\{u_{n_k}(z_0)\}$ that converges to some limit l, possibly, infinite. Taking a closed disk $\overline{D}(z_0, r)$ in Ω, by Harnack's inequality applied to these functions, we have

$$\frac{1}{c} u_{n_k}(z_0) \leq u_{n_k}(z) \leq c u_{n_k}(z_0)$$

for all $z \in \overline{D}(z_0, r/2)$. If $l < \infty$, then the sequence $\{u_{n_k}\}$ is bounded in a neighborhood of z_0 and hence normal there. Otherwise, if $l = \infty$, then $u_{n_k} \to \infty$ uniformly in $\overline{D}(z_0, r)$, so that in either case, $\mathcal{H}^+$ is a normal family. □

7.50 Remark. It is clear that if $u(z) = \alpha_0 \in \mathbb{R}$ for all $u \in \mathcal{H}^+$, then the family is also compact.

Harnack principle

Intuitively, the next result is obvious in 1-dimension. A monotonically increasing/decreasing sequence of straight lines (which represent harmonic functions) defined in a domain either converges to another straight line in the domain or carries on to $\pm\infty$. Moreover, if the sequence converges at one point, then it converges throughout the domain.

7.51 Harnack principle. *Let $\{u_n\}$ be a sequence of harmonic functions on a domain Ω with $u_n \le u_{n+1}$ for $n = 1, 2, 3, \ldots$. Then either $u_n \to \infty$ uniformly on compact subsets of Ω, or u_n converges to a harmonic function uniformly on compact subsets of Ω. Analogously, for a decreasing sequence of harmonic functions.*

Proof. A normal family argument makes light work of the traditionally more technical proof. Indeed, setting $v_n = u_n - u_1$, the sequence of positive harmonic functions $\{v_n\}$ forms a normal family in Ω. Since $\{v_n\}$ is increasing, it must either converge or diverge to $+\infty$ at each point of Ω. In view of the normality of the sequence $\{v_n\}$, the convergence is uniform on compact subsets either to a harmonic function or to $+\infty$. Then the same holds for $\{u_n\}$, and the result is proved, and likewise for a decreasing sequence. □

We can replace the property of monotonicity by the omission of a single value. This is the analogue of the Montel theorem for analytic functions.

7.52 Theorem. *If a family $\mathcal{H}$ of harmonic functions on a domain Ω omits a single value $a \in \mathbb{R}$, then it is normal.*

To see this, note that for each $u \in \mathcal{H}$, we have either $u(z) > a$ or $u(z) < a$ *for all* $z \in \Omega$, since if it were the case that $u(z_1) > a$ and $u(z_2) < a$, then there would be a point z on any curve joining z_1 to z_2 at which $u(z) = a$. As a consequence, we can divide the family $\mathcal{H}$ into two disjoint subfamilies

$$\mathcal{H}_1 = \{u \in \mathcal{H} : u(z) > a\}, \quad \mathcal{H}_2 = \{u \in \mathcal{H} : u(z) < a\},$$

both of which are normal, establishing the normality of $\mathcal{H}$.

The Laplace equation is a subclass of the more general class of elliptic partial equations. The above theorem has been extended to this more general category by A. Beardon.[24]

7.53 Exercise. Show that the family $\mathcal{H}$ of harmonic functions that do not take any value in the interval $(0, 1)$ is normal.

We will require the following, which is the harmonic analogue of the analytic case (Corollary 4.8).

---

24 Montel's theorem for subharmonic functions and solutions of partial differential equations, *Proc. Camb. Phil. Soc.* 69 (1971), 123–150.

7.54 Proposition. *A normal family of harmonic functions that is bounded at a point is locally bounded.*

Proof. Exercise. □

Schottky theorem for harmonic functions

In view of the above proposition, we have the harmonic version of Schottky's theorem. The previous version (Chapter 4) was for analytic functions omitting the values 0 and 1, which allowed the use of a normal family argument. In the harmonic version, we assume more generally that the family is normal.

7.55 Theorem. *Let*

$$\mathcal{H} = \{u \text{ harmonic in } |z| < R : u(0) = a_0\}$$

be a normal family. Then for each $0 < \delta < 1$, *there is a constant* $M = M(a_0, \delta)$ *such that*

$$|u(z)| \le M(a_0, \delta)$$

for $|z| \le \delta R$ *and all* $u \in \mathcal{H}$.

The proof (exercise) follows as in the analytic case by means of the preceding proposition.

Julia theorem for harmonic functions

We next prove an analogue of the Julia theorem for analytic functions in a sector emanating from an essential singularity. This result is a consequence of the preceding Schottky theorem and demonstrates the somewhat remarkable behavior of a harmonic function in the complex plane even though we are not dealing with any type of singularity in this case.

7.56 Theorem.[25] *Let* $u(z)$ *be a nonconstant harmonic function in* $\mathbb{C}$. *Then there is at least one ray* $\arg z = \phi$ *emanating from the origin such that in each sector* $\phi - \varepsilon < \arg z < \phi - \varepsilon$, *the function* $u(z)$ *assumes every real value infinitely often.*

Proof. For $z \in U : |z| < 1$ with $z = x + iy$, define the sequence of harmonic functions

$$u_n(x, y) = u(2^n x, 2^n y)$$

25 P. Montel, Sur quelques familles de fonctions harmoniques, *Fund. Math.* 25 (1935), 388–407.

for $n = 1, 2, 3, \ldots$. We claim that the family of functions $\{u_n\}$ does *not* form a normal family in U. If it were a normal family in U, then together with the fact that

$$u_n(0) = u(0) = a_0,$$

the preceding harmonic version of Schottky's theorem would imply that

$$|u_n(z)| \leq M(a_0, 1/2)$$

in the disk $|z| \leq 1/2$. This in turn would mean that the function u is bounded in $\mathbb{C}$, which cannot be the case as u is nonconstant. We infer that there is a point (an *irregular point*) in U at which the family of harmonic functions $\{u_n\}$ is not normal.

If say $z = 0$ were the only irregular point in U, then by normality there would be a subsequence of $\{u_n\}$ that converges uniformly on the compact set $|z| = r < 1$ and thus on the disk $|z| \leq \rho < r$ by Poisson's formula. We may conclude from this argument that an irregular point cannot be isolated.

As a consequence, there is an irregular point $z_0 \neq 0$ such that in any arbitrarily small disk with center z_0, the family $\{u_n\}$ is not normal. Therefore $\{u_n\}$ attains every value in a suitably small disk $D_0 : |z - z_0| < r$ for *infinitely* many values of n.

If we define a sequence of homothetic disks

$$D_n = \{|z - 2^n z_0|\} < 2^n r$$

for $n = 1, 2, 3, \ldots$, then we have $u_n(D_0) = u(D_n)$. It follows that the function u takes every real value in the collection of disks $\{D_n\}$ infinitely many times, proving the theorem. □

8 Subharmonic/superharmonic functions

Subharmonic and superharmonic functions are important for many reasons and particularly since they arise naturally in complex function theory and potential theory. In one dimension, harmonic functions are represented by straight lines, and subharmonic (superharmonic) functions are convex (concave) functions that lie below (above) them, and the same analogy applies for subharmonic (superharmonic) functions in the complex plane.

8.1 Definition. A function $v : \Omega \to \mathbb{R} \cup \{-\infty\}$ is *subharmonic* on a domain $\Omega \subseteq \mathbb{C}$ if:

(i) v is upper semicontinuous (u. s. c.) in Ω, that is, for each $z_0 \in \Omega$ and each $\eta > v(z_0)$, there exists $\delta = \delta(z_0, \eta) > 0$ such that $v(z) < \eta$ whenever $|z - z_0| < \delta$. In other words, the sets where $v(z_0) < \eta$ are open. Equivalently,

$$\overline{\lim_{z \to z_0}}\, v(z) \le v(z_0).$$

(ii) For any subdomain $D \subset \Omega$ and harmonic function h in D that is continuous on $\overline{D} \subset \Omega$ with $v \le h$ on ∂D, $v(z) \le h(z)$ for all $z \in D$.

8.2 Remarks. Although a subharmonic function is allowed to take the value $-\infty$, the value $+\infty$ cannot be in the range of the function. Property (ii) is essentially how the function gets its name. Moreover, it follows by property (i) that if $v(z_0) = -\infty$, then $\lim_{z \to z_0} v(z) = -\infty$. The definition allows for the function $v \equiv -\infty$ to be subharmonic although some authors do not allow this by definition. In what follows, we *assume that $v \not\equiv -\infty$ unless it is convenient to assume otherwise.*

Analogously, a function v is *superharmonic* in a domain Ω if $-v$ is subharmonic in Ω. Thus a superharmonic function can take the value $+\infty$ but not the value $-\infty$.

As is the case of continuous functions, we have the following:

8.3 Proposition. *An upper semicontinuous function v on a domain Ω attains it maximum on any compact set $K \subset \Omega$.*

In fact, let

$$\sup_{z \in K} v(z) = M,$$

and let $\{z_n\}$ be a sequence in K such that $\lim_{n\to\infty} v(z_n) = M$. Since K is compact, there is a convergent subsequence $\{z_{n_k}\}$ such that $\lim_{k\to\infty} z_{n_k} = z_0 \in K$. Then by property (i)

$$M = \overline{\lim_{z_{n_k} \to z_0}}\, v(z_{n_k}) \le v(z_0) \le M,$$

so that $v(z_0) = M < \infty$.

https://doi.org/10.1515/9783110757828-008

Every upper semicontinuous function v has the following property, which we will exploit further.

8.4 Lemma. *If v is upper semicontinuous on any set $E \subset \Omega$, then v is the limit of a nonincreasing sequence of continuous functions ϕ_n on E, that is, $\phi_n(z) \searrow v(z)$ as $n \to \infty$ for all $z \in E$.*

For the very technical proof, see Hayman and Kennedy (1976, Theorem 1.4).

For a subharmonic function v on Ω, we will exploit this as follows. For some disk D with $\overline{D} \subset \Omega$, there is a sequence of continuous functions ϕ_n on ∂D such that $\phi_n \searrow v$. Solving the Dirichlet problem in D for each continuous boundary value ϕ_n gives a corresponding sequence of harmonic functions h_n in D such that $h_n|\partial D = \phi_n|\partial D$ and $h_n \geq h_{n+1} \geq \dots$. By Harnack's principle, $h_n \searrow h$ for some harmonic function h in D, and $h_n \geq v$ on ∂D implies by property (ii) that $h_n \geq v$ in D, and hence $h \geq v$ in D.

Submean value properties

As we might expect, a subharmonic function takes a value at the center of each disk that is below its mean value on the boundary of the disk.

8.5 Theorem. *If v is subharmonic in a domain Ω and $v(z_0) > -\infty$, then v satisfies the submean-value properties:*

(i)

$$v(z_0) \leq \frac{1}{2\pi} \int_0^{2\pi} v(z_0 + re^{i\theta})d\theta$$

(ii)

$$v(z_0) \leq \frac{1}{\pi r^2} \int_0^r \int_0^{2\pi} v(z_0 + \rho e^{i\theta})\rho d\rho d\theta$$

in any disk $\{|z - z_0| \leq r\} \subset \Omega$. If v is superharmonic in Ω and $v(z_0) < +\infty$, then the reverse inequalities hold.

Proof. For the disk $|z - z_0| \leq r$, consider a nonincreasing sequence of harmonic functions $h_n \searrow h$ in $|z-z_0| < r$ with $h_n \geq v$ on $|z-z_0| = r$. Then by the monotonic convergence theorem

$$-\infty < v(z_0) \leq h(z_0) = \lim_{n\to\infty} h_n(z_0) = \frac{1}{2\pi} \int_0^{2\pi} \lim_{n\to\infty} h_n(z_0 + re^{i\theta})d\theta = \frac{1}{2\pi} \int_0^{2\pi} v(z_0 + re^{i\theta})d\theta,$$

and the function v is Lebesgue-integrable on $[0, 2\pi]$, completing the proof of (i).

The second part follows from (i) by integrating $\rho d\rho$ on both sides of the inequality from 0 to r. Note that the integrability over the disk $|z - z_0| \le r$ means that $v(z) = -\infty$ can possibly hold on at most a set of two-dimensional Lebesgue measure zero. Moreover, it is known that $v(z_0) > -\infty$ at a single point $z_0 \in \Omega$ implies that $v(z) > -\infty$ on a dense subset of Ω (cf. Radó 1971, p. 4). □

8.6 Corollary. *A function v that is both subharmonic and superharmonic satisfies the mean value property*

$$v(z_0) = \frac{1}{2\pi}\int_0^{2\pi} v(z_0 + re^{i\theta})d\theta,$$

and since v is continuous, it is harmonic.[1]

The above mean-value theorems in fact characterize subharmonic functions in the following sense.

8.7 Theorem. *Let v be a function that is upper semicontinuous in a domain Ω such that $-\infty \le v < +\infty$, $v \not\equiv -\infty$, and such that for each point $z_0 \in \Omega$ with $-\infty < v(z_0)$, there is a disk $|z - z_0| < \rho$ in Ω for which the integral mean satisfies*

$$v(z_0) \le \frac{1}{2\pi}\int_0^{2\pi} v(z_0 + re^{i\theta})d\theta$$

for $0 < r < \rho$. Then v is subharmonic in Ω.

Proof. Let h be a harmonic function in a disk that is continuous on $\overline{D} \subset \Omega$ such that $v \le h$ on ∂D. We must show that $v(z) \le h(z)$ for all $z \in D$, so let us assume on the contrary that $v(z_0) > h(z_0)$ for some $z_0 \in D$. Then the upper semicontinuous function $u = v - h$ has a positive maximum value m in D. Since the set S of all points $z \in D$ for which $u(z) = m$ forms a closed subset of D, there is a point $z_0' \in S$ such that z_0' minimizes the distance to $\partial\Omega$. Then for all sufficiently small circles $|z - z_0'| = r$ contained in D, we have $u(z) < m$ on some arc $c_r \subseteq \{|z - z_0'| = r\}$ that lies *outside of* S. For the integral mean of u, we have

$$\frac{1}{2\pi}\int_0^{2\pi} v(z_0' + re^{i\theta})d\theta - h(z_0') = \frac{1}{2\pi}\int_0^{2\pi} u(z_0' + re^{i\theta})d\theta < m = v(z_0') - h(z_0'),$$

implying

1 By Corollary 7.8.

$$v(z_0') > \frac{1}{2\pi}\int_0^{2\pi} v(z_0' + re^{i\theta})d\theta$$

for all sufficiently small values of r, contradicting the hypothesis. □

A variation of the above argument establishes the following maximum principle.

8.8 Proposition. *If v is subharmonic in Ω, then v has no maximum in Ω unless $v \equiv$ constant.*

Indeed, if $v(z_0) = m$ is maximum for $z_0 \in \Omega$, then

$$S = \{z \in \Omega : v(z) = m\}$$

is a closed set in Ω as above. For any $z_1 \in S$, take a disk $D(z_1, r)$ with $\overline{D}(z_1, r) \subset \Omega$. We claim that $D(z_1, r) \subset S$. For if not, then there is a point $z_2 \in D(z_1, r)$ that is not in S and $|z_1 - z_2| = \rho < r$. Since $v(z_2) < m$, we have that $v(z) < m$ and an arc $c_\rho \subseteq \{|z_1 - z_2| = \rho\}$, again by the upper semicontinuity of v. Then

$$m = v(z_1) \le \frac{1}{2\pi}\int_0^{2\pi} v(z_1 + \rho e^{i\theta})d\theta < m,$$

a contradiction, proving that $D(z_1, r) \subset S$, and hence S is an open set in Ω. Since Ω is connected, either $S = \emptyset$ or $S = \Omega$, establishing the result.

As is the case with harmonic functions we maintain the following:

8.9 Proposition. *If $\{v_n\}$ is a sequence of subharmonic functions converging uniformly on compact subsets of a domain Ω to a function v, then v is subharmonic in Ω.*

Proof. We can readily verify that the function v is upper semicontinuous, and, moreover, for $z_0 \in \Omega$ with $-\infty < v(z_0)$,

$$v(z_0) = \lim_{n\to\infty} v_n(z_0) \le \lim_{n\to\infty} \frac{1}{2\pi}\int_0^{2\pi} v_n(z_0 + re^{i\theta})d\theta = \frac{1}{2\pi}\int_0^{2\pi} v(z_0 + re^{i\theta})d\theta$$

for all sufficiently small values of r, so that by the preceding theorem, v is subharmonic. □

Let us denote the *integral mean value* of $v(z_0)$ by

$$L(v, z_0, r) = \frac{1}{2\pi}\int_0^{2\pi} v(z_0 + re^{i\theta})d\theta,$$

which is a nondecreasing function of r. To see this, suppose that $r_1 < r_2$ and $\overline{D}(z_0, r_2) \subset \Omega$. In view of Lemma 8.4, there is a nonincreasing sequence of harmonic functions $\{h_n\}$ that are continuous on $\overline{D}(z_0, r_2)$ with $h_n \searrow v$ on $|z - z_0| = r_2$, and thus $h_n \geq v$ in $D(z_0, r_2)$. It follows that

$$L(h_n, z_0, r_2) = h_n(z_0) = L(h_n, z_0, r_1) \geq L(v, z_0, r_1)$$

for all n. Since $h_n \searrow v$ on $|z - z_0| = r_2$, upon letting $n \to \infty$, we conclude that $L(v, z_0, r_2) \geq L(v, z_0, r_1)$, as desired.

Growth rate/integral means

To determine the rate of growth of $L(v, z_0, r)$, it suffices to consider our domain as an annular region $\mathcal{A} : \rho < |z - z_0| < R$. For a harmonic function u in $\mathcal{A}$ and $\rho < r_1 < r_2 < R$, by Green's second identity we have

$$\int_{|z-z_0|=r_1} \frac{\partial u}{\partial n} ds = \int_{|z-z_0|=r_2} \frac{\partial u}{\partial n} ds,$$

and hence for any $r_1 \leq r \leq r_2$ and $C_r : |z - z_0| = r$,

$$c = r \int_{C_r} \frac{\partial u}{\partial r} d\theta = r \frac{\partial}{\partial r} \int_0^{2\pi} u(z_0 + re^{i\theta}) d\theta,$$

where c is some constant. Therefore, on each circle $C_r : |z - z_0| = r$, we have

$$L(u, z_0, r) = a \log r + b.$$

For a subharmonic function v in $\mathcal{A}$, $v \not\equiv -\infty$, since the Dirichlet problem is solvable for the closed annulus $r_1 \leq r \leq r_2$, we can find a nonincreasing sequence of harmonic functions $\{h_n\}$ in $r_1 < |z - z_0| < r_2$ having continuous boundary values with $h_n \searrow v$ on the boundary $r = r_1, r_2$.

Then by the above in the closed annular region $r_1 \leq r \leq r_2$, we have

$$L(h_n, z_0, r) = a_n \log r + b_n \tag{8.1}$$

for $n = 1, 2, 3, \ldots$.

Now by Harnack's principle the functions $h_n \searrow h$ uniformly on the compact subsets C_r $(r_1 < r < r_2)$, and h is harmonic in $r_1 < r < r_2$ with

$$\lim_{n\to\infty} (a_n \log r + b_n) = L(h, z_0, r)$$

for all $r_1 < r < r_2$. Taking two values of r in the preceding equation and subtracting the results, we find that the limits $\lim_{n\to\infty} a_n = a$ and $\lim_{n\to\infty} b_n = b$ exist and are finite with $L(h, z_0, r) = a \log r + b$.

Now for $r = r_1$, by (8.1) we have

$$\lim_{n\to\infty} L(h_n, z_0, r_1) = \lim_{n\to\infty} (a_n \log r_1 + b_n) = a \log r_1 + b, \tag{8.2}$$

and likewise for $r = r_2$. By the max/min principle the integral means $L(h_n, z_0, r_1)$ are decreasing, and we have demonstrated that the sequence of integral means is bounded below. This allows us to take the limit inside the integral on the left side of (8.2) to conclude that $L(v, z_0, r_1)$ is finite. Indeed, $L(v, z_0, r)$ exists on any circle in the annular region $\mathcal{A}$.

Finally, note that for $r_1 < r < r_2$, since $h_n \geq v$, we have

$$L(v, z_0, r) \leq \lim_{n\to\infty} L(h_n, z_0, r) = a \log r + b$$

with equality at the endpoints:

$$L(v, z_0, r_1) = a \log r_1 + b, \quad L(v, z_0, r_2) = a \log r_2 + b.$$

Thus we have proved the theorem of Riesz (1926).

8.10 Theorem. *If v is subharmonic in a domain Ω containing the annular region* $r_1 \leq r \leq r_2$, *then the integral mean of v is a convex function of* $\log r$.

We also have a subharmonic areal mean value as in Theorem 7.1b given by

$$A(v, z_0, r) = \frac{1}{\pi r^2} \int_0^r \int_0^{2\pi} v(z_0 + \rho e^{i\theta}) \rho d\rho d\theta.$$

The relationship between the circular and areal means is somewhat counterintuitive.

8.11 Proposition. *For v subharmonic in a domain Ω, on any disk contained in Ω,*

$$A(v, z_0, r) \leq L(v, z_0, r).^2$$

Proof. For any disk $D(z_0, r)$ with $\overline{D}(z_0, r) \subset \Omega$, note that for $0 \leq \rho \leq r$,

$$A(v, z_0, r) = \frac{2}{r^2} \int_0^r L(v, z_0, \rho) \rho d\rho. \tag{8.3}$$

Moreover, $L(v, z_0, \rho)$ is a continuous function of ρ, and if we let $F(\rho) = L(v, z_0, \rho) \cdot \rho$, then its Riemann sum from 0 to r is given by

2 If v is continuous, then the converse is also true, that is, $A(v, z_0, r) \leq L(v, z_0, r)$. Cf. Beckenbach, E. F. and Radó, T., Subharmonic functions and surfaces of negative curvature, *Trans. Amer. Math. Soc.* 35 (1933), 662–674.

$$\sum_{k=1}^{n} L\left(v, z_0, \frac{k}{n}r\right)\left(\frac{k}{n}r\right)\left(\frac{r}{n}\right) = r^2 \sum_{k=1}^{n} L\left(v, z_0, \frac{k}{n}r\right)\left(\frac{k}{n^2}\right).$$

Therefore by (8.3) each areal mean can be written as

$$A(v, z_0, r) = \lim_{n\to\infty} 2\sum_{k=1}^{n} L\left(v, z_0, \frac{k}{n}r\right)\left(\frac{k}{n^2}\right).$$

As $L(v, z_0, \rho)$ is a nondecreasing function of ρ, (noted previously) we obtain

$$A(v, z_0, r) \leq 2L(v, z_0, r) \cdot \lim_{n\to\infty} \sum_{k=1}^{n} \frac{k}{n^2} = L(v, z_0, r),$$[3]

proving the result. □

Laplacian

As a harmonic function is characterized by its Laplacian being equal to zero, there is a similar characterization of subharmonic functions via its Laplacian.

8.12 Proposition. *$v \in C^2(\Omega)$ is subharmonic in Ω if and only if $\Delta v \geq 0$.*

Proof. Suppose that $\Delta v > 0$. Let $\overline{D} \subset \Omega$, and let h be a harmonic function in D that is continuous on $\overline{D}$ and such that $v \leq h$ on ∂D. To show that $v \leq h$ in D, suppose on the contrary that $v(z_0) > h(z_0)$ for some $z_0 \in D$. Then the subharmonic function $u = v - h$ attains a maximum at some $z_1 \in D$, implying that $\Delta u \leq 0$ at z_1. However, as the Laplace operator is linear, $0 < \Delta v = \Delta u + \Delta h = \Delta u$ at the point z_1, a contradiction. Thus $v \leq h$ in D, and v is subharmonic.

It remains to show that if $\Delta v \geq 0$, then v is subharmonic. Consider the function $v' = v + \alpha(x^2 + y^2)$ for $\alpha > 0$, which is subharmonic and satisfies $\Delta v' > 0$. Then v' is subharmonic, and $\alpha \to 0$ implies $v' \to v$ locally uniformly in Ω, implying that v is subharmonic by Proposition 8.9.

For the converse, suppose that v is subharmonic and that $\Delta v < 0$ at some point $z_0 \in D$, which implies that v is superharmonic in a neighborhood of z_0 by the analogue of the preceding argument. Therefore v is subharmonic and superharmonic in this neighborhood and so harmonic there, i. e., $\Delta v = 0$ in the neighborhood, a contradiction. □

8.13 Examples.

(i) Green's function $G(z, z_0)$ for a domain Ω defined in Chapter 7 is superharmonic in Ω as it is harmonic in Ω except for a logarithmic singularity at the pole z_0.

(ii) If u is a nonnegative solution to the elliptic partial differential equation

$$\Delta u = Pu$$

---

3 The proof of the theorem is from: Montel, P., Sur les fonctions convexes et les fonctions sousharmoniques, *J. Math. Pures Appl., Ser.* 9, 7 (1928), 29–60.

with $P \geq 0$, then u is subharmonic.[4] Similarly, any solution u to the equation $\Delta u = P$ is subharmonic for $P \geq 0$.

(iii) If u is harmonic, then $|u|^p$ is subharmonic for $p \geq 1$. Indeed, for $p = 1$, the result is obvious by the mean value property. For $p > 1$, by Hölder's inequality

$$|u(z_0)| \leq \frac{1}{2\pi}\left(\int_0^{2\pi} |u(z_0 + re^{i\theta})|^p \, d\theta\right)^{\frac{1}{p}} \cdot (2\pi)^{\frac{1}{q}},$$

where $\frac{1}{p} + \frac{1}{q} = 1$. Raising both sides to the power p yields

$$|u(z_0)|^p \leq \frac{1}{2\pi} \int_0^{2\pi} |u(z_0 + re^{i\theta})|^p \, d\theta,$$

as desired. Likewise, if f is analytic, then $|f|^p$ is subharmonic for $p > 1$, and:

(iv) If $f(z)$ is analytic in a domain Ω, then $|f(z)|$ is subharmonic, which follows from the mean value property for analytic functions.

(v) If $f(z)$ is analytic in a domain Ω, then $\log|f(z)|$ is harmonic wherever $f(z) \neq 0$. With $\log|f(z_0)| = -\infty$ whenever $f(z_0) = 0$, clearly, the submean value property is satisfied at z_0. As a consequence, $\log|f(z)|$ is subharmonic in Ω.

(vi) If v_1, v_2 are subharmonic, then so is $v_1 + v_2$ or any finite sum for that matter.

(vii) If v_1 and v_2 are subharmonic, then so is $v = \max(v_1, v_2)$. It is clear that condition (i) holds, so now take a point z_0 and a circle $C_r : |z - z_0| = r$ in a domain Ω. As v_1, v_2 are upper semicontinuous, they are both bounded above on C_r by some constant m, and, consequently, $v_1 \leq v \leq m$ on C_r. As both v_1 and m are integrable around C_r with respect to θ, so is v. Assuming that $v(z_0) = v_1(z_0)$, we have

$$v(z_0) \leq \frac{1}{2\pi} \int_0^{2\pi} v_1(z_0 + re^{i\theta}) d\theta \leq \frac{1}{2\pi} \int_0^{2\pi} v(z_0 + re^{i\theta}) d\theta,$$

and v is subharmonic. Likewise, if $v_1, v_2, \ldots, v_n$ are subharmonic, then $v = \max(v_1, v_2, \ldots, v_n)$ is subharmonic.

(viii) The class of functions $v \geq 0$ such that $u = \log v$ is subharmonic in a domain Ω was introduced by T. Radó as the class *PL*. At the zeros of v, we take u to be $-\infty$. The modulus of an analytic function is of class *PL* by example (v). Moreover, if v is of class *PL*, then v itself must be subharmonic. Indeed, the subharmonicity of $\log v$ means that $v \not\equiv -\infty$ and that v is upper semicontinuous. To demonstrate the submean value property, we fortunately have the pleasing inequality

4 These are known as *P-harmonic functions*.

$$\frac{1}{2\pi}\int_0^{2\pi} \log f(\theta)d\theta \le \log \frac{1}{2\pi}\int_0^{2\pi} f(\theta)d\theta$$

for Lebesgue integrable functions.[5] Since $u = \log v$ is subharmonic, from the preceding inequality we obtain

$$v(z_0) = e^{u(z_0)} \le e^{L(u,z_0,r)} \le L(v, z_0, r),$$

and thus v is subharmonic.

(ix) If $f : \Omega_1 \to \Omega_2$ is conformal and v is subharmonic on Ω_2, then $v \circ f$ is subharmonic on Ω_1. Compare with Lemma 7.10. The proof is left as an exercise.

Another approach to some of the above examples is via the following useful result.

8.14 Jensen Inequality.[6] *Let f be a real-valued μ-integrable function on a finite measure space X, and let I be an interval such that $f(X) \subset I$. If ϕ is a convex function on I and $\phi \circ f$ is integrable, then*

$$\phi\left(\frac{1}{\mu(X)}\int_X f d\mu\right) \le \frac{1}{\mu(X)}\int_X (\phi \circ f) d\mu.$$

8.15 Example. Let ϕ be a convex increasing function on an interval I, and let v be subharmonic in a domain Ω with $v(z) \in I$. Then the composition $\phi \circ v$ is subharmonic in Ω.

Proof. As ϕ is continuous, $\phi \circ v$ remains upper semicontinuous. Furthermore, since v is subharmonic and ϕ is an increasing function,

$$(\phi \circ v)(z_0) \le \phi\left(\frac{1}{2\pi}\int_0^{2\pi} v(z_0 + re^{i\theta})d\theta\right) \le \frac{1}{2\pi}\int_0^{2\pi} (\phi \circ v)(z_0 + re^{i\theta})d\theta,$$

where the last inequality follows by the Jensen inequality. Hence $\phi \circ v$ is subharmonic by Theorem 8.7. □

Observe that Example (iii) immediately follows since the function $|x|^p$ is convex and increasing for $p > 1$.

8.16 Corollary. *If v is subharmonic, then $e^{\alpha v}$ is subharmonic for $\alpha > 0$.*

---

5 A proof can be found in Riesz, F., Sur les valeurs moyennes des fonctions, *J. London Math. Soc.* 5 (1930), 120–121.

6 Sur les fonctions convexes et les inégalités entre les valeurs moyennes, *Acta Math.* 30 (1906), 175–193.

Growth rate/subharmonic functions

Let v be subharmonic in $\mathbb{C}$ and consider a closed annulus $\mathcal{A}$: $r_1 \le r \le r_2$, setting $m(r) = \max_{|z|=r} v(z)$, which exists by Proposition 8.3. Then the function

$$h(z) = \frac{\log r_2 - \log r}{\log r_2 - \log r_1} m(r_1) + \frac{\log r - \log r_1}{\log r_2 - \log r_1} m(r_2) \tag{8.4}$$

is harmonic in $\mathcal{A}$ and satisfies $v(z) \le h(z)$ whenever $|z| = r_i$, $i = 1, 2$. Hence $v(z) \le h(z)$ for all $z \in \mathcal{A}$. This is the Hadamard three-circles theorem for subharmonic functions, and equality (8.4) means that $m(r)$ is a convex function of $\log r$.

Furthermore, if we now assume that v that is subharmonic in $\mathbb{C}$ and satisfies the growth condition $\underline{\lim}_{r\to\infty}(m(r)/\log r) = 0$, then letting $r_2 \to \infty$ in (8.4), we obtain

$$v(z) \le m(r_1)$$

for $0 < r_1 < |z|$. Since v is upper semicontinuous, we obtain

$$v(z) \le \overline{\lim_{r_1\to 0}}\, m(r_1) = v(0)$$

for all $z \in \mathbb{C}$, implying that v is constant by the maximum principle. Thus we have established the following:

8.17 Theorem. *A subharmonic function v in $\mathbb{C}$ satisfying*

$$\underline{\lim}_{r\to\infty} \frac{m(r)}{\log r} = 0,$$

where $m(r) = \max_{|z|=r} v(z)$, reduces to a constant. In particular, if $v(z)$ is bounded above in $\mathbb{C}$, then $v(z)$ reduces to a constant.

Growth rate/superharmonic functions

For positive superharmonic functions in the unit disk, as was the case for their harmonic counterparts, there are also bounds on just how rapidly such functions can approach the value zero without collapsing to zero.

8.18 Theorem.[7] *If $s(z)$ is positive superharmonic in U and*

$$\underline{\lim}_{z\to\zeta_0} \frac{s(z)}{1-|z|} = 0$$

for some point $\zeta_0 \in \partial U$, then $s \equiv 0$.

7 Ü. Kuran and J. L. Schiff, A uniqueness theorem for nonnegative superharmonic functions in planar domains, *J. Math. Anal. Appl.* 93 (1) (1983), 195–205.

Proof. For positive superharmonic functions, we use the Riesz–Herglotz representation[8]

$$s(z) = \iint_U G(z,\alpha)d\upsilon(\alpha) + \int_{\partial U} K(z,\zeta)d\mu(\zeta),$$

where $G(z,\alpha) = \log\left|\frac{1-\bar{\alpha}z}{z-\alpha}\right|$ is Green's function for U with pole at α, $K(z,\zeta)$ is the Poisson kernel, and υ and μ are nonnegative Borel measures. The second integral is a positive harmonic function $h(z)$, and the hypothesis implies that $h(z) \equiv 0$ by Proposition 7.40.

For the first integral, by Fatou's lemma we have

$$\begin{aligned} 0 &= \lim_{z\to\zeta_0} \frac{1}{1-|z|} \iint_U G(z,\alpha)d\upsilon(\alpha) \\ &\geq \iint_U \lim_{z\to\zeta_0} \frac{1}{1-|z|} \log\left|\frac{1-\bar{\alpha}z}{z-\alpha}\right| d\upsilon(\alpha). \end{aligned}$$

Considering the integrand of the integral on the right, from the growth estimate of Corollary 5.3 we obtain

$$\log\left|\frac{1-\bar{\alpha}z}{z-\alpha}\right| \geq \frac{(1-|\alpha|)(1-|z|)}{1+|z|},$$

and we infer that

$$\lim_{z\to\zeta_0} \frac{1}{1-|z|} \log\left|\frac{1-\bar{\alpha}z}{z-\alpha}\right| \geq \frac{1-|\alpha|}{2} > 0.$$

This implies that $\upsilon \equiv 0$, and therefore $s \equiv 0$. □

Poisson extension

For an upper semicontinuous function $v(z)$ in a domain Ω, we next need to extend the Poisson integral to any disk $\overline{D}(z,\rho) \subseteq \Omega$ using the boundary values of v. Again by considering a sequence of continuous functions $\phi_n \searrow v$ on $\zeta = z_0 + \rho e^{i\phi}$ we have

$$\begin{aligned} P_v(z) &= \lim_{n\to\infty} \frac{1}{2\pi} \int_0^{2\pi} \frac{\rho^2 - |z-z_0|^2}{|\zeta - z|^2} \phi_n(\zeta)d\phi \\ &= \frac{1}{2\pi} \int_0^{2\pi} \frac{\rho^2 - |z-z_0|^2}{|\zeta - z|^2} v(\zeta)d\phi \end{aligned}$$

8 See Helms (1969, p. 116).

by the monotone convergence theorem. Thus P_v is harmonic in the disk, where $v(z) \leq P_v(z)$. We know from Chapter 7 that when v is continuous at $\zeta \in \partial\Omega$, we have

$$\lim_{z\to\zeta} P_v(z) = v(\zeta).$$

In the present case,

$$\overline{\lim_{z\to\zeta}}\, P_v(z) \leq v(\zeta)$$

for all $\zeta \in \partial\Omega$. We omit the technical details.[9]

8.19 Theorem. *Let v be subharmonic in a domain Ω, and let D be an open disk with $\overline{D} \subset \Omega$. Then the function w defined by*

$$w(z) = \begin{cases} P_v(z) & z \in D \\ v(z) & z \in D' = \Omega - D, \end{cases}$$

is subharmonic in Ω.

Proof. To verify the upper semicontinuity, note that for any $\zeta \in \partial D$ and $z \in D'$,

$$\overline{\lim_{z\to\zeta}}\, w(z) = \overline{\lim_{z\to\zeta}}\, v(z) \leq v(\zeta) = w(\zeta)$$

and for $z \in D$,

$$\overline{\lim_{z\to\zeta}}\, w(z) = \overline{\lim_{z\to\zeta}}\, P_v(z) \leq v(\zeta) = w(\zeta).$$

To establish the subharmonicity of w, take $\zeta \in \partial D$ such that for all $r > 0$ sufficiently small,

$$w(\zeta) = v(\zeta) \leq \frac{1}{2\pi}\int_0^{2\pi} v(\zeta + re^{i\theta})d\theta \leq \frac{1}{2\pi}\int_0^{2\pi} w(\zeta + re^{i\theta})d\theta,$$

and now w satisfies the submean value property in all of D. □

Perron method

The following method initiated by Oskar Perron (1880–1975) is a very versatile method for solving the Dirichlet problem.[10] For a bounded region $\Omega \subset \mathbb{C}$ whose boundary con-

---

9 These can be found in Helms (1969, p. 24).

10 Eine neue Behandlung der ersten Randwertaufgabe für $\Delta u = 0$, *Math. Zeit.* 18 (1923), 42–54.

sists of finitely many Jordan curves and boundary values f (not necessarily continuous), the method determines an associated harmonic function H_f^Ω so that when the classical Dirichlet problem has a solution, it is identically equal to H_f^Ω. Other methods developed by Norbert Weiner and Marcel Brelot produce the same function H_f^Ω, which is known as the *generalized solution to the Dirichlet problem.*

8.20 Theorem. *Consider the family $\mathfrak{H} = \mathfrak{H}_f$ of subharmonic functions v defined on a bounded domain $\Omega \subset \mathbb{C}$ such that for any real-valued function f (which can take infinite values) defined on $\partial\Omega$, the following condition holds:*

$$\overline{\lim_{z\to\zeta}}\, v(z) \le f(\zeta)$$

for all $\zeta \in \partial\Omega$. Then the function

$$H_f^\Omega(z) = \sup_{v\in\mathfrak{H}} v(z)$$

is either harmonic in Ω or identically $\pm\infty$ in Ω.

Proof. Here if we allow $v \equiv -\infty$ to be subharmonic then the family $\mathfrak{H}$ is always nonempty, and if $\mathfrak{H}$ only consists of this single function, then we set $H_f^\Omega(z) = -\infty$. Otherwise, take a small open disk D with $\overline{D} \subset \Omega$ such that for a fixed $z_0 \in D$, there is a sequence $v_n \in \mathfrak{H}$ with

$$v_n(z_0) \nearrow H_f^\Omega(z_0)$$

as $n \to \infty$. Define $V_n = \max(v_1, v_2, \dots, v_n) \in \mathfrak{H}$ and let

$$w_n(z) = \begin{cases} P_{V_n}(z), & z \in D, \\ V_n(z), & z \in D' = \Omega - D. \end{cases}$$

As we have seen by Theorem 8.19, w_n is subharmonic in Ω, and for all $\zeta \in \partial\Omega$,

$$\overline{\lim_{z\to\zeta}}\, w_n(z) = \overline{\lim_{z\to\zeta}}\, V_n(z) \le f(\zeta),$$

so that $w_n \in \mathfrak{H}$. Moreover, $V_n \le V_{n+1}$ implies that $w_n \le w_{n+1}$, and then we obtain

$$v_n(z_0) \le V_n(z_0) \le P_{V_n}(z_0) = w_n(z_0) \le H_f^\Omega(z_0).$$

Firstly, we assume that $H_f^\Omega(z_0) = \infty$, so that by Harnack's principle $\infty = \lim_{n\to\infty} w_n(z) \le H_f^\Omega(z)$ in D. On the other hand, if $H_f^\Omega(z_0) < \infty$, then $w(z) = \lim_{n\to\infty} w_n(z)$ is a harmonic function in D such that $w(z_0) = H_f^\Omega(z_0)$ and $w(z) \le H_f^\Omega(z)$ for all $z \in D$.

To show that $w = H_f^\Omega$ in D, we take any other point $z_1 \in D$ and a sequence $v_n'(z_1) \nearrow H_f^\Omega(z_1)$ as $n \to \infty$. Now for the clever part, we first take $V_n' = \max(v_n, v_n')$, and then proceeding just as above, let $\hat{V}_n = \max(V_1', V_2', \dots, V_n')$. Thus, for $W_n(z) = P_{\hat{V}_n}(z)$ in D (and $= \hat{V}_n$ in D'), we obtain $\lim_{n\to\infty} W_n(z) = W(z)$, which is harmonic in D or $+\infty$, and

$$w(z) \le W(z) \le H_f^{\Omega}(z)$$

with $W(z_1) = H_f^{\Omega}(z_1)$. Furthermore, $w(z_0) = H_f^{\Omega}(z_0)$ implies that $w(z_0) = W(z_0)$, but then the harmonic function $W - w$ in D attains a zero minimum at the interior point z_0, implying that $W = w$ in D, and so $w(z_1) = H_f^{\Omega}(z_1)$. We conclude that, $w = H_f^{\Omega}$ in D.

Consequently, taking an arbitrary open disk D, either H_f^{Ω} is a harmonic function in D or equals ∞ in D. A connectedness argument then proves the theorem. □

8.21 Remark. If the Dirichlet problem happens to have a solution h in Ω for continuous boundary values f, then certainly $h \in \mathfrak{H}$, and so $h \le H_f^{\Omega}$. On the other hand, for each $v \in \mathfrak{H}$,

$$\overline{\lim_{z\to\zeta}}\, v(z) \le f(\zeta) = h(\zeta)$$

at all points $\zeta \in \partial\Omega$ means that for any point ζ on the boundary and any $\varepsilon > 0$, there is an open disk D_ζ about the point such that $v(z) < h(z) + \varepsilon$ for all $z \in D_\zeta \cap \Omega$. Now for $u = v - h$, if there is a point $z_0 \in \Omega$ with $u(z_0) = m > 0$, once again consider the set $S = \{z \in \Omega : u(z) = m\}$. In view of the boundary condition, every point $\zeta \in \partial\Omega$ has a disk D_ζ about it with $u(z) < m/2$ on each $D_\zeta \cap \Omega$. Thus assuming m to be the maximum of u, since S is a closed subset of Ω, we can find a point $z_1 \in S$ that minimizes the distance to $\partial\Omega$, and the arc argument of Theorem 8.7 on a sufficiently small disk about z_1 gives a contradiction, so that $u \le 0$. Hence $v(z) \le h(z)$ for all $z \in \Omega$ and all $v \in \mathfrak{H}$ implies $H_f^{\Omega} \le h$, and, consequently, $H_f^{\Omega} = h$.

8.22 Corollary. *If v is subharmonic in a bounded domain Ω and*

$$\overline{\lim_{z\to\zeta}}\, v(z) \le h(\zeta)$$

for h harmonic in Ω and continuous on $\overline{\Omega}$, then $v(z) \le h(z)$ in Ω. Moreover, $v(z) < h(z)$ unless $v(z) \equiv h(z)$.

The latter statement is true since $v(z) - h(z)$ cannot attain a maximum in Ω unless it is identically constant.

Barrier/regular points

Now that we have associated the harmonic function H_f^{Ω} with given boundary values in a seemingly natural way, we must investigate the boundary behavior of this harmonic function and therefore the nature of the boundary comes into play. To this end, we require the following type of functions. Without loss of much generality, we will henceforth consider only functions f that are bounded on the boundary of some bounded open set in the complex plane.

8.23 Definition. A subharmonic function $\mathfrak{b}(z)$ defined in Ω is called a *barrier* at a point $\zeta_0 \in \partial\Omega$ if

(i) $\lim_{z\to\zeta_0} \mathfrak{b}(z) = 0$ and

(ii) $\overline{\lim}_{z\to\zeta} \mathfrak{b}(z) < 0$ for all $\zeta \in \partial\Omega$, $\zeta \neq \zeta_0$.

Points on the boundary that have a barrier are known as *regular* (for the Dirichlet problem).

8.24 Remark. From the preceding corollary we see that $\mathfrak{b}(z) < 0$ in Ω. For any disk D_{ζ_0} about a point $\zeta_0 \in \partial\Omega$, condition (ii) implies $\sup\{\mathfrak{b}(z) \in \Omega \cap (D_{\zeta_0})'\} = -m < 0$. Without loss of generality, we can normalize our barrier by letting $-m = -1$ and subsequently taking $\max(\mathfrak{b}(z), -1)$, which is also a barrier at ζ_0 and will again be denoted by $\mathfrak{b}(z)$, which now equals -1 outside of $D_{\zeta_0} \cap \Omega$. Equivalently, a barrier can be defined in terms of superharmonic functions (cf. Helms 1969).

Here is the reason we want a barrier at a boundary point. All the properties of the barrier come to the fore.

8.25 Theorem. *If ζ_0 is a regular point and f is bounded on $\partial\Omega$, then*

$$\underline{\lim}_{\zeta\to\zeta_0} f(\zeta) \leq \underline{\lim}_{z\to\zeta_0} H_f^{\Omega}(z) \leq \overline{\lim}_{z\to\zeta_0} H_f^{\Omega}(z) \leq \overline{\lim}_{\zeta\to\zeta_0} f(\zeta). \tag{8.5}$$

Proof. Let us start with the first inequality of (8.5) by setting $\alpha = \underline{\lim}_{\zeta\to\zeta_0} f(\zeta)$ for $\zeta \in \partial\Omega$. Then given $\varepsilon > 0$, there is a closed disk $\mathcal{N}$ of ζ_0 such that $f(\zeta) > \alpha - \varepsilon$ for $\zeta \in \mathcal{N}$. Note that for $|f(\zeta)| \leq M$, the function

$$\phi(z) = (\alpha + M)\mathfrak{b}(z) + (\alpha - \varepsilon)$$

is subharmonic in Ω. Moreover, since $\overline{\lim}_{z\to\zeta} \mathfrak{b}(z) \leq 0$ for any $\zeta \in \mathcal{N}$, we have

$$\overline{\lim}_{z\to\zeta} \phi(z) \leq \alpha - \varepsilon < f(\zeta).$$

On the other hand, for $\zeta \in \mathcal{N}'$, since $\mathfrak{b}(z) \equiv -1$ outside the closed set $\mathcal{N}$,

$$\overline{\lim}_{z\to\zeta} \phi(\zeta) = -M - \varepsilon < f(\zeta).$$

It follows that our function $\phi(z)$ belongs to the family $\mathfrak{H}$, and hence $\phi(z) \leq H_f^{\Omega}(z)$.

Finally, since $\lim_{z\to\zeta_0} \mathfrak{b}(z) = 0$, we conclude that

$$\underline{\lim}_{z\to\zeta_0} H_f^\Omega(z) \geq \lim_{z\to\zeta_0} \phi(z) = \alpha - \varepsilon.$$

As $\varepsilon > 0$ was arbitrary, $\underline{\lim}_{\zeta\to\zeta_0} f(\zeta) \leq \underline{\lim}_{z\to\zeta_0} H_f^\Omega(z)$ as desired.

For the second inequality of (8.5), let us set $\beta = \overline{\lim}_{\zeta\to\zeta_0} f(\zeta)$, so that given some $\varepsilon > 0$, there is a closed neighborhood $\mathcal{N}$ of ζ_0 in which $f(\zeta) < \beta + \varepsilon$ for $\zeta \in \mathcal{N}$. In this instance we consider an analogous function to $\phi(z)$,

$$\psi(z) = (M - \beta)\mathfrak{b}(z) + (v - \beta) \tag{8.6}$$

for an arbitrary $v \in \mathfrak{H}$. Note that $\psi(z)$ is subharmonic in Ω, and since $\overline{\lim}_{z\to\zeta}\mathfrak{b}(z) \leq 0$ for $\zeta \in \mathcal{N}$,

$$\overline{\lim}_{z\to\zeta} \psi(z) \leq \overline{\lim}_{z\to\zeta} v(z) - \beta \leq f(\zeta) - \beta < \varepsilon.$$

Moreover, as $\mathfrak{b}(z) \equiv -1$ outside $\mathcal{N}$ and since $v \in \mathfrak{H}$ implies that $\overline{\lim}_{z\to\zeta} v(z) \leq M$, we again obtain that $\overline{\lim}_{z\to\zeta}\psi(z) < \varepsilon$ for $\zeta \in \mathcal{N}'$.[11] As a consequence, $\overline{\lim}_{z\to\zeta}\psi(z) < \varepsilon$ for all $\partial\Omega$, and hence $\psi(z) < \varepsilon$ for all $z \in \Omega$ and $v \in \mathfrak{H}$. Therefore, after solving for v in equation (8.6), we deduce that

$$H_f^\Omega(z) \leq \beta - (M - \beta)\mathfrak{b}(z) + \varepsilon,$$

and hence

$$\overline{\lim}_{z\to\zeta_0} H_f^\Omega(z) < \beta + \varepsilon.$$

We conclude that

$$\overline{\lim}_{z\to\zeta_0} H_f^\Omega(z) \leq \overline{\lim}_{\zeta\to\zeta_0} f(\zeta),$$

proving the theorem. □

The upshot of all this is that at points of continuity, we have exactly what we want:

8.26 Corollary. *If the boundary function f is continuous at the regular point ζ_0, then*

$$\lim_{z\to\zeta_0} H_f^\Omega(z) = f(\zeta_0).$$

8.27 Corollary. *If the boundary function has continuous boundary values f and each point of the boundary is regular, then the Dirichlet problem has a unique solution.*

---

11 In fact, in this case, $\overline{\lim}_{z\to\zeta}\psi(z) \leq 0$.

Interestingly, the converse is also true.

8.28 Corollary. *If the Dirichlet problem has a solution for every continuous function on the boundary, then every point of the boundary is regular.*

This is so as it is always possible to find a continuous function that is zero at a boundary point and negative at all other boundary points. Then the solution to the Dirichlet problem is the requisite barrier at that point.

Harmonic measure extension

At this juncture, let us utilize the Perron method to extend the notion of harmonic measure as initiated by Nevanlinna in Chapter 7 in a natural way from arcs to Borel sets on $\partial\Omega$. See Garnett (1986) for further details.

8.29 Definition. Let E be a Borel set on $\partial\Omega$ with the characteristic function χ_E of E, and let $\mathfrak{H}_{\chi_E}$ be the Perron family of subharmonic functions v in Ω satisfying

$$\overline{\lim_{z\to\zeta}}\, v(z) \le \chi_E(\zeta)$$

for all $\zeta \in \partial\Omega$. Then the *harmonic measure* of E with respect to the domain Ω is the bounded harmonic function

$$\omega(z, E, \Omega) \equiv H^{\Omega}_{\chi_E}(z) = \sup_{v\in\mathfrak{H}} v(z)$$

in Ω satisfying $0 \le \omega(z, E, \Omega) \le 1$.

It is clear that when E is an arc on $\partial\Omega$, the two definitions are the same. As in the case of arcs, the harmonic measure is conformally invariant under a conformal mapping of the domain.

As we have previously encountered in Chapter 7 another harmonic measure with regard to the Riesz representation theorem, let us apply the Perron method to establish that our two notions of harmonic measure are in fact the same, that is, the measure $\mu^{\Omega}_z(\zeta)$ from the RRT for functions $f(\zeta)$ continuous on $\partial\Omega$

$$H^{\Omega}_f(z) = \int_{\Gamma} f(\zeta)\,d\mu^{\Omega}_z(\zeta)$$

is the same as the harmonic measure defined by $\omega(z, E, \Omega) = H^{\Omega}_{\chi_E}$ in Chapter 7.

To this end, for a bounded domain Ω, let E be a closed set on $\partial\Omega$. Our proof uses the family $\mathfrak{H}$ of Theorem 8.20. Take a sequence of continuous functions $\{f_n\}$ on $\partial\Omega$ such that $f_n(\zeta) \searrow \chi_E(\zeta)$. Then $H^{\Omega}_{f_n}$ is a decreasing sequence of harmonic functions, so that $H^{\Omega}_{f_n} \searrow H^{\Omega}$ and H^{Ω} is harmonic in Ω by Harnack's principle. It follows that

$$H^{\Omega}(z) = \lim_{n\to\infty} \int_{\partial\Omega} f_n(\zeta) d\mu_z(\zeta) = \int_E \lim_{n\to\infty} f_n(\zeta) d\mu_z(\zeta) = \mu_z^{\Omega}(E).$$

Moreover, for all n,

$$\overline{\lim_{z\to\zeta}} H^{\Omega}(z) \le \overline{\lim_{z\to\zeta}} H_{f_n}^{\Omega}(z) = f_n(\zeta),$$

whence $\overline{\lim}_{z\to\zeta} H^{\Omega}(z) \le \chi_E(\zeta)$. Thus H^{Ω} is belongs to the Perron family $\mathfrak{H}_{\chi_E}$, which implies that $H^{\Omega}(z) \le H_{\chi_E}^{\Omega}(z)$.

Moreover, every function $v \in \mathfrak{H}_{\chi_E}$ also belongs to the family $\mathfrak{H}_{f_n}$ for each n, implying that $v \le H_{f_n}^{\Omega}$ for each n. We conclude that

$$H_{\chi_E}^{\Omega}(z) \le \lim_{n\to\infty} H_{f_n}^{\Omega}(z) = H^{\Omega}(z),$$

and hence $\omega(z, E, \Omega) = H_{\chi_E}^{\Omega}(z) = H^{\Omega}(z) = \mu_z^{\Omega}(E)$, which was to be proved.

Next, let $E \subseteq \partial\Omega$ be any Borel set and note that

$$\mu_z^{\Omega}(E) = \sup\{\mu_z^{\Omega}(K) : K \text{ closed}, K \subseteq E\} \le H_{\chi_E}^{\Omega}(z),$$

where we have used the fact that $f \le g$ implies $H_f^{\Omega}(z) \le H_g^{\Omega}(z)$. Finally, since $\mathfrak{H}_f \cup \mathfrak{H}_g \subseteq \mathfrak{H}_{f+g}$, setting $\partial\Omega - E = E'$, we have

$$\mu_z^{\Omega}(E) = 1 - \mu_z^{\Omega}(E') \ge 1 - H_{\chi_{E'}}^{\Omega}(z) \ge H_{\chi_E}^{\Omega}(z),$$

and we conclude that $\omega(z, E, \Omega) = \mu_z^{\Omega}(E)$ for all Borel sets on $\partial\Omega$.

This means that we can write

$$H_f^{\Omega}(z) = \int_{\partial\Omega} f(\zeta) d\omega(z, \zeta)$$

for every $f(\zeta)$ continuous on $\partial\Omega$. In the case the domain Ω is the open unit disk, the above formulation reduces to the Poisson integral.

9 Iteration of rational mappings

The modern study of the iteration of rational functions began with the work of Gaston Julia[1] and Pierre Fatou[2] early in the 20th century basing their studies on the prior work of Montel regarding normal families. No doubt, a great impetus for this large outpouring of research was the *Grand Prix des Sciences Mathématiques* to be awarded in 1918 for work in this subject area, and although Julia entered (and won), Fatou did not. The subject occurs implicitly in the well-known Newton–Raphson method of the 17th century.[3] It was the development of the theory of normal families that allowed an elegant separation of iterations that at a point behaved wildly from those that had a more regular development. One striking feature of the iteration of rational functions is that very simple functions can have extraordinarily complex dynamics, which makes them so interesting to study. Unlike Julia and Fatou, we now have computers to calculate the behavior of the iterates. If only they could see now what beautiful complexity they had unleashed, as for example, displayed in the book by Peitgen and Richter (1986).

Rational functions

We consider the quotient of two polynomials considered as mappings from $\hat{\mathbb{C}}$ to $\hat{\mathbb{C}}$,

$$R(z) = \frac{P(z)}{Q(z)} = \frac{a_0 + a_1 z + \cdots + a_p z^p}{b_0 + b_1 z + \cdots + b_q z^q},$$

where the coefficients are complex numbers, and neither polynomial is identically zero, nor do they have common zeros. Associated with such rational functions is the number

$$d = \deg(R) = \max\{\deg P, \deg Q\}.$$

An interesting fact of rational functions is that the number of roots of $R(z) = a$ in $\hat{\mathbb{C}}$ equals the number of poles of $R(z)$ in $\hat{\mathbb{C}}$ (both counted according to multiplicity), and it equals the degree d (exercise).

A further fact that will be useful is that *a meromorphic function $f(z)$ on $\hat{\mathbb{C}}$ is a rational function*. To see this, note that the number of poles of $f(z)$ is finite by the compactness of $\hat{\mathbb{C}}$ and let us assume in the most general case that $b_\infty = \infty$ is also a pole. Enumerating the finite poles and the one at infinity, we have the set $\{b_1, b_2, \ldots, b_n, b_\infty\}$.

---

1 Mémoire sur l'iteration des fractions rationnelles, *J. Math. Pures Appl.* (8) 1 (1918), 47–245.

2 Sur les équations fonctionnelles, *Bull. Soc. Math. France*, 47 (1919), 161–271; 48 (1920), 33–94 and 208–314. "Nous avons fait exclusivement usage dans nos recherches des propositions de M. Montel."

3 Indeed, Ernst Schröder published in 1870/71 a work in this regard associated with the polynomial $p(z) = z^2 - 1$. The Newton–Raphson method is also known as Newton's method and referred to by A. Cayley as the Newton–Fourier method in the context of polynomials of a complex variable. See Figure 9.7.

https://doi.org/10.1515/9783110757828-009

At each finite pole b_j, let $P_j(\frac{1}{z-b_j})$ be the principal part of $f(z)$, each of which represents a rational function. For such values of j, the point at infinity is a removable singularity since $P_j(\frac{1}{z-b_j}) \to 0$ as $z \to \infty$.

At b_∞, letting $w = 1/z$, the principal part of $f(w)$ at $w = 0$ is of the form

$$P_\infty\left(\frac{1}{w}\right) = \frac{k_m}{w^m} + \frac{k_{m-1}}{w^{m-1}} + \cdots + \frac{k_1}{w},$$

for some m, so that $P_\infty(z)$ is a polynomial. As a consequence, the function

$$G(z) = f(z) - \sum_{j=1}^{n} P_j\left(\frac{1}{z-b_j}\right) - P_\infty(z)$$

is analytic in $\hat{\mathbb{C}}$ as all the singularities are removable. Since $G(z)$ is continuous on the compact set $\hat{\mathbb{C}}$ and hence bounded on $\mathbb{C}$, an application of Liouville's theorem implies that $G = c$. Thus $f = \sum_{j=1}^{n} P_j + P_\infty + c$, proving that f is rational.

Orbits

Starting with some initial point z_0, we will be concerned with the successive iterations of this point by the rational function $R(z)$, that is,

$$R^n(z_0) = R \circ R \circ \cdots \circ R(z_0)$$

(n-fold composition), and we write $z_n = R^n(z_0)$, $n = 1, 2, 3, \dots$, with $z_0 = R^0(z_0)$. The result is a sequence of successive iterations

$$Or^+(z_0) = \{z_0, z_1, z_2, \dots\}$$

in $\hat{\mathbb{C}}$ known as the *forward orbit* of z_0. Analogously, the *backward orbit* of z_0 is

$$Or^-(z_0) = \{z \in \hat{\mathbb{C}} : R^k(z) = z_0 \text{ for some } k = 0, 1, 2, \dots\}.$$

A useful notion from the linear algebra of matrices is the following.

Conjugates

9.1 Definition. Rational functions $R(z)$ and $S(z)$ are said to be *conjugate* if there exists a Möbius transformation $M(z) = \frac{az+b}{cz+d}$, $ad - bc \neq 0$, such that $S = M \circ R \circ M^{-1}$.

Note that the conjugate of a rational function is again a rational function of the same degree. Moreover, $S^n = M \circ R^n \circ M^{-1}$ (why?), that is, the iterates are conjugate as well. Therefore their dynamical developments are "equivalent." So, for example, the quadratic polynomial

$$R(z) = az^2 + bz + c$$

satisfies the conjugacy relation $S = M \circ R \circ M^{-1}$, where $S(z) = z^2 + d$ and $M(z) = az + b/2$, and d is the constant $ac + b/2 - b^2/4$.[4] Hence, to study the dynamical nature of a general quadratic $R(z)$, it suffices to study the case of $S(z) = z^2 +$ constant.

If $R(z_0) = z_0$, then z_0 is a *fixed point* of $R(z)$, and the iterations remain stationary. If a sequence of iterates $\{z_n\}$ of z_0 converges to some point ζ, then

$$R(\zeta) = R\Big(\lim_{n\to\infty} z_n\Big) = \lim_{n\to\infty} R(z_n) = \lim_{n\to\infty} z_{n+1} = \zeta$$

by the continuity of $R(z)$, implying that ζ is necessarily a fixed point of $R(z)$. Note that z_0 is a fixed point of $R(z)$ if and only if $M(z_0)$ is a fixed point of the conjugate $S(z) = M \circ R \circ M^{-1}(z)$.

If $R^n(z_0) = z_0$ for some positive integer n, that is, z_0 is a fixed point of R^n, and $R^m(z_0) \neq z_0$ for $0 < m < n$, then z_0 is a *periodic point of period* n or a *fixed point of order* n. Hence the orbit consists of n distinct points

$$\gamma = Or^+(z_0) = \{z_0, z_1, \dots, z_{n-1}\}$$

where $z_k = R^k(z_0)$, $k = 0, 1, 2, \dots, n-1$. In this instance, $Or^+(z_0)$ forms an *n-cycle* or a *periodic orbit*. A point z_0 is called *preperiodic* if z_0 is not periodic but some iterate $R^m(z_0)$ is. If $Or^+(z_0) = \{z_0, z_1, z_2, \dots\}$ is infinite, then the point z_0 is a *wandering point.*

9.2 Example. $R(z) = z^2$. Then there are exactly three fixed points of order one, namely, $0, 1$, and ∞. Any point z_0 with $0 < |z_0| < 1$ or $|z_0| > 1$ is a wandering point since $Or^+(z_0)$ is a sequence of points converging to the origin or to ∞, respectively. The point $z_0 = -1$ is preperiodic since its orbit is $\{-1, 1, 1, \dots\}$. Finally, the points $e^{2\pi i/3}$ and $e^{4\pi i/3}$ on the unit circle $|z| = 1$ form a 2-cycle.

Fixed points and their nature form a fundamental part of the theory, but not all fixed points are created equal. The distinguishing factor is the derivative of $(R^n)'(z_0)$ and the reasons for that will become more apparent further. Let $z_0 \in \mathbb{C}$ be a fixed point of order n, and let $\lambda = (R^n)'(z_0)$. This gives rise to the following classification:
(i) If $\lambda = 0$, then $Or^+(z_0)$ is *superattracting*;
(ii) If $0 < |\lambda| < 1$, then $Or^+(z_0)$ is *attracting*;
(iii) If $|\lambda| = 1$, then $Or^+(z_0)$ is *indifferent*;
(iv) If $|\lambda| > 1$, then $Or^+(z_0)$ is *repelling*.

If $z_0 = \infty$, then we take $\lambda = 1/(R^n)'(\infty)$.

4 The reader is invited to check that the author has got this right.

By repeated applications of the chain-rule we obtain (when all terms are finite)

$$\lambda = R'(z_0) \cdot R'(z_1) \cdots R'(z_{n-1}),$$

which means that each point z_k of $Or^+(z_0)$ has the same value $\lambda = (R^n)'(z_k)$ (exercise), and thus each z_k has the same classification as z_0. Thus the *eigenvalue* λ is an invariant of the orbit of z_0.

9.3 Example.

(i) $R(z) = z^2 - z$. Then 0, 2, and ∞ are fixed points of $R(z)$. Moreover, $|R'(0)| = 1$ and $|R'(2)| = 3$, making $z = 0$ an indifferent fixed point and $z = 2$ a repelling fixed point. Also, $\lambda = 0$ for $z = \infty$, making $z = \infty$ a superattracting fixed point.

(ii) $R(z) = z^2 - 1$. Then the points 0 and -1 form a 2-cycle, and since

$$|\lambda| = |R'(0) \cdot R'(-1)| = 0,$$

the 2-cycle is superattracting.

(iii) $R(z) = \frac{1}{z^2}$. Then the points 0 and ∞ are attractors of period 2, and $(R^2)'(0) = (R^2)'(\infty) = 0$, so that the 2-cycle is superattracting.

Iteration of Möbius transformations

Since the Möbius transformations are rational functions of degree 1, we first consider their iterations before moving on to more general considerations. Recall from Chapter 1 that the Möbius transformations $w = T(z) \in \text{Möb}(\hat{\mathbb{C}})$, i. e.,

$$T(z) = \frac{az + b}{cz + d},$$

$ad - bc = 1$, can be classified according to the square of their trace values $\tau = a + d$, where the fixed points of the transformations are given by the roots

$$\xi_1, \xi_2 = \frac{(a - d) \pm \sqrt{(a + d)^2 - 4}}{2c}.$$

Let us now examine the iterations of each class.

One fixed point (parabolic case: $\tau^2 = 4$)

With one fixed point, $\xi = \xi_1 = \xi_2$. If $c = 0$, then $T(z) = \frac{a}{d}z + \frac{b}{d}$, and $\xi = \infty$, so without loss of generality, in this case, let us assume that $a = d = 1$. Hence $T(z) = z + b$, $b \neq 0$. It follows that $T^n(z) = z + nb \to \infty$ as $n \to \infty$.

On the other hand, if $c \neq 0$, then $\xi = (a - d)/2c \neq \infty$, and there is a Möbius transformation $w = M(z) = 1/(z - \xi)$ that maps ξ to ∞. Then the conjugate $S = M \circ T \circ M^{-1}$ has a single fixed point at ∞ such that $S^n(w) \to \infty$. Since $T^n = M^{-1} \circ S^n \circ M$, we see

Figure 9.1: A Parabolic transformation with a single fixed-point attractor (center). Courtesy José Ibrahim Villanueva Gutiérrez.

that $T^n(z) \to M^{-1}(\infty) = \xi$ as $n \to \infty$. We conclude that in either case, $T^n(z) \to \xi$ as $n \to \infty$ for all $z \in \mathbb{C}$, and T is a *parabolic* transformation. See Figure 9.1.

Two fixed points

Denote the fixed points of $T(z)$ by ξ_1, ξ_2 and assume that $c \neq 0$, so that both points are finite. Then the Möbius transformation

$$w = M(z) = \frac{z - \xi_1}{z - \xi_2}$$

maps ξ_1 and ξ_2 to 0 and ∞, respectively.[5] It follows that the conjugate mapping $S = M \circ T \circ M^{-1}$ fixes the points 0 and ∞. Then we can write $S(w) = kw$, which satisfies $S^n(w) = k^n w$. Clearly, the behavior of S^n as $n \to \infty$ at all $z \neq 0, \infty$ depends crucially on the value of k.

As $S(w) = M \circ T \circ M^{-1}(w) = kw$, we obtain the *normal form* of the transformation $T(z)$

$$\frac{T(z) - \xi_1}{T(z) - \xi_2} = k\frac{z - \xi_1}{z - \xi_2}. \tag{9.1}$$

5 Of course, we could just as well write

$$w = M(z) = \frac{z - \xi_2}{z - \xi_1}.$$

We can deduce something interesting about the value of the *multiplier* k by using a cross-ratio[6] since the points ξ_1, ξ_2, and ∞ are mapped by $T(z)$, respectively, to ξ_1, ξ_2, and a/c. The cross-ratio in this instance becomes

$$\frac{(w-\xi_1)(\frac{a}{c}-\xi_2)}{(w-\xi_2)(\frac{a}{c}-\xi_1)} = \frac{z-\xi_1}{z-\xi_2},$$

implying that

$$k = \frac{(\frac{a}{c}-\xi_1)}{(\frac{a}{c}-\xi_2)}.$$

9.4 Exercise. In the preceding case, using the fact that $\xi_1+\xi_2 = \frac{a-d}{c}$, $\xi_1\xi_2 = -\frac{b}{c}$, $\xi_1^2+\xi_2^2 = \frac{(a-d)^2+2bc}{c^2}$, and $ad-bc=1$, show that

$$k + \frac{1}{k} = (a+d)^2 - 2 = \tau^2 - 2.$$

Returning to equality (9.1), we deduce that $|T'(\xi_1)| = |k|$ and $|T'(\xi_2)| = 1/|k|$. It follows that if (i) $|k| < 1$, then ξ_1 is an attractor, and ξ_2 is a repeller, and (ii) if $|k| > 1$, then the roles of ξ_1 and ξ_2 are reversed. In other words, the iterates T^n converge to

Figure 9.2: Iterations of a loxodromic transformation with two finite fixed points, one an attractor and the other a repeller. Courtesy José Ibrahim Villanueva Gutiérrez.

6 Here for convenience, we use an equivalent cross-ratio to that given in (1.12) of the Möbius transformation $w = T(z)$ that maps the points $w_i = T(z_i)$, $i = 1, 2, 3$, all of which are finite, namely

$$\frac{(w-w_1)(w_2-w_3)}{(w-w_2)(w_1-w_3)} = \frac{(z-z_1)(z_2-z_3)}{(z-z_2)(z_1-z_3)}$$

with the right-hand side becoming $(z-z_1)/(z-z_2)$ when $z_3 = \infty$ as in the present case.

Figure 9.3: The iterates of a hyperbolic transformation traverse circular orbits away from the repeller toward the attractor. Courtesy José Ibrahim Villanueva Gutiérrez.

one of the two fixed points, respectively, with the second point being a repeller. If k is complex, then this corresponds to a *loxodromic* transformation: $\tau^2 \in \mathbb{C}$. The iterates move along S-shaped orbits away from the repeller toward the attractor. See Figure 9.2.

If (iii) k is real, $k > 0$, $k \neq 1$, then the iterates move in circular orbits away from the repeller toward the attractor. This is the *hyperbolic* case (τ real, $\tau^2 > 4$). See Figure 9.3.

In the case (iv) $|k| = 1$, $k \neq 1$, since a Möbius transformation maps circles to circles (straight lines are circles of infinite radius), there are two possibilities:

(a) k is an nth root of unity, and therefore the iterates T^n cycle through a finite set of points on a circle, or
(b) the iterates form a dense subset of a circle. In either case, T is an *elliptic* transformation (τ real, $0 \leq \tau^2 < 4$), and iterations move in fixed circular orbits about one or the other of the fixed points. See Figure 9.4.

In all the above cases of two fixed points, we have stipulated that $c \neq 0$. If $c = 0$, then the fixed points are $\xi_1 = \frac{b}{d-a}$ and $\xi_2 = \infty$. The development engenders the same classifications as in the preceding cases (exercise).

Next, for our subsequent considerations, *we will assume that* $d = \deg(R) \geq 2$.

Julia and Fatou sets

We now define some of the most fascinating sets in mathematics. Our discussion of normal families in Chapter 4 will now play a major role.

Figure 9.4: Iterations of an elliptic transformation with two fixed points. All iterations remain on circles whose radius is that of the seed point. Courtesy José Ibrahim Villanueva Gutiérrez.

9.5 Definition. For a rational mapping $R(z)$, the *Fatou set* (*stable set*) is

$$F_R = \{\zeta \in \hat{\mathbb{C}} : \{R^n(z)\}, n = 0, 1, 2, \ldots, \text{ is a normal family at } \zeta\},$$

and the *Julia set* is $J_R = \hat{\mathbb{C}} - F_R$.[7]

Recall that a family is normal at a point if it is normal in a neighborhood of the point, so that the Fatou set is open, the Julia set is closed, and the connected components of F_R are domains of maximum normality of the family $\{R^n(z)\}$.

9.6 Example. Given $R(z) = z^d$, $d \geq 2$, the iterates of any z_0 with $|z_0| < 1$ satisfy $R^n(z) \to 0$ uniformly on compact subsets of the unit disk $U : |z| < 1$, and, furthermore, $R^n(z) \to \infty$ uniformly on compact subsets of $U' : |z| > 1$. Moreover, for any z_0 with $|z_0| = 1$, there is no neighborhood of z_0 in which the family of iterates $\{R^n(z)\}$ constitutes a normal family. We conclude that $F_R = U \cup U'$ and that $J_R = \{|z| = 1\}$.

The significance of the derivative $(R^n)'(z_0)$ will now become more apparent.

9.7 Theorem. *If $Or^+(z_0)$ is a (super)attracting periodic orbit, then $Or^+(z_0) \subseteq F_R$. If $Or^+(z_0)$ is a repelling periodic orbit, then $Or^+(z_0) \subseteq J_R$.*

Proof. Without loss of generality, we assume that z_0 is an attracting fixed point (of order one) and finite. Then for all points z in some sufficiently small disk $D(z_0, r)$, we have

$$\left| \frac{R(z) - R(z_0)}{z - z_0} \right| < \varepsilon < 1,$$

7 '…définissons comme l'ensemble des points où les itérées ne forment pas une suite normale…' Fatou op. cit., p. 163.

so that $|R(z)-z_0| < \varepsilon|z-z_0|$. Consequently, $|R^n(z)-z_0| < \varepsilon^n r$ for $n = 1, 2, 3, \dots$, implying that the family $\{R^n(z)\}$ is uniformly bounded in $D(z_0, r)$ and hence normal at z_0. We conclude that $z_0 \in F_R$ and, as a matter of fact, $D(z_0, r) \subseteq F_R$.

The second part of the theorem is proved analogously. □

As a consequence, since J_R is closed, we have

$$\overline{\{\text{repelling periodic points}\}} \subseteq J_R.$$

There is the equality in the above relation, which we mention further on.

9.8 Exercise. Prove that if ζ is a repelling fixed point of $R(z)$, then the iterates z_n can converge to ζ only if $z_n = \zeta$ for all n sufficiently large.

Note that in the first part of the preceding theorem, for any $z \in D(z_0, r)$, $R^n(z) \to z_0$, $n = 1, 2, 3, \dots$. Moreover, in the example $R(z) = z^2$, there were two open sets U and U' in which the iterates tended to an attracting fixed point. This suggests the following:

9.9 Definition. If z_0 is an attracting fixed point in $\hat{\mathbb{C}}$, then the *basin (domain) of attraction* is the set

$$A(z_0) = \{z \in \hat{\mathbb{C}} : R^n(z) \to z_0 \text{ as } n \to \infty\}.$$

It consists of all the points z whose forward orbits $Or^+(z)$ approach z_0 and clearly includes all the points of $Or^-(z_0)$. The *immediate basin of attraction* is denoted by $A^*(z_0)$ and is the connected component of $A(z_0)$ that contains the point z_0.

In the case of a cycle γ of period n, each of the fixed points of period n, $z_k = R^k(z_0)$, $k = 0, 1, 2, \dots, n-1$, has a basin of attraction for the iterates R^n, and the basin of attraction of the n-cycle $A(\gamma)$ is the union of these basins.[8] The *immediate basin of attraction* of γ, denoted by $A^*(\gamma)$, is the union of the corresponding immediate basins of attraction of all the points $z_0, z_1, \dots, z_{n-1}$ in the cycle.

Exceptional points

One of the most salient features of the theory of normal families is the fundamental normality test 4.35, namely that a family of meromorphic functions defined on a domain $\Omega \subseteq \hat{\mathbb{C}}$ that omits three distinct values is normal.[9] This result can be applied in relation to a Julia set.

---

8 The Newton–Raphson rational function associated with $p(z) = z^2 - 1$ is $R(z) = \frac{z^2+1}{2z}$. In the aforementioned work, E. Schröder showed that the points ± 1 are superattracting fixed points of $R(z)$ and that their respective basins of attraction are half-planes $x > 0$ and $x < 0$, respectively.

9 See also the comments after Theorem 4.29.

9.10 Proposition. *For each point $\zeta \in J_R$ and any neighborhood Δ of ζ, then*

$$E_\Delta = \hat{\mathbb{C}} - \bigcup_{n>0} R^n(\Delta)$$

contains at most two points (exceptional points).

For example, if $R(z) = a_n z^n + a_{n-1} z^{n-1} + \cdots + a_0$, $n \geq 2$, then $z = \infty$ is a superattracting fixed point belonging to the Fatou set. Therefore the Julia set is contained in $\mathbb{C}$, so that

$$\bigcup_{n>0} R^n(\mathbb{C}) = \mathbb{C},$$

implying that $z = \infty$ is the sole exceptional point.

9.11 Definition. Given $\zeta \in J_R$, the *set of exceptional points* is

$$E_\zeta = \bigcup E_\Delta,$$

where the union is taken over all neighborhoods Δ of ζ.

As we have seen, the cardinality of E_ζ satisfies $0 \leq \overline{\overline{E_\zeta}} \leq 2$, and for all sufficiently small neighborhoods Δ, E_ζ is independent of Δ.

It is interesting that the nature of the set E_ζ allows for the classification of the rational function $R(z)$ in the following manner.

9.12 Theorem. *Let $\zeta \in J_R$ with $E_\zeta \neq \emptyset$.*
(i) *If $\overline{\overline{E_\zeta}} = 1$, then $R(z)$ is conjugate to a polynomial;*
(ii) *If $\overline{\overline{E_\zeta}} = 2$ then $R(z)$ is conjugate to the mapping $z \to z^{\pm d}$ with $d = \deg(R)$.*

Proof. It is evident that $R^{-1}(E_\zeta) = E_\zeta$ with the set E_ζ consisting of either one fixed point, or two fixed points, or a 2-cycle. Thus we have three cases to consider.
(i) If $E_\zeta = \{a\}$, then we can find a Möbius transformation $M(z)$ such that $M(a) = \infty$. Then the conjugate mapping

$$P(z) = M \circ R \circ M^{-1}(z)$$

is a rational function satisfying $P(\infty) = \infty$ but having no poles in $\mathbb{C}$. Therefore $P(z)$ reduces to a polynomial.
(ii) If $E_\zeta = \{a, b\}$, then again choose a Möbius transformation with $M(a) = \infty$ and $M(b) = 0$. For the conjugate $P(z) = M \circ R \circ M^{-1}(z)$, we have two possibilities:
(*) If a and b both are fixed points of $R(z)$, then 0 and ∞ are the respective fixed points of $P(z)$. Then as in (i), $P(z)$ is a polynomial. Moreover, $P^{-1}(0) = \{0\}$ (why?), so that the origin is a zero of multiplicity $d = \deg(R)$ and $P(z) = cz^d$.

By taking an expansion or contraction we can conclude that $R(z)$ is conjugate to $Q(z) = z^d$.

(**) If a and b are 2-cycles of $R(z)$, then the two points 0 and ∞ are a 2-cycle of $P(z)$. With analogous reasoning $R(z)$ is conjugate to $Q(z) = z^{-d}$ (exercise). □

At first glance, it might seem that the set E_ζ could depend on the point ζ of the Julia set. In fact, it does not, and, interestingly, any point of E_ζ belongs to the Fatou set.

9.13 Corollary.
(i) *E_ζ is independent of the choice of $\zeta \in J_R$.*
(ii) *$E_\zeta \subseteq F_R$.*

Proof. (i) For $E_\zeta \neq \emptyset$ and $\zeta \in J_R$, by the theorem $R(z)$ is conjugate either to a polynomial or to the mapping $Q(z) = z^{\pm d}$. In the first instance, taking the domain $\Delta = \mathbb{C}$ implies that $E_\zeta = E_\Delta$, which is also valid in the second case taking $\Delta = \hat{\mathbb{C}} - \{0, \infty\}$.[10]

To prove (ii), simply note that if either $\overline{\overline{E_\zeta}} = 1$ or $\overline{\overline{E_\zeta}} = 2$, then E_ζ consists either of a single superattracting fixed point or of two superattracting fixed points or a superattracting 2-cycle, respectively. In all cases, the result is a consequence of Theorem 9.7. □

Since the set E_ζ only depends on the rational function $R(z)$, we set $E_\zeta = E_R$.

We have seen that for $R(z) = z^d$, $d \geq 2$, the Julia set is simply the unit circle $|z| = 1$. However, with the simplest of modifications, say $R(z) = z^2 + c$, the Julia set can turn out to be exquisitely intricate (see Figure 9.5). So let us now turn to examining the properties and structure of J_R more closely.

Basic properties of J_R

9.14 Theorem. *For the Julia set of $R(z)$:*
(i) *$R(J_R) = J_R$, forward invariance; $J_R = R^{-1}(J_R)$, backward invariance;*[11]
(ii) *$J_R \neq \emptyset$, and it is equal to its set of accumulation points and hence is a perfect set;*
(iii) *If J_R contains an interior point, then $J_R = \hat{\mathbb{C}}$;*
(iv) *For any $a \in J_R$, $J_R = \overline{\bigcup_{n\geq 0} R^{-n}(a)}$;*
(v) *Given any attracting fixed point z_0 of $R(z)$, we have $A(z_0) \subseteq F_R$ and $\partial A(z_0) = J_R$.*[12]

---

10 To simplify matters, we need not distinguish between $R(z)$ and its conjugate.

11 A set that is both forward invariant and backward invariant is *completely invariant*. Since $R(z)$ is surjective, backward invariance is the same as complete invariance for $R(z)$.

12 Note that this means that if $R(z)$ has multiple attracting fixed points, say α, β, γ, then $J_R = \partial A(\alpha) \cup \partial A(\beta) \cup \partial A(\gamma)$.

Figure 9.5: The Julia set of the function $R(z) = z^2 + c$ with $c = 0.26006 + 0.00178i$. The set is actually connected, which is discussed in the sequel. Courtesy José Ibrahim Villanueva Gutiérrez.

Proof. To establish (i), it is perhaps simpler to consider the complementary Fatou set F_R. Since each $R(z)$ is a continuous open mapping, open neighborhoods are mapped to open neighborhoods by both R^{-1} and R. As a consequence, given any $z \in F_R$, we have $R(z) \in F_R$ and $R^{-1}(z) \subseteq F_R$. These observations are sufficient to show that (exercise)

$$R(F_R) = F_R = R^{-1}(F_R),$$

and thusly for J_R.

(ii) To show that J_R is nonempty, let us assume on the contrary that $J_R = \emptyset$. This means that $\{R^n\}$ is normal on the Riemann sphere $\hat{\mathbb{C}}$ and there is a subsequence $\{R^{n_k}\}$ that converges spherically uniformly on $\hat{\mathbb{C}}$ to a limit function $S(z)$. Then $S(z)$ is meromorphic in $\hat{\mathbb{C}}$ (or $\equiv \infty$) by Corollary 3.18 and hence a rational function of degree δ, including the case where $S(z)$ may be a finite or identically infinite constant. If $S(z)$ is not constant, then let a be an arbitrary point, and if $S(z) \equiv c$, then take $a \neq c$. Since the equation $S(z) - a = 0$ has exactly δ roots, we conclude that $R^{n_k} - a = 0$ also has δ roots *for all k* sufficiently large (why?). However, $\deg(R^{n_k}) \to \infty$ as $k \to \infty$, a contradiction proving that $J_R \neq \emptyset$.

We proceed to show that J_R is equal to the set of all its accumulation points. First, we establish that for any $a \in J_R$, there is $b \in J_R$ such that $a \in Or^+(b)$ but $b \notin Or^+(a)$. Indeed, if a is a nonperiodic point, then we can simply choose any $b \in R^{-1}(a)$ (why?).

However, if $a \in J_R$ is a periodic point with period n, then consider the mapping $S(z) = R^n(z)$ and the equation

$$S(z) = a. \tag{9.2}$$

Suppose for a moment that $S^{-1}(a) = \{a\}$, in which case we choose a Möbius transformation M such that $M(a) = \infty$. Then the conjugate

$$P(z) = M \circ S \circ M^{-1}(z)$$

has no poles in $\mathbb{C}$, and only $P(\infty) = \infty$. Therefore $P(z)$ is a polynomial with attracting fixed point $z_\infty = \infty$, implying that $z_\infty \in F_P$ or, in other words, $a \in F_S$. This however contradicts the fact that $a \in J_R$. Hence we can conclude that there is another point $b \in S^{-1}(a)$, that is, $R^n(b) = a$, and once again, $a \in Or^+(b)$. Furthermore, observe that the point a is the only solution of equation (9.2) that is a member of $Or^+(a)$, which implies that $b \notin Or^+(a)$, as stated above.

We are now in a position to prove that J_R is a perfect set. Take an arbitrary point $a \in J_R$ and a neighborhood Δ_a of a. By the preceding there is another point $b \in J_R$ such that $a \in Or^+(b)$ but $b \notin Or^+(a)$. Moreover, no point of J_R can be an exceptional point (since $E_R \subseteq F_R$ by Corollary 9.13), implying that $b \notin E_R$. This means that $b \in R^m(\Delta_a)$ for some positive integer m, and hence there is $c \in \Delta_a$ such that $R^m(c) = b$. As $b \notin Or^+(a)$, it follows that $c \neq a$. On the other hand, since J_R is backward invariant and $b \in J_R$, we conclude that $c \in J_R$, which means that a is an accumulation of point of J_R, as desired.[13]

(iii) Assume that $a \in J_R$ is an interior point, and let $\Delta_a \subseteq J_R$ be an open disk about a. Since $\{R^n(z)\}$ is not a normal family in Δ_a, we know that $\bigcup_{n>0} R^n(\Delta_a)$ can omit at most two points in $\hat{\mathbb{C}}$. The forward invariance property of the Julia set implies that $\bigcup_{n>0} R^n(\Delta_a) \subseteq J_R$, and since J_R is a closed set, we conclude that $J_R = \hat{\mathbb{C}}$.

(iv) Take some point $a \in J_R$ and a neighborhood Δ_b of some $b \in J_R$. As we saw earlier, $a = R^m(c)$ for some $c \in \Delta_b$, $m \geq 1$, since $a \notin E_R$. Therefore $c \in \bigcup_{n\geq 0} R^{-n}(a)$. Since the neighborhood Δ_b can be arbitrarily small, we conclude that b is an accumulation point of the set $\bigcup_{n\geq 0} R^{-n}(a)$. As b was an arbitrary point of the Julia set, we obtain $J_R \subseteq \overline{\bigcup_{n\geq 0} R^{-n}(a)}$. To establish the reverse inclusion, note that by the backward invariance of J_R we have $\bigcup_{n\geq 0} R^{-n}(a) \subseteq J_R$. Since J_R is a closed set, $\overline{\bigcup_{n\geq 0} R^{-n}(a)} \subseteq J_R$, providing the desired equality.

(v) Note that if z_0 is an attracting fixed point, then $z_0 \in F_R$ by Theorem 9.7. Since F_R is an open set, take a disk $D(z_0, r) \subseteq F_R$ and any point $z \in A(z_0)$. Then for all n sufficiently large, the iterates $R^n(z)$ are in the disk $D(z_0, r)$ which means that $z \in F_R$ in view of the backward invariance of F_R. Hence $A(z_0) \subseteq F_R$, as desired.

---

13 Then, trivially, J_R contains no isolated points.

Furthermore, take a point $a \in J_R$; by the last inclusion $a \in A(z_0)'$ so that any neighborhood Δ_a of a satisfies $\Delta_a \cap A(z_0)' \neq \emptyset$, implying that the sequence $\{R^n(\Delta_a)\}$ can omit at most two points. We conclude that the iterations of some points of Δ_a belong to the set $A(z_0)$, which means that $\Delta_a \cap A(z_0) \neq \emptyset$ (why?), and thus $J_R \subseteq \partial A(z_0)$. Finally, if $a \in \partial A(z_0)$, any neighborhood Δ_a of a intersects both $A(z_0)$ and $A(z_0)'$. Then $\{R^n\}$ cannot be a normal family in Δ_a, implying that $a \in J_R$, proving that $\partial A(z_0) = J_R$ and thus concluding the proof of the theorem. □

It was demonstrated in 1918 that (iii) actually does occur, in particular, that for the rational function

$$R(z) = \frac{(z^2+1)^2}{4z(z^2-1)}, \tag{9.3}$$

$J_R = \hat{\mathbb{C}}$.[14]

It is interesting that the proof involves the Weierstrass $\wp$-function of Chapter 3 and the duplication formula it satisfies,[15] namely

$$\wp(2w) = \frac{(\wp^2(w)+1)^2}{4\wp(w)(\wp^2(w)-1)}.$$

Given a point $z_0 \in \mathbb{C}$, suppose that w_0 is a solution to the equation $\wp(w) = z_0$. For any arbitrarily small neighborhood $N(z_0)$ about z_0, there is a neighborhood $N(w_0)$ of w_0 such that $\wp(N(w_0)) = N(z_0)$. Then for any $z \in N(z_0)$, we infer from (9.3) that $R(z) = R(\wp(w)) = \wp(2w)$ and, in general,

$$R^n(z) = R^n(\wp(w)) = \wp(2^n w)$$

for $n = 1, 2, 3, \ldots$; that is, the values of the iterates $R^n(z)$ for $z \in N(z_0)$ take the values of $\wp(w)$ in an increasing sequence of homothetic neighborhoods $N^n(w_0)$ obtained from $N(w_0)$ by the multiplicative factor of 2^n. With increasing n, these neighborhoods cover an increasing number of period parallelograms, which means that for each $z \in N(z_0) = \wp(N(w_0))$, the iterates $R^n(z) = \zeta$ take each value ζ an ever-increasing number of times. This being the case, suppose now that there is a subsequence $R^{n_k}(z)$ that converges spherically uniformly on compact subsets of $N(z_0)$ to a rational function $S(z)$ (or to $\equiv \infty$), that we have noted previously. As in the proof of part (ii) of Theorem 9.14, we can conclude that for all k sufficiently large, the equations $S(z) = \zeta$ and $R^{n_k}(z) = \zeta$ have the same number of roots in $N(z_0)$. This is clearly a contradiction, which implies that

14 S. Lattès, Sur l'itération des substitutions rationnelles et les fonctions de Poincaré, *C. R. Acad. Sci. Paris*, 166 (1918), 26–28.

15 See J. V. Armitage and W. F. Eberlein, *Elliptic Functions*, LMS Student Texts 67, Cambridge University Press, 2006.

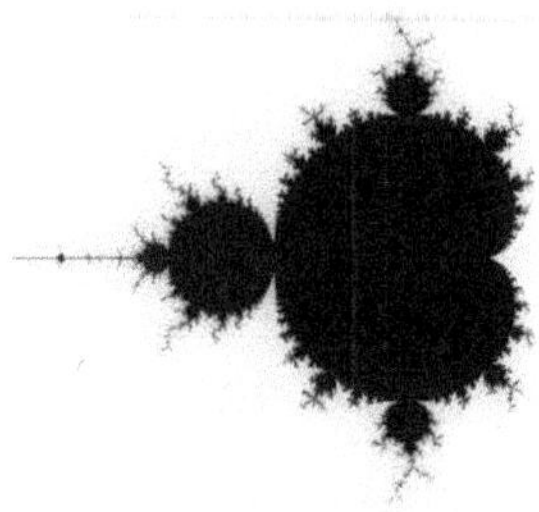

Figure 9.6: The famous Mandelbrot set is an atlas of the values of the parameter c of the quadratic polynomial $R_c(z) = z^2 + c$, for which the corresponding Julia set is connected. Equivalently, for each value of c in the Mandelbrot set, $R_c^n(0) \not\to \infty$ as $n \to \infty$. Gray regions are indicative of relative escape times to infinity. Courtesy José Ibrahim Villanueva Gutiérrez.

the sequence $\{R^n(z)\}$ cannot be normal in $N(z_0)$. As $z_0 \in \mathbb{C}$ was arbitrary, we conclude that $J_R = \hat{\mathbb{C}}$.[16]

Mandelbrot set

We have not said anything about the connectedness or otherwise of the Julia set, because it can be either connected or indeed totally disconnected. As it turns out, in the quadratic case for $R_c(z) = z^2 + c$, the Julia set is connected whenever c belongs to the *Mandelbrot set*,[17] which is the set of complex values c for which $R_c^n(0) \not\to \infty$ as $n \to \infty$ (Figure 9.6). Thus the Julia set in Figure 9.5 is connected as the parameter c lies within the Mandelbrot set.[18] For values of $c \in \mathbb{C}$ in the complement of the Mandelbrot set, J_R is totally disconnected. The connectedness of the Mandelbrot set itself was established by A. Douady and J. Hubbard.[19]

16 In more recent times the rational function $R(z) = \left(\frac{z-2}{z}\right)^2$ was also shown to have $J_R = \hat{\mathbb{C}}$ in R. Mañé, P. Sad, D. Sullivan, On the dynamics of rational maps, *Ann. Sci. École Norm. Sup.* 16 (1983) 193–217.

17 Attributed to B. Mandelbrot (1924–2010), but the set had appeared in the work by J. P. Matelski and R. W. Brooks (1978) prior to Mandelbrot's publication in 1980.

18 For the value $c = 0.26006 + 0.00178i$, which was used to generate the Julia set of Figure 9.5, we can determine that the iterations $R_c^n(0)$ have a periodic orbit via a brief computer program. We leave it to the reader to determine the period.

19 Itération des polynômes quadratiques complexes, *C. R. Acad. Sci. Paris* 294 (1982), 123–126.

Critical points

To discuss the Julia set a bit further, we first want to examine the relationship between J_R and the periodic points, and so we will need the following notion. Recall that if $R(z)$ is of degree $d \geq 2$, then it represents a d-fold mapping of $\hat{\mathbb{C}}$ onto itself.

9.15 Definition. A point $c \in \hat{\mathbb{C}}$ is a *critical point* of the rational function $R(z)$ if $R(z)$ is not one-to-one in any neighborhood of c, and if $v \in \hat{\mathbb{C}}$ satisfies $R(c) = v$ for some critical point c, then v is called a *critical value*.

Let C_R = {critical points} and V_R = {critical values}. Clearly, $R(C_R) = V_R$. Furthermore, if $z \in \mathbb{C}$ and $R'(z) \neq 0, \infty$, then $R(z)$ is one-to-one in a neighborhood of z, which means that $\overline{\overline{C_R}} < \infty$. As we might expect, the cardinality of C_R depends on the degree of $R(z)$, and in fact $\overline{\overline{C_R}} \leq 2d - 2$.[20]

In the case where $w = R(z)$ is *not* a critical value, its inverse $R^{-1}(w)$ consists of d distinct points z_i, $i = 1, 2, \ldots, d$. This means that for $i = 1, 2, \ldots, d$, there is a neighborhood N_i of z_i that is mapped by $R(z)$ one-to-one and onto some neighborhood of N_w of w with d inverse branches $R_i^{-1}(N_w) \to N_i$, $i = 1, 2, \ldots, d$. This leads to the following:

9.16 Theorem. $J_R \subseteq \overline{\{periodic\ points\}}$.

Proof. We make use of the set

$$K_R = J_R - \{\text{critical values of } R^2\},$$

which differs from J_R by only a finite set of points. Thus we need only demonstrate that $K_R \subseteq \overline{\{\text{periodic points}\}}$ since J_R is a perfect set. The ingenious reason for taking R^2 will now become apparent. For any point $w \in K_R$ and open neighborhood W of w, it follows that $R^{-2}(w)$ consists of at least four distinct points since $d \geq 2$ with at least three of them a_1, a_2, a_3 distinct from w. In view of our preceding discussion, there exist pairwise disjoint neighborhoods N_1, N_2, N_3 of a_1, a_2, a_3, respectively, such that

$$R^2(z) : N_i \to W' \subseteq W$$

is a homeomorphism for $i = 1, 2, 3$.

Consider now the inverse mapping of $R^2 | N_i$, which we denote by $S_i : W' \to N_i$, $i = 1, 2, 3$. If for all $\zeta \in W'$,

$$R^n(\zeta) \neq S_i(\zeta),$$

---

20 This result is a consequence of the Riemann–Hurwitz theorem, where $d = \deg(R)$. See: P. Blanchard, Complex analytic dynamics on the Riemann sphere, *Bull. Amer. Soc.*, 11 (1) (1984), 85–141.

for $i = 1, 2, 3$ and $n = 1, 2, 3, \ldots$, then $\{R^n\}$ would omit three values and hence be normal in W' by the FNT, a contradiction since $w \in J$. Hence there must be $\zeta' \in W'$, $m \geq 1$, and $1 \leq j \leq 3$ such that $R^m(\zeta') - S_j(\zeta')$. Then an application of the mapping R^2 gives

$$R^{m+2}(\zeta') = R^2(S_j(\zeta')) = \zeta',$$

which means that $\zeta' \in W' \subseteq W$ is a periodic point of R. Since $w \in K_R$ was arbitrary, we have the desired conclusion $K_R \subseteq \overline{\{\text{periodic points}\}}$. □

Therefore, at this juncture, we have

$$\overline{\{\text{repelling periodic points}\}} \subseteq J_R \subseteq \overline{\{\text{periodic points}\}}.$$

The difference between the periodic and repelling periodic points is the attracting and indifferent fixed points, and we can say something significant regarding the number of attracting fixed points.

9.17 Theorem. *The number of attracting cycles of a rational mapping $R(z)$ with* $\deg(R) = d \geq 2$ *is less than or equal to* $2d - 2$.

Proof. Recall that the number $2d-2$ is the upper bound of the number of critical points, and we will show that for each attracting cycle γ, the immediate basin of attraction $A^*(\gamma)$ contains at least one critical value. To this end, let us first consider the case where a is an attracting fixed point (order 1) and assume that no critical value lies in $A^*(a)$. Let $N_a \subseteq A^*(a)$ be an open neighborhood of a and consider a branch R_*^{-1} of the inverse $R^{-1}(z)$ satisfying $R^{-1}(a) = a$. By the monodromy theorem the branch R_*^{-1} can be extended to a meromorphic function in all N_a, since by assumption there are no critical values in N_a arising from critical points in $A^*(a)$. As a consequence, $R_*^{-1}(N_a) \subseteq A^*(a)$. In the same fashion, we can define a sequence of functions

$$R_*^{-n}(z) = R_*^{-1}(R_*^{-(n-1)}(z)), \quad z \in N_a,$$

for $n = 2, 3, 4, \ldots$, and once again, we obtain $R_*^{-n}(z) \subseteq A^*(a)$. We have thus derived a sequence of meromorphic functions $\{R_*^{-n}\}$ defined on N_a such that $R_*^{-n}(N_a) \cap J_R = \emptyset$. This implies that $\{R_*^{-n}\}$ is normal in N_a, which contradicts the fact that a must be a repelling fixed point of R_*^{-1}. We conclude that $A^*(a)$ contains at least one critical value arising from critical points in $A^*(a)$. Thus the number of attracting fixed points is less than or equal to $2d - 2$.

In the general case of an attracting cycle $\gamma = \{a, a_1, \ldots, a_m\}$, we obtain the functions $R_*^{-n}(z)$ derived from

$$R_*^{-1}(a) = a_m, R_*^{-2}(a) = a_{m-1}, \ldots, R_*^{-m}(a) = a$$

and proceed as in the previous argument. We leave the details as an exercise. □

No wandering domains

Although much attention was paid to characterizing the Julia set, we state here a remarkable result regarding the nature of the Fatou set. It is evident that if C is a component of the Fatou set, then the iterates $R^n(C)$ are also components of the Fatou set. Can these iterates wander indefinitely all over the complex plane?

9.18 No wandering domains theorem.[21] *Let $R(z)$ be a rational map of degree $d \geq 2$. Then every component of the Fatou set is eventually periodic, that is, if C is a component of the Fatou set, then for some $0 < m < n$, $R^n(C) = R^m(C)$.*

The proof is beyond the scope of this book. As it turns out for entire maps, the statement is false since the function $f(z) = z + \sin(2\pi z)$ has wandering Fatou components.

Newton–Raphson method

Let us now revisit the Newton–Raphson method regarding rational functions. The root finding method is given by the iterative sequence

$$z_{n+1} = z_n - \frac{f(z_n)}{f'(z_n)},$$

which under suitable assumptions produces successively better approximations to the roots of the equation $f(z) = 0$, starting with an appropriate initial value for z_0.

In the case of quadratic equations $P(z) = az^2 + bz + c$, the situation is not so interesting. Cayley (1879) established that if the (distinct) roots are z_1 and z_2 in the complex plane and L is the perpendicular bisector of the line segment joining z_1 to z_2, then by Newton's method, L divides the complex plane into two half-planes that are the two basins of attraction $A(z_1)$ and $A(z_2)$. This is just what Ernst Schröder found in the aforementioned study of $p(z) = z^2 - 1$, where $A(1) = \{z : R(z) > 0\}$ and $A(-1) = \{z : R(z) < 0\}$. The Julia set is the imaginary axis, which is $\partial A(1) = \partial A(-1)$. Cayley hoped to extend his results to the case of cubic polynomials but admitted that this proved too difficult.[22]

Indeed, it is! Only with the advent of modern computers could the basins of attraction for cubic polynomials and those of higher degree ever be realized together with their separating boundary the Julia set. For a simple cubic polynomial like $f(z) = z^3 - 1$,

---

21 D. Sullivan, Quasiconformal homeomorphisms and dynamics. I. Solution of the Fatou–Julia problem on wandering domains, *Ann. Math.* 122 (1985) (3) 401–418.

22 "J'espère appliquer cette théorie au cas d'une équation cubique, mais les calculs sont beaucoup plus difficiles."

Figure 9.7: The Newton–Raphson method for the equation $f(z) = z^3 - 1$ leads to the rational function $R(z) = \frac{2z^3+1}{3z^2}$. The Julia set for $R(z)$ is in white bordering the basins of attraction for the three roots (indicated by small black circles) of $f(z)$: 1 (red), $e^{i2\pi/3}$ (yellow), and $e^{i4\pi/3}$ (blue). Courtesy José Ibrahim Villanueva Gutiérrez.

the basins are remarkably complex structures exhibiting the property of self-similarity at smaller and smaller scales. See Figure 9.7.

10 Analytic number theory

Entre deux vérités du domaine réel, le chemin le plus facile et le plus court passe bien souvent par le domaine complexe[1]...

Paul Painlevé

In this chapter, we consider some applications of complex analysis to deal with issues in number theory. Of course, the subject is vast, so we touch on a few of the most salient highlights. Many of the topics have had their origins defined for real variables but take on a new life in the complex domain.

Gamma function

This function originates with Euler's generalization of the notion of the factorial function $n!$. Since its creation in 1730, it has found applications in various scientific fields, and we will see it again in the discussion of the Riemann zeta function.

For a complex variable $s = \sigma + i\tau$, the *gamma function* is defined by the integral

$$\Gamma(s) = \int_0^{\infty} t^{s-1}e^{-t}dt, \tag{10.1}$$

which is analytic for $\sigma = \mathrm{Re}(s) > 0$.[2] It represents the Mellin transform of the function e^{-t}.[3] It is clear that $\Gamma(1) = 1$.

---

1 Between two truths of the real domain, the easiest and shortest path very often passes through the complex domain.

2 We use the complex variable s instead of our more customary complex variable z, because the master Riemann used the variable s in his seminal work regarding the Riemann zeta function, and this has become somewhat conventional. Euler's original formulation was given by the integral

$$\int_0^1 (-\log t)^{x-1}dt$$

for $x > 0$, which after a change of variable yields the conventional form using Legendre's notation

$$\Gamma(x) = \int_0^{\infty} t^{x-1}e^{-t}dt.$$

3 The Mellin transform of a function $f(t)$ is defined as $\mathcal{M}(f(t)) = \Phi(s) = \int_0^{\infty} t^{s-1}f(t)dt$ for $s \in \mathbb{C}$. It is named after Hjalmar Mellin (1854–1933). Under suitable conditions, its inverse is given by

$$\mathcal{M}^{-1}\big(\Phi(s)\big) = f(t) = \frac{1}{2\pi i}\int_{x-i\infty}^{x+i\infty} t^{-s}\Phi(s)ds.$$

See later regarding the Laplace transform, where the integration is taken over an infinite vertical line.

https://doi.org/10.1515/9783110757828-010

In fact, considering the tail of the integral $\int_a^\infty t^{s-1}e^{-t}dt$ for $a > 0$ and s constrained to a compact subset of $\mathrm{Re}(s) > 0$,

$$\left|\int_a^\infty t^{s-1}e^{-t}dt - \int_a^n t^{s-1}e^{-t}dt\right| \leq \left|\int_n^\infty t^{s-1}e^{-t}dt\right| \leq \int_n^\infty t^{\sigma-1}e^{-t}dt < C\int_n^\infty e^{-\frac{t}{2}}dt = Ce^{-\frac{n}{2}} \to 0$$

as $n \to \infty$, which implies the uniform convergence of the integral $\int_a^\infty t^{s-1}e^{-t}dt$ on compact subsets of the right half-plane.

Defining

$$F_n(s) = \int_{\frac{1}{n}}^\infty t^{s-1}e^{-t}dt \quad (\mathrm{Re}(s) > 0),$$

it follows by Corollary 1.4 of Chapter 1 that each $F_n(s)$ is analytic for $\mathrm{Re}(s) > 0$. We now show that the sequence $\{F_n(s)\}$ in fact converges *uniformly* to $\Gamma(s)$, which implies that $\Gamma(s)$ is analytic by the Weierstrass Theorem 1.27. Indeed, it suffices to show that the sequence is uniformly Cauchy.

To this end, take $\sigma \geq \alpha > 0$, and we only need to consider the values $0 < t \leq 1$, where our integration will take place. Then for $n > m$,

$$|F_n(s) - F_m(s)| = \left|\int_{\frac{1}{n}}^{\frac{1}{m}} t^{s-1}e^{-t}dt\right| \leq \int_{\frac{1}{n}}^{\frac{1}{m}} t^{\sigma-1}dt \leq \int_{1/n}^{1/m} t^{\alpha-1}dt = \frac{1}{\alpha}\left(\frac{1}{m^\alpha} - \frac{1}{n^\alpha}\right).$$

As the last quantity can be made arbitrarily small for m, n sufficiently large, we conclude that the sequence $\{F_n(s)\}$ is uniformly convergent on compact subsets of $\mathrm{Re}(s) > 0$, and thus $\Gamma(s)$ is analytic.

Just as for real variables, integration by parts applied to $\Gamma(s)$ yields

$$\Gamma(s+1) = -t^s e^{-t}\big|_0^\infty + s\int_0^\infty t^{s-1}e^{-t}dt = s\Gamma(s) \quad (\mathrm{Re}(s) > 0). \tag{10.2}$$

Since $\Gamma(1) = 1$, we have the familiar formula

$$\Gamma(n+1) = n!$$

by induction on $n = 1, 2, 3, \ldots$. Thus $\Gamma(s)$ is viewed as a generalization of the factorial to the half-plane $\mathrm{Re}(s) > 0$. Furthermore, equation (10.2) yields the following elegant extension of the gamma function.

10.1 Theorem. *$\Gamma(s)$ can be extended to a meromorphic function in $\mathbb{C}$ with simple poles at zero and the negative integers.*

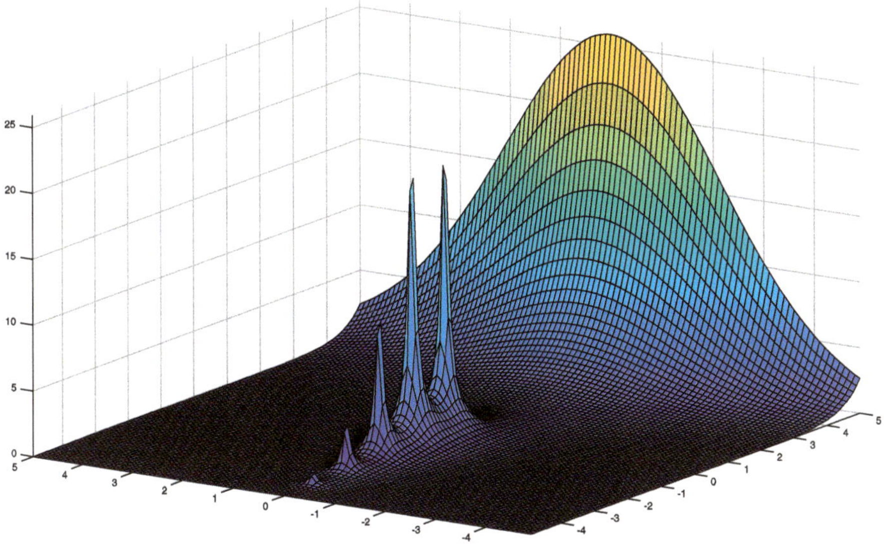

Figure 10.1: Graph of $|\Gamma(s)|$ showing the poles at the origin and the negative integers. Courtesy Chris King.

Proof. For each integer $n \geq 1$ and $\mathrm{Re}(s) > 0$, $(n-1)$ applications of equation (10.2) give

$$\Gamma(s) = \frac{\Gamma(s+n)}{s(s+1)\cdots(s+n-1)}. \tag{10.3}$$

Since the function $\Gamma(s+n)$ is analytic in the half-plane $\mathrm{Re}(s) > -n$, we conclude that $\Gamma(s)$ has a meromorphic extension to $\mathrm{Re}(s) > -n$ given by equation (10.3) with simple poles at $0, -1, -2, \ldots, -(n-1)$. As n is arbitrary, the proof is complete! See Figure 10.1. □

10.2 Corollary. *The residue of* $\Gamma(s)$ *at each simple pole* $s = -n$ *is given by*

$$\mathrm{Res}(-n) = \frac{(-1)^n}{n!}.$$

Indeed, by direct computation via (10.2)

$$\mathrm{Res}(-n) = \lim_{s \to -n} (s+n)\Gamma(s) = \frac{\Gamma(s+n+1)}{s(s+1)\cdots(s+n-1)} = \frac{(-1)^n}{n!},$$

which is also valid for $n = 0$.

10.3 Example. Note that for positive integers $n > m$,

$$\frac{d^m}{dx^m} z^n = \frac{n!}{(n-m)!} z^{n-m}$$

so that, at least formally, we can write

$$\frac{d^\mu}{dx^\mu} z^n = \frac{\Gamma(n+1)}{\Gamma(n-\mu+1)} z^{n-\mu},$$

which is a *fractional derivative*. More generally, if $f(z) - \sum_{n=0}^{\infty} a_n z^n$ is analytic, then the fractional derivative is given by

$$f^\mu(z) = \sum_{n=0}^{\infty} a_n \frac{\Gamma(n+1)}{\Gamma(n-\mu+1)} z^{n-\mu}, \quad z \neq 0,\ \mu \in R,$$

and $z^{n-\mu}$ is the principal branch of $e^{(n-\mu)\ln z}$, $-\pi < \operatorname{Im}(\log z) \leq \pi$, where it is also possible to consider $\mu \in \mathbb{C}$. However, we will not pursue this topic further. See Miller and Ross (1993) for more details about fractional derivatives.

The Gauss version of the gamma function is the *pi-function* $\Pi(s)$ given by

$$\Pi(s) = \Gamma(s+1) = \int_0^\infty t^s e^{-t}\,dt.$$

This led Gauss to the relation

$$\Gamma(s) = \lim_{n\to\infty} \frac{n! n^s}{s(s+1)\cdots(s+n)}, \tag{10.4}$$

which we will now prove as it is useful for establishing another characterization of the gamma function.[4]

First, we need to establish a lemma for the gamma function, which is not unexpected given the definition of the exponential function.

10.4 Lemma. *For* $\operatorname{Re}(s) > 0$, *the gamma function can be expressed as*

$$\Gamma(s) = \lim_{n\to\infty} \int_0^n \left(1 - \frac{t}{n}\right)^n t^{s-1}\,dt.$$

Proof. Define

$$\gamma_n(s) = \int_0^n \left(1 - \frac{t}{n}\right)^n t^{s-1}\,dt,$$

so that

$$\Gamma(s) - \gamma_n(s) = \int_0^n \left(e^{-t} - \left(1 - \frac{t}{n}\right)^n\right) t^{s-1}\,dt + \int_n^\infty e^{-t} t^{s-1}\,dt. \tag{10.5}$$

---

4 Of course, Euler had his own characterization of the gamma function for $s \neq 0, -1, -2, \ldots$:

$$\Gamma(s) = \frac{1}{s} \prod_{n=1}^{\infty} \frac{\left(1+\frac{1}{n}\right)^s}{\left(1+\frac{s}{n}\right)}.$$

As the integral for the gamma function converges for $\mathrm{Re}(s) > 0$, the second integral converges to zero as $n \to \infty$. As for the first integral, taking $0 \le x \le 1$, note that $(1+x) \le e^x \le \frac{1}{1-x}$, which gives

$$\left(1 - \frac{t}{n}\right)^n \le e^{-t} \le \left(1 + \frac{t}{n}\right)^{-n}$$

for $x = t/n$ and $0 \le t \le n$. Therefore, for the integrand of the first integral in (10.5),

$$\begin{aligned} 0 \le e^{-t} - \left(1 - \frac{t}{n}\right)^n &= e^{-t}\left(1 - e^t\left(1 - \frac{t}{n}\right)^n\right) \\ &\le e^{-t}\left(1 - \left(1 + \frac{t}{n}\right)^n\left(1 - \frac{t}{n}\right)^n\right) \\ &= e^{-t}\left(1 - \left(1 - \frac{t^2}{n^2}\right)^n\right). \end{aligned} \tag{10.6}$$

If $0 \le y \le 1$ and $ny < 1$, then we have $(1-y)^n \ge 1 - ny$, which holds trivially if $ny \ge 1$. Setting $y = t^2/n^2$ gives

$$1 - \left(1 - \frac{t^2}{n^2}\right)^n \le \frac{t^2}{n}$$

for $0 \le t \le n$. Hence, it follows from (10.6) that

$$0 \le e^{-t} - \left(1 - \frac{t}{n}\right)^n \le \frac{t^2}{n} e^{-t}.$$

We conclude that for $\mathrm{Re}(s) = \sigma > 0$,

$$\left| \int_0^n \left(e^{-t} - \left(1 - \frac{t}{n}\right)^n\right) t^{s-1} dt \right| \le \frac{1}{n} \int_0^n t^{\sigma-1} t^2 e^{-t} dt < \frac{1}{n}\Gamma(\sigma + 2) \to 0$$

as $n \to \infty$, proving the lemma. □

10.5 Theorem (Euler 1729, Gauss 1811). *For all $s \in \mathbb{C}$, $s \ne 0, -1, -2, \ldots$ the gamma function can be expressed as*

$$\Gamma(s) = \lim_{n\to\infty} \frac{n! n^s}{s(s+1)\cdots(s+n)}.$$

Proof. Using the substitution $\frac{t}{n} = v$ and integrating by parts, we have

$$\int_0^n \left(1 - \frac{t}{n}\right)^n t^{s-1} dt = n^s \int_0^1 (1-v)^n v^{s-1} dv$$

$$= n^s\left(\left((1-v)^n\frac{v^s}{s}\right)\Bigg|_0^1 + \frac{n}{s}\int_0^1 (1-v)^{n-1}v^s dv\right)$$

$$= n^s\left(\frac{n}{s}\int_0^1 (1-v)^{n-1}v^s dv\right). \tag{10.7}$$

To treat this last integral, repeated integration by parts leads to (exercise)

$$\int_0^1 (1-v)^{n-1}v^s dv = \frac{(n-1)(n-2)\cdots 1}{(s+1)(s+2)\cdots(s+n-1)}\int_0^1 v^{s+n-1}dv,$$

which coupled with (10.7) implies that

$$\int_0^n \left(1-\frac{t}{n}\right)^n t^{s-1}dt = \frac{n!n^s}{s(s+1)\cdots(s+n)}.$$

In view of the preceding lemma, taking the limit as $n \to \infty$ of both sides yields the result. □

This leads to a further characterization of the gamma function in terms of a Weierstrass product.

10.6 Theorem (Weierstrass 1856). *For all $s \in \mathbb{C}$, $s \neq 0, -1, -2, \ldots$,*

$$\frac{1}{\Gamma(s)} = se^{\gamma s}\prod_{n=1}^{\infty}\left(1+\frac{s}{n}\right)e^{-s/n},$$

where γ is the Euler–Mascheroni constant of Example 2.5.

Proof. By the preceding theorem

$$\begin{aligned}
\Gamma(s) &= \lim_{n\to\infty}\frac{n!n^s}{s(s+1)(s+2)\cdots(s+n)} \\
&= \lim_{n\to\infty}\frac{e^{s\log n}}{s(1+s)(1+\frac{s}{2})\cdots(1+\frac{s}{n})} \\
&= \lim_{n\to\infty}\frac{e^{s(\log n-1-\frac{1}{2}-\cdots-\frac{1}{n})}}{s}\,\frac{e^{s(1+\frac{1}{2}+\cdots+\frac{1}{n})}}{(1+s)(1+\frac{s}{2})\cdots(1+\frac{s}{n})} \\
&= \frac{e^{-\gamma s}}{s}\lim_{n\to\infty}\prod_{k=1}^{n}\left(1+\frac{s}{k}\right)^{-1}e^{s/k}.
\end{aligned} \tag{10.8}$$

Since the canonical product in (10.8) is nonzero in view of Example 2.5, the result follows by inverting the last expression. □

The *Euler reflection formula* is the following beautiful expression relating the gamma function to trigonometric functions.[5]

10.7 Proposition. *For all* $s \in \mathbb{C}$, $s \neq 0, \pm 1, \pm 2, \dots$,

$$\Gamma(s)\Gamma(1-s) = \frac{\pi}{\sin \pi s}.$$

Proof. By Theorem 10.5

$$\begin{aligned}\Gamma(s)\Gamma(-s) &= \lim_{n\to\infty} \frac{n!n^s}{s(s+1)\cdots(s+n)} \frac{n!n^{-s}}{-s(-s+1)\cdots(-s+n)}\\ &= \lim_{n\to\infty} \frac{1}{-s^2 \prod_{k=1}^{n}(1+\frac{s}{n})(1-\frac{s}{n})}\\ &= -\frac{\pi}{s\sin \pi s}\end{aligned}$$

by the Euler product for the sine in Example 2.7 of Chapter 2. Since $\Gamma(s+1) = s\Gamma(s)$ (for s as above we can establish this via (10.8) – exercise), implying that

$$\Gamma(s)\Gamma(1-s) = \Gamma(s)(-s)\Gamma(-s) = \frac{\pi}{\sin \pi s},$$

as desired. Note that the formula holds for all s where both sides become infinite. □

A trivial but noteworthy consequence of the preceding result is that for $s = \frac{1}{2}$,

$$\Gamma\left(\frac{1}{2}\right) = \sqrt{\pi}.$$

Observe that here we link a generalization of the notion of factorial with the ratio of the circumference of a circle to its diameter. This illustrates that something quite extraordinary is going on in the netherworld of complex analysis.

Furthermore,

$$\frac{\sqrt{\pi}}{2} = \frac{1}{2}\int_0^\infty e^{-t}t^{-1/2}dt,$$

and with the substitution $t = x^2$, via symmetry of the resulting function, we obtain

$$\sqrt{\pi} = \int_{-\infty}^{\infty} e^{-x^2}dx. \tag{10.9}$$

5 The Euler reflection formula can also be derived from the canonical product $P(z)$ and Proposition 2.6 of Chapter 2 coupled with (2.9). Exercise.

The bell-shaped curve

$$f(x) = \frac{1}{\sqrt{2\pi}} e^{-\frac{1}{2}x^2}$$

is the *probability density function* of a *standard normal distribution*. From the preceding we conclude that

$$\frac{1}{\sqrt{2\pi}} \int_{-\infty}^{\infty} e^{-\frac{1}{2}x^2} dx = 1.$$

Beta function

We now derive a function of two complex variables from the gamma function introduced by Euler and studied by Legendre. The resulting beta function appears in statistics, computing, and particle physics, including string theory.[6]

For $\mathrm{Re}(z) > 0$ and $\mathrm{Re}(w) > 0$, consider the product

$$\Gamma(z)\Gamma(w) = \int_0^{\infty} e^{-\tau}\tau^{z-1}d\tau \cdot \int_0^{\infty} e^{-t}t^{w-1}dt = \int_0^{\infty}\int_0^{\infty} e^{-\tau-t}\tau^{z-1}t^{w-1}d\tau dt.$$

Substituting $\tau = y^2$ and $t = x^2$ yields

$$\Gamma(z)\Gamma(w) = 4\int_0^{\infty}\int_0^{\infty} e^{-(x^2+y^2)}y^{2z-1}x^{2w-1}dxdy.$$

Using polar coordinates $x = r\cos\theta$, $y = r\sin\theta$, we get

$$\begin{aligned}
\Gamma(z)\Gamma(w) &= 4\int_0^{\infty}\int_0^{\pi/2} e^{-r^2}r^{2(z+w)-1}\sin^{2z-1}\theta\cos^{2w-1}\theta drd\theta \\
&= \left(2\int_0^{\infty} e^{-r^2}r^{2(z+w)-1}dr\right)\left(2\int_0^{\pi/2}\sin^{2z-1}\theta\cos^{2w-1}\theta d\theta\right) \\
&= \Gamma(z+w)\left(\int_0^{\pi/2} 2\sin^{2z-1}\theta\cos^{2w-1}\theta d\theta\right). \qquad (10.10)
\end{aligned}$$

6 Quite remarkably, two centuries later, Euler's beta function was used to describe interactions in elementary particle physics. The *dual resonance model* of Gabriele Veneziano (in work between 1968–1973) only made physical sense with one-dimensional strings instead of zero-dimensional points, that led to the birth of string theory, which is now one of the leading theories attempting to describe the workings of the Universe at its smallest scale. See Riddhi D., Beta function and its applications, http://sces.phys.utk.edu/~moreo/mm08/Riddi.pdf

Define the *beta function* (*Euler integral of the first kind*) as the latter integral:

$$B(z,w) = 2\int_0^{\pi/2} \sin^{2z-1}\theta\cos^{2w-1}\theta d\theta = 2\int_0^{\pi/2} \cos^{2z-1}\theta\sin^{2w-1}\theta d\theta = B(w,z)$$

$$= \int_0^1 y^{z-1}(1-y)^{w-1}dy, \tag{10.11}$$

where the last integral equals the first one via the substitution $y = \sin^2\theta$. Thus by equation (10.10) we have established *Euler's formula for the beta function*:

$$B(z,w) = B(w,z) = \frac{\Gamma(z)\Gamma(w)}{\Gamma(z+w)}.$$

So, for example,

$$B\left(\frac{1}{2},\frac{1}{2}\right) = \pi,$$

and for $0 < \mathrm{Re}(z) < 1$ (and actually valid for $z \neq 0, \pm1, \pm2, \dots$)

$$B(z,1-z) = \frac{\pi}{\sin \pi z}$$

by the Euler reflection formula (10.7). For positive integers m, n, we have

$$B(m,n) = \frac{(m-1)!(n-1)!}{(m+n-1)!}.$$

Furthermore, formula (10.10) with $z = 1/2$ and $w = 3/2$ gives

$$\Gamma\left(\frac{3}{2}\right) = \frac{2}{\sqrt{\pi}}\int_0^{\pi/2}\cos^2\theta d\theta = \frac{\sqrt{\pi}}{2}.$$

This is a particular case of the following formula for $\Gamma(z+\frac{1}{2})$.

10.8 Legendre duplication formula (1809). For $\mathrm{Re}(z) > 0$ and $\mathrm{Re}(w) > 0$,

$$\Gamma(2z) = \frac{2^{2z-1}}{\sqrt{\pi}}\Gamma(z)\Gamma\left(z+\frac{1}{2}\right).$$

Proof. From Euler's formula for the beta function and representation (10.11) we have

$$B(z,w) = \frac{\Gamma(z)\Gamma(w)}{\Gamma(z+w)} = \int_0^1 y^{z-1}(1-y)^{w-1}dy.$$

Letting $w = z$, this becomes

$$\frac{\Gamma^2(z)}{\Gamma(2z)} = \int_0^1 y^{z-1}(1-y)^{z-1}dy.$$

The substitution $y = \frac{1+x}{2}$ gives

$$\frac{\Gamma^2(z)}{\Gamma(2z)} = \frac{1}{2}\int_{-1}^{1}\left(\frac{1+x}{2}\right)^{z-1}\left(\frac{1-x}{2}\right)^{z-1}dx = \frac{1}{2^{2z-1}}\int_{-1}^{1}(1-x^2)^{z-1}dx = \frac{2}{2^{2z-1}}\int_0^1(1-x^2)^{z-1}dx,$$

that is,

$$2^{2z-1}\Gamma^2(z) = 2\Gamma(2z)\int_0^1(1-x^2)^{z-1}dx. \tag{10.12}$$

Again by (10.11),

$$B\left(\frac{1}{2}, z\right) = \int_0^1 y^{-1/2}(1-y)^{z-1}dy,$$

and by the substitution $y = t^2$

$$B\left(\frac{1}{2}, z\right) = 2\int_0^1(1-t^2)^{z-1}dt.$$

As a consequence, from (10.12) and $\Gamma(\frac{1}{2}) = \sqrt{\pi}$ we infer that

$$2^{2z-1}\Gamma^2(z) = \Gamma(2z)B\left(\frac{1}{2}, z\right) = \Gamma(2z)\sqrt{\pi}\frac{\Gamma(z)}{\Gamma(z+\frac{1}{2})},$$

which establishes the Legendre formula. □

An immediate consequence is the following for n a nonegative integer:

10.9 Corollary.

$$\Gamma\left(n+\frac{1}{2}\right) = \frac{\Gamma(\frac{1}{2})(2n)!}{2^{2n}n!}.$$

10.10 Exercise. The *Wallis integrals* are defined as

$$W_n = \int_0^{\frac{\pi}{2}} \sin^n\theta d\theta = \int_0^{\frac{\pi}{2}} \cos^n\theta d\theta.$$

Show that
(a) $W_n = \frac{1}{2}B(\frac{n+1}{2}, \frac{1}{2})$,
(b) $W_{2n+1} = \frac{2^{2n} n!^2}{(2n+1)!}$,
(c) $W_{2n} = \frac{(2n)!}{2^{2n} n!^2} \cdot \frac{\pi}{2}$.

Riemann zeta function

The famous Riemann zeta function has at its source the ordinary p-series from calculus

$$\frac{1}{1^p} + \frac{1}{2^p} + \frac{1}{3^p} + \frac{1}{4^p} + \frac{1}{5^p} + \cdots,$$

which converges to some finite value for $p > 1$. At $p = 1$, we obtain the harmonic series that diverges to infinity. For $p = 2$, the series converges to $\frac{\pi^2}{6}$, for $p = 4$, the series converges to $\frac{\pi^4}{90}$, and for $p = 6$, the sum is $\frac{\pi^6}{945}$.[7]

The first step is replacing the power p with a complex number power s (which is the traditional letter). So, the series now reads

$$\zeta(s) = \sum_{n=1}^{\infty} \frac{1}{n^s}. \tag{10.13}$$

Here $n^s = e^{s \log n}$, $n = 1, 2, 3, \ldots$, and each term of the series is an entire function. For this series to converge, we must have $\mathrm{Re}(s) > 1$, which is analogous to $p > 1$.[8]

In fact, for any $\varepsilon > 0$ and $\sigma = \mathrm{Re}(s) \geq 1 + \varepsilon$,

$$\left|\frac{1}{n^s}\right| = \frac{1}{n^\sigma} \leq \frac{1}{n^{1+\varepsilon}}.$$

By the Weierstrass M-test we conclude that the series in (10.13) converges absolutely and uniformly in $\sigma \geq 1+\varepsilon$, and hence $\zeta(s)$ represents an analytic function in $\mathrm{Re}(s) > 1$. It is interesting that in view of the absolute convergence for $\sigma \geq 2$,

$$|\zeta(s)| \leq \sum_{n=1}^{\infty} \frac{1}{n^\sigma} \leq \sum_{n=1}^{\infty} \frac{1}{n^2} = \frac{\pi^2}{6},$$

which is the value of $\zeta(s)$ at $s = 2$.

A closely related function is the *Dirichlet eta function* (*alternating zeta function*) given by

---

7 These values for the p-series can be established via Fourier series for an appropriate function.

8 The series representation in (10.13) is a particular case of a *Dirichlet series* $\sum_{n=1}^{\infty} \frac{a_n}{n^s}$ with complex s and a_n.

$$\eta(s) = \sum_{n=1}^{\infty} \frac{(-1)^{n-1}}{n^s},$$

which converges for $\mathrm{Re}(s) > 0$. To see this, we prove a more general result about Dirichlet series of the form $\sum_{n=1}^{\infty} \frac{a_n}{n^s}$ for $\sigma = \mathrm{Re}(s) \geq \varepsilon > 0$ and $S_n = a_1 + a_2 + \cdots + a_n$ satisfying $|S_n| \leq C < \infty$. To this end, note that for $m > n \geq n_0$, we can write (exercise)

$$\sum_{k=n+1}^{m} \frac{a_k}{k^s} = \sum_{k=n+1}^{m} \frac{S_k - S_{k-1}}{k^s} = \frac{S_m}{(m+1)^s} - \frac{S_n}{(n+1)^s} + \sum_{k=n+1}^{m} S_k\left(\frac{1}{k^s} - \frac{1}{(k+1)^s}\right).$$

Therefore

$$\begin{aligned}
\left|\sum_{k=n+1}^{m} \frac{a_k}{k^s}\right| &\leq \frac{C}{m^\sigma} + \frac{C}{n^\sigma} + \sum_{k=n+1}^{m} |S_k| \left|\frac{1}{k^s} - \frac{1}{(k+1)^s}\right| \\
&\leq \frac{2C}{n^\varepsilon} + C \sum_{k=n+1}^{m} \left|\int_k^{k+1} \frac{s}{x^{s+1}} dx\right| \\
&\leq \frac{2C}{n^\varepsilon} + C|s| \sum_{k=n+1}^{m} \frac{1}{k^{1+\sigma}} \\
&\leq \frac{2C}{n^\varepsilon} + C|s| \int_n^m \frac{1}{x^{1+\varepsilon}} dx \\
&\leq \frac{2C}{n^\varepsilon} + C|s| \frac{1}{\varepsilon n^\varepsilon} \leq C\left(2 + \frac{|s|}{\varepsilon}\right)\frac{1}{n_0^\varepsilon}.
\end{aligned}$$

We conclude that the series $\sum_{n=1}^{\infty} \frac{a_n}{n^s}$ under the above conditions converges uniformly on compact subsets of $\mathrm{Re}(s) > 0$ and hence defines an analytic function. In particular, the result holds for the Dirichlet eta series where $a_n = (-1)^{n-1}$ and so $|S_n| \leq 1$.

The eta function is related to the zeta function by the following formula for $\mathrm{Re}(s) > 1$:

$$\eta(s) = (1 - 2^{1-s})\zeta(s).$$

Indeed, since the absolute value of the terms of $\eta(s)$ coincide with those of $\zeta(s)$, the series for $\eta(s)$ converges absolutely for $\mathrm{Re}(s) > 1$ and hence converges. Thus, rearranging the terms, we have

$$\begin{aligned}
\eta(s) &= 1^{-s} - 2^{-s} + 3^{-s} - 4^{-s} + \cdots \\
&= (1^{-s} + 2^{-s} + 3^{-s} + 4^{-s} + \cdots) - 2(2^{-s} + 4^{-s} + 6^{-s} \ldots) \\
&= \zeta(s) - 2 \cdot 2^{-s}(1^{-s} + 2^{-s} + 3^{-s} + \cdots) \\
&= (1 - 2^{1-s})\zeta(s),
\end{aligned}$$

establishing the result.

Note that the zeros of $\eta(s)$ occur at all the zeros of $\zeta(s)$, including all the points where $s = 1 - 2^{1-s} = 0$, that is, where $s = 1 - \frac{2k\pi i}{\ln 2}$, for all integers $k \neq 0$. These latter are arrayed equidistantly along the line $\mathrm{Re}(s) = 1$.

Riemann functional equation

We now proceed to establish the values of the Riemann zeta function in the entire complex plane via a remarkable functional equation established by Riemann (1859). We start with the gamma function and make the substitution $t = nx$, so that

$$\Gamma(s) = \int_0^\infty t^{s-1} e^{-t} dt = n^s \int_0^\infty x^{s-1} e^{-nx} dx,$$

that is,

$$\frac{\Gamma(s)}{n^s} = \int_0^\infty x^{s-1} e^{-nx} dx.$$

As a consequence,

$$\Gamma(s)\zeta(s) = \sum_{n=1}^\infty \int_0^\infty x^{s-1} e^{-nx} dx = \int_0^\infty x^{s-1} \sum_{n=1}^\infty e^{-nx} dx,$$

where reversing the order of summation and integration is a consequence of the convergence of $\sum \int_0^\infty |x^{s-1} e^{-nx}| dx$. Therefore we have established a wonderful relationship between the gamma function and zeta function:[9]

$$\Gamma(s)\zeta(s) = \int_0^\infty \frac{x^{s-1}}{e^x - 1} dx \quad (\mathrm{Re}(s) > 1). \tag{10.14}$$

To treat the preceding integral and avoid the issue of integrability at the origin, let Γ be the contour depicted in Figure 10.2 ensuring that the circular curve γ_r is sufficiently small so as to exclude any of the poles at $\xi = \pm 2n\pi i$ for $n \neq 0$. Consider the integral[10]

9 Note that this formulation is the *Mellin transform* of the function $f(x) = \frac{1}{e^x - 1}$. For $s = 2$, we have

$$\frac{\pi^2}{6} = \int_0^\infty \frac{x}{e^x - 1} dx.$$

10 Taking the principal branch of $\log z$, note that on $[r, \infty)$, $z^{s-1} = e^{(s-1)(\log|z| + i \operatorname{Arg} z)}$, where $\operatorname{Arg} z = 0$ on the upper edge of the branch cut, and $\operatorname{Arg} z = 2\pi$ on the lower edge.

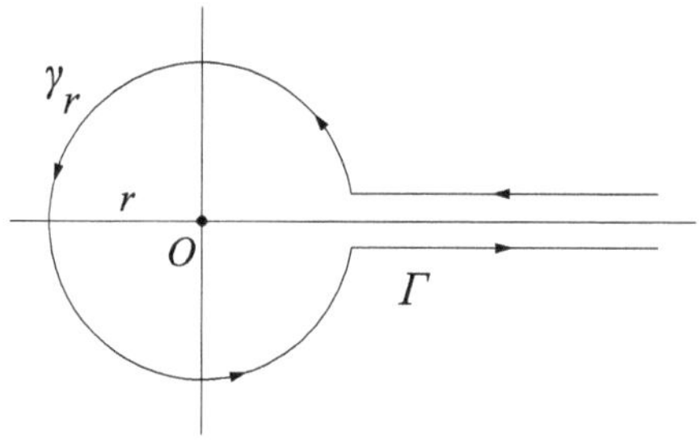

Figure 10.2: The contour Γ for the evaluation of the integral $I(s)$. Image courtesy Katy Metcalf.

$$I(s) = \int_{\Gamma} \frac{z^{s-1}}{e^z - 1} dz = -\int_r^{\infty} \frac{x^{s-1}}{e^x - 1} dx + \int_{\gamma_r} \frac{z^{s-1}}{e^z - 1} dz + \int_r^{\infty} \frac{x^{s-1} e^{2\pi i s}}{e^x - 1} dx. \tag{10.15}$$

It is left as an exercise to prove that this definition of $I(s)$ is *independent of the choice of* $r > 0$ by taking $0 < r' < r$ and showing that the integral over the contour $\Gamma - \Gamma'$ is zero by Cauchy's theorem.[11] Moreover, $I(s)$ is an entire function.[12]

For the integral over γ_r and $\sigma = \mathrm{Re}(s) > 1$, we can write $e^z - 1 = zg(z)$ for $g(z)$ analytic and $g(0) = 1$. Then for z on γ_r and all $r > 0$ sufficiently small, we will have $|g(z)| \geq 1/2$, implying

$$\frac{1}{|e^z - 1|} \leq \frac{2}{r}.$$

It follows that[13] as $r \to 0$

$$\left| \int_{\gamma_r} \frac{z^{s-1}}{e^z - 1} dz \right| \leq M \int_0^{2\pi} r^{\sigma-1} d\theta = M r^{\sigma-1} \to 0.$$

By (10.14) and (10.15) we conclude that for $\mathrm{Re}(s) > 1$,

$$I(s) = (e^{2\pi i s} - 1) \int_0^{\infty} \frac{x^{s-1}}{e^x - 1} dx = (e^{2\pi i s} - 1)\Gamma(s)\zeta(s). \tag{10.16}$$

Now $I(s)$ is an entire function with the simple zeros of $(e^{2\pi i s} - 1)$ at the points at $s = 0, -1, -2, \ldots$, exactly cancelling the simple poles of $\Gamma(s)$. Since we already know

---

11 The contour Γ' is the curve corresponding to the circle of radius r'.

12 "The contour integral is in fact absolutely convergent for any $s \in \mathbb{C}$ and from the usual argument involving the Cauchy, Fubini, and Morera theorems we see that this integral depends holomorphically on s." (Exercise). From: https://terrytao.wordpress.com/2014/12/15/254a-supplement-3-the-gamma-function-and-the-functional-equation-optional/

13 Observe that for $s = \sigma + i\tau$, $z = re^{i\theta}$, then $|z^{s-1}| = r^{\sigma-1} e^{-\tau\theta} = O(r^{\sigma-1})$.

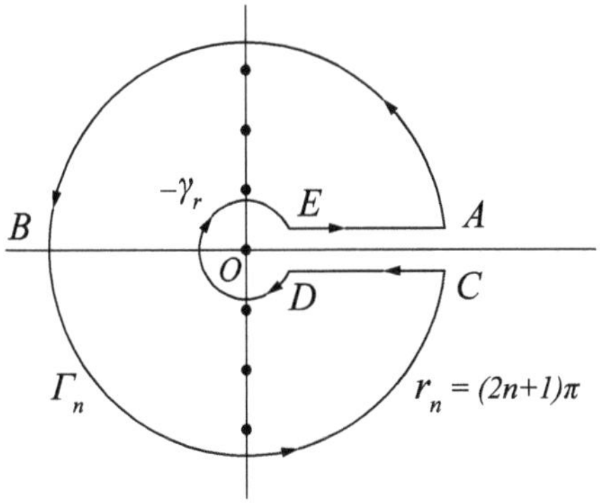

Figure 10.3: The contour ABCDEA enclosing the poles with a branch cut along the positive real axis. Image courtesy Katy Metcalf.

that $\zeta(s)$ is analytic for $\mathrm{Re}(s) > 1$, *we use* (10.16) *to provide an analytic continuation of* $\zeta(s)$ *to* $\mathrm{Re}(s) \le 1$ except when $(e^{2\pi is} - 1)\Gamma(s) = 0$ in this half-plane, which is only at the point $s = 1$. This sole point is a simple pole of $\zeta(s)$ with

$$\mathrm{Res}(s = 1) = \lim_{s\to 1}(s - 1)\frac{I(s)}{(e^{2\pi is} - 1)\Gamma(s)} = \frac{I(1)}{2\pi i}.$$

From the definition of $I(s)$ we have for $s = 1$,

$$I(1) = \int_{\gamma_r} \frac{1}{e^z - 1}dz = 2\pi i$$

by the residue theorem in this instance applied to the integrand. It follows that $\mathrm{Res}(s = 1) = 1$ at the simple pole of $\zeta(s)$, which is analytic at every other point of $\mathbb{C}$, which we state formally as the following:

10.11 Theorem. *The Riemann zeta function* $\zeta(s)$ *is an analytic function in the complex plane except for a simple pole at the point* $s = 1$ *with* $\mathrm{Res}(s = 1) = 1$. *Moreover, equality* (10.16) *holds for all* $s \in \mathbb{C}$.

To establish the celebrated functional equation, let us take a branch cut along the positive real axis and integrate a branch of our function $z^{s-1}/(e^z - 1)$ over contour *ABCDEA* as in Figure 10.3, with the radius of Γ_n being $r_n = (2n + 1)\pi$, $n = 1, 2, 3, \ldots$, and with fixed radius $0 < r < 2\pi$ of γ_r (see Figure 10.3) and $\sigma = \mathrm{Re}(s) < 0$. The latter condition is only relevant near the end of our deliberations.

In this instance, we have simple poles in the annular region enclosed by the contour ABCDEA at the points $\xi = \pm 2k\pi i$, $k = 1, 2, \ldots, n$, indicated by the dots in Figure 10.3. A direct calculation, which we leave to the reader, gives

$$\mathrm{Res}(2k\pi i) = -i(2k\pi)^{s-1}e^{\frac{i\pi s}{2}}, \quad \mathrm{Res}(-2k\pi i) = i(2k\pi)^{s-1}e^{3i\pi s/2}.$$

As a consequence of the residue theorem,

$$\int_{ABCDEA} \frac{z^{s-1}}{e^z - 1} dz = (2\pi)^s e^{\frac{i\pi s}{2}} \sum_{k=1}^{n} \frac{1}{k^{1-s}} - (2\pi)^s e^{\frac{3i\pi s}{2}} \sum_{k=1}^{n} \frac{1}{k^{1-s}}$$

$$= -2i(2\pi)^s e^{i\pi s} \sin\left(\frac{\pi s}{2}\right) \sum_{k=1}^{n} \frac{1}{k^{1-s}}. \tag{10.17}$$

As is normally done with contour integrals, we now show that $\int_{\mathrm{ABC}=\Gamma_n} \to 0$ as $n \to \infty$. To this end, we first show the following:

10.12 Lemma. *The function* $f(z) = \frac{1}{e^z-1}$ *is uniformly bounded for* $z \in \Gamma_n$ *with radius* $r_n = (2n+1)\pi$, $n = 1, 2, 3, \ldots$.

Proof. To achieve our result, we need to find a lower found for $|e^z - 1|$.

Case (i) $\mathrm{Re}(z) \geq 1$: In this instance, $|e^z - 1| \geq e - 1$.

Case (ii) $\mathrm{Re}(z) \leq -1$: Then $|e^z - 1| \geq 1 - \frac{1}{e}$.

Case (iii) $-1 < \mathrm{Re}(z) < 1$: Since the function $f(z) = \frac{1}{e^z-1}$ is $2\pi i$-periodic, we need only consider its behavior in the strip

$$S = \{z = x + iy : -1 < x < 1, 2n\pi \leq y < 2(n+1)\pi\}$$

as in Figure 10.4.

Let us avoid the pole at any point $\xi = 2n\pi i$ by taking a circle of radius $r = 1$ about it. Then the function $f(z) = \frac{1}{e^z-1}$ is analytic in the shaded region R of S in Figure 10.4, external to any such circle about a pole and hence bounded there, say $|f(z)| \leq M$ for $z \in R$. Therefore for $z \in \Gamma_n \cap R$, we have $|f(z)| \leq M$, and by periodicity and (i) and (ii) we conclude that $f(z)$ is uniformly bounded for all $z \in \Gamma_n$ for $n = 1, 2, 3, \ldots$, proving the lemma. □

Rewriting equality (10.17) as

$$\int_{\Gamma_n} \frac{z^{s-1}}{e^z - 1} dz + \int_{CD} \frac{x^{s-1} e^{2\pi i s}}{e^x - 1} dx + \int_{-\gamma_r} \frac{z^{s-1}}{e^z - 1} dz + \int_{EA} \frac{x^{s-1}}{e^x - 1} dx$$

$$= -2i(2\pi)^s e^{i\pi s} \sin\left(\frac{\pi s}{2}\right) \sum_{k=1}^{n} \frac{1}{k^{1-s}}, \tag{10.18}$$

we see that the very first term vanishes since

$$\left| \int_{\Gamma_n} \frac{z^{s-1}}{e^z - 1} dz \right| \leq C[(2n+1)\pi]^\sigma \to 0$$

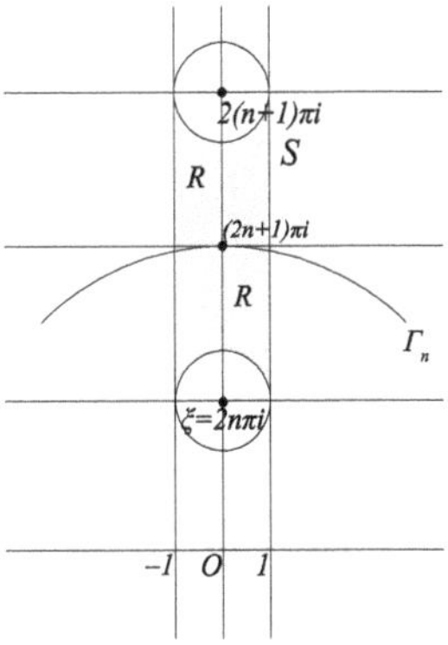

Figure 10.4: A strip S containing the region R for $-1 < \operatorname{Re}(z) < 1$, which includes an arc of the curve Γ_n. Image not to scale. Courtesy Katy Metcalf.

as $n \to \infty$, invoking the condition $\sigma = \operatorname{Re}(s) < 0$. As well, the last term of (10.18) converges to $-2i(2\pi)^s e^{i\pi s} \sin \frac{\pi s}{2} \zeta(1-s)$ as $n \to \infty$. Regarding the convergence of the left-hand side of (10.18) and taking the limit as $n \to \infty$ for the remaining terms,

$$\int_{CD} \frac{x^{s-1} e^{2\pi i s}}{e^x - 1} dx + \int_{-\gamma_r} \frac{z^{s-1}}{e^z - 1} dz + \int_{EA} \frac{x^{s-1}}{e^x - 1} dx \to -I(s)$$

in view of (10.15), since the corresponding direction of integration in this instance is the opposite to that of the previous case. Thus via (10.16), (10.18), and the preceding considerations, we arrive at

$$I(s) = (e^{2\pi i s} - 1)\Gamma(s)\zeta(s) = 2i(2\pi)^s e^{i\pi s} \sin\left(\frac{\pi s}{2}\right)\zeta(1-s) \tag{10.19}$$

for $\operatorname{Re}(s) < 0$. By the identity theorem equality (10.19) remains valid for all $s \in \mathbb{C}$. We have nearly reached our goal. Since $e^{2\pi i s} - 1 = 2ie^{i\pi s} \sin \pi s$ and by Proposition 10.7 we have $\Gamma(s) = \pi/\Gamma(1-s)\sin \pi s$, we conclude from equality (10.19) that

$$\zeta(s) = 2^s \pi^{s-1} \sin\left(\frac{\pi s}{2}\right)\Gamma(1-s)\zeta(1-s)$$

for all $s \in \mathbb{C}$. Moreover, this equation demonstrates how the zeta function takes values in the left-half plane $\operatorname{Re}(s) < 0$ directly from the values at its reflected counterpart $1-s$ in the half-plane $\operatorname{Re}(s) > 1$. We restate this absolutely magnificent result.

10.13 Riemann functional equation. *For all complex values $s \in \mathbb{C}$, we have*[14]

$$\zeta(s) = 2^s \pi^{s-1} \sin\left(\frac{\pi s}{2}\right)\Gamma(1-s)\zeta(1-s),$$

where $\zeta(s)$ is analytic in $\mathbb{C} - \{1\}$ with a simple pole at $s = 1$ and simple zeros at the points $-2, -4, -6, \ldots$.

Observe that on the right-hand side of the Riemann functional equation, at $s = 2n$, for $n = 1, 2, 3, \ldots$, the factor $\sin(\frac{\pi s}{2})$ has simple zeros that cancel the simple poles of $\Gamma(1-s)$ at those points. We have already seen the values of $\zeta(s)$ at $s = 2, 4$, and 6, namely $\frac{\pi^2}{6}$, $\frac{\pi^4}{90}$, and $\frac{\pi^6}{90}$, respectively. A simple calculation shows that $\zeta(0) = -\frac{1}{2}$. At $s = 2n + 1$, $n = 0, 1, 2, \ldots$, the factor $\zeta(1 - s)$ must have simple zeros that cancel the remaining poles of $\Gamma(1 - s)$ except, of course, for the pole at $s = 1$, which is shared by both sides. Furthermore, we find that $\zeta(s)$ has simple zeros at $s = -2n$, $n = 1, 2, \ldots$, and these are called the *trivial zeros* of the zeta function and are of no particular interest. On the other hand, there are other zeros of the Riemann zeta function at infinitely many other points in the complex plane. Riemann himself calculated a few of these zeros and noted that they all have real part $\sigma = 1/2$. See Figures 10.5 and 10.6 for the location of the first few of these "nontrivial" zeros. To keep track of the known (nontrivial) zeros, we write them as

$$s_n = \frac{1}{2} + it_n,$$

where the real part is $\frac{1}{2}$, and the imaginary part will be some decimal number that increases the more zeros we calculate. Due to the symmetric nature of the Riemann zeta function, each of these zeros has a counterpart with negative imaginary part, but since we know that they are always there, we need only focus on those with positive imaginary parts. To visualize the Riemann zeta function $\zeta(s)$ we consider the values $|\zeta(s)|$ for $s \in \mathbb{C}$. A portion of the real values near the *critical line* $x = 1/2$ is depicted in Figure 10.5 with the simple pole at $s = 1$. Note the zeros where the function dips down to the complex plane along the critical line as determined in Figure 10.6.

---

14 There are various other formulations of this result and Riemann's original one in his landmark paper, On the number of prime numbers less than a given quantity, *Monatsberichte der Berliner Akademie, 1859*, is of a slightly different form. His proof also used the Euler product formula and an identity for Jacobi theta functions, whereas our proof is more elementary but retains the flavor of the original.

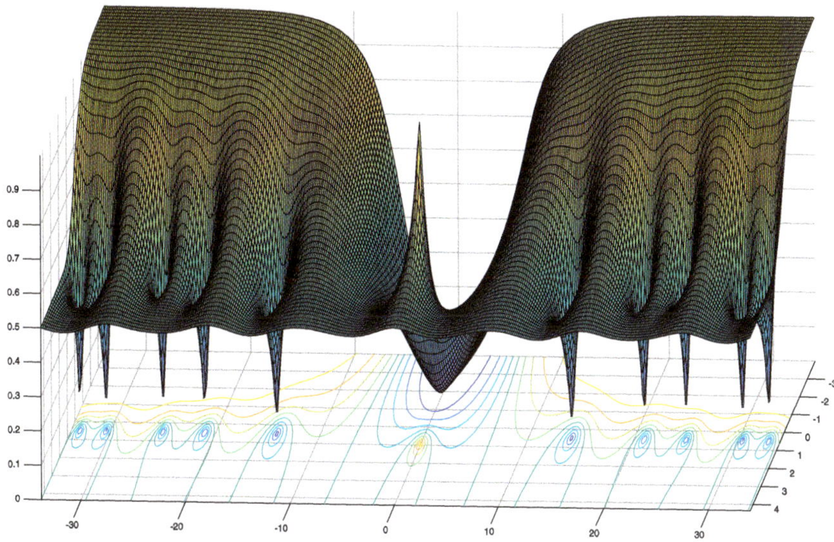

Figure 10.5: A plot of a small region for the Riemann zeta function. The vertical axis represents $|\zeta(s)|$ for $-3 \leq x \leq 4$ and $-30 \leq y \leq 30$. The imaginary axis runs horizontally. The sharp upward peak is at the point $s = 1$, where the function $\zeta(s)$ has a pole. Courtesy Chris King.

Riemann hypothesis

Riemann's great conjecture, the Riemann hypothesis (RH), which the Clay Mathematics Institute has put US$1 million price on its resolution, is that *all* the (nontrivial) zeros s_n of the zeta function lie on the critical line $\sigma = 1/2$ (Figure 10.6). According to the 20th century mathematical giant David Hilbert, "*If I were to awaken after having slept for a thousand years, my first question would be: Has the Riemann hypothesis been proven?*" Indeed, the conjecture is without doubt one of the most important unresolved questions in mathematics, certainly one of the most famous, and since it is still unresolved since 1859, one of the most difficult. See the fine account by Derbyshire (2003).

Prime number theorem

Although it seems rather obscure, RH has deep connections with the distribution of prime numbers, as well as connections with quantum mechanics. A lot of historical work has gone into understanding the distribution of prime numbers, which culminated in the prime number theorem:[15]

$$\pi(x) \sim \frac{x}{\ln x},$$

15 Conjectured by Adrien-Marie Legendre in 1798, but of course also known to Gauss.

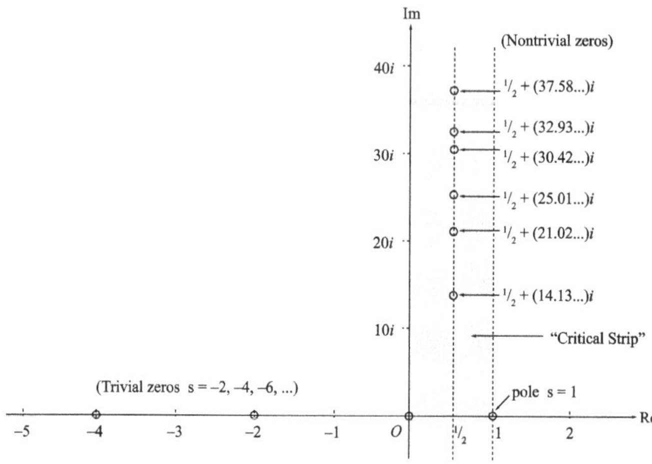

Figure 10.6: A depiction of the first few trivial and nontrivial zeros of the Riemann zeta function, with the latter all on the vertical line $\sigma = 1/2$. Courtesy Katy Metcalf/Springer.

where $\pi(x)$ is the number of primes less than or equal to x. In other words,

$$\lim_{x\to\infty} \frac{\pi(x)}{x/\ln x} = 1.$$

The first formal proofs were given by Jacques Hadamard and Charles de la Vallée Poussin in 1896. As it turns out, there is an explicit formula due to Riemann for determining the number of prime numbers up to any given number and measuring their distribution. However, to utilize the formula, we have to take an infinite sum over the nontrivial zeros of $\zeta(s)$ all of which must lie on the critical line $x = 1/2$. Some progress over the years has been made by various mathematicians. The distinguished English mathematician G. H. Hardy (1877–1947) proved that there are infinitely many zeros on the critical line. On the other hand, the mathematical scientist Andrew Odlyzko using a computer showed that the first ten trillion zeros of the zeta function were all on the critical line. Of course, to disprove RH, we only have to find a single zero not on the critical line. Thus far (2022), it has been shown that at least two-fifths of the nontrivial zeros lie on the critical line.[16]

Relation to quantum mechanics

Another approach to probing the validity of RH lies in the realm of quantum mechanics. Evidence suggests a correspondence between the distribution of the zeros of the

16 J. Conrey, At least two fifths of the zeros of the Riemann zeta function lie on the critical line, *Bull. Amer. Math. Soc.* 20 (1989), 79–81.

Riemann zeta function along the critical line and the distribution of the eigenvalues of a random Hermitian matrix (the Hilbert–Pólya conjecture). Such matrices are directly related to quantum mechanical states so that if a quantum system can be found such that the nontrivial zeros s_n of $\zeta(s)$ correspond to the energy levels E_n of the system with

$$s_n = \frac{1}{2} + iE_n,$$

then RH would be proved. The reason is because the energy levels E_n are all actual real numbers, and therefore all the zeros s_n must necessarily lie on $\mathrm{Re}(s) = 1/2$. Alternatively, there are various other mathematical results equivalent to the Riemann hypothesis, which, if proven, would establish the validity of RH. In this regard, we will discuss the Mertens conjecture in the sequel. Let us digress first and put the value $s = -1$ into the Riemann zeta function given by Theorem 10.13:

$$\zeta(-1) = 2^{-1}\pi^{-2}\sin\left(\frac{-\pi}{2}\right)\Gamma(2)\zeta(2).$$

Since $\sin(\frac{-\pi}{2}) = -1$, $\Gamma(2) = 1$, and $\zeta(2) = \pi^2/6$, we find that

$$\zeta(-1) = \frac{1}{2}\frac{1}{\pi^2}(-1)(1)\left(\frac{\pi^2}{6}\right) = -1/12.$$

Next, suppose that we consider the original definition (10.13) of the Riemann zeta function, that is,

$$\zeta(s) = \frac{1}{1^s} + \frac{1}{2^s} + \frac{1}{3^s} + \frac{1}{4^s} + \frac{1}{5^s} + \cdots,$$

which was only valid for $\mathrm{Re}(s) > 1$. Throwing all caution to the wind, let us fudge matters just a little and put the value $s = -1$ into this formula. Then we arrive at the ridiculously absurd conclusion that

$$1 + 2 + 3 + \cdots = -1/12.$$

On the other hand, the genius Indian mathematician Srinivasa Ramanujan using a technique known as Ramanujan summation arrived at exactly the same result.[17]

17 The *Ramanujan sum* of the series can be derived from the *Ramanujan summation formula*

$$1 + 2 + 3 + \cdots = \sum_{n=1}^{\infty} f(n) = -\frac{1}{2}f(0) - \sum_{n=1}^{\infty} \frac{B_{2n}}{(2n)!} f^{(2n-1)}(0) = -\frac{1}{12}$$

with $f(x) = x$, the Bernoulli numbers B_{2n}, and $B_2 = 1/6$. From: Bruce C. Berndt, *Ramanujan's Notebooks Part 1*, Springer-Verlag, 1985, pp. 134–135.

Moreover, this result also appears in a book on string theory,[18] one of the foremost theories attempting to bridge the gap between quantum mechanics and the theory of relativity. Shakespeare's Hamlet comes to mind: "There are more things in Heaven and Earth, Horatio, that are dreamt of in your philosophy."

Euler product formula

There is a beautiful connection between the Riemann zeta function and the set of all prime numbers. It is a wonderful equality known as the *Euler product formula*:

$$\zeta(s) = \prod_{p \text{ prime}} \frac{1}{1-p^{-s}} = \left(\frac{1}{1-2^{-s}}\right)\left(\frac{1}{1-3^{-s}}\right)\left(\frac{1}{1-5^{-s}}\right)\cdots\left(\frac{1}{1-p^{-s}}\right)\cdots$$

with the infinite product taken over all primes $2, 3, 5, 7, 11, \ldots$ in the denominators, and $\mathrm{Re}(s) > 1$, where the product converges absolutely. The proof going back to Euler (who used positive integer values for s) is as follows. Consider

$$\zeta(s) = \frac{1}{1^s} + \frac{1}{2^s} + \frac{1}{3^s} + \frac{1}{4^s} + \frac{1}{5^s} + \cdots, \tag{10.20}$$

which converges absolutely and uniformly for $\mathrm{Re}(s) \geq \alpha > 1$, and likewise for the series of all the primes p_n since $\sum \frac{1}{|p_n^s|} \leq \sum \frac{1}{|n^s|}$ as all the terms of the first series are in common with the second series.[19] Next, multiply expression (10.20) by $1/2^s$, giving

$$\frac{1}{2^s}\zeta(s) = \frac{1}{2^s} + \frac{1}{4^s} + \frac{1}{6^s} + \frac{1}{8^s} + \frac{1}{10^s} + \cdots.$$

Subtraction yields

$$\left(1 - \frac{1}{2^s}\right)\zeta(s) = \frac{1}{1^s} + \frac{1}{3^s} + \frac{1}{5^s} + \frac{1}{7^s} + \frac{1}{9^s} + \frac{1}{11^s} + \cdots,$$

with all the even terms removed. Multiplying the last expression with $1/3^s$ and subtracting the result from it, we obtain

$$\left(1 - \frac{1}{3^s}\right)\left(1 - \frac{1}{2^s}\right)\zeta(s) = \frac{1}{1^s} + \frac{1}{5^s} + \frac{1}{7^s} + \frac{1}{11^s} + \cdots,$$

where all multiples of 3 are eliminated. We continue in this fashion, multiplying the expression previously obtained by $1/p_n^s$, where p_n increasingly runs through all the

18 J. Polchinski, *String Theory*, Cambridge University Press, 1998, p. 43.

19 Interestingly the sum $\sum \frac{1}{p_n}$ diverges, as was shown by Euler in spite of the increasing sparseness of the primes.

primes, thus sieving out all the multiples of $2, 3, \ldots, n$, so that by the nth stage we have

$$\left(1-\frac{1}{p_n^s}\right)\cdots\left(1-\frac{1}{3^s}\right)\left(1-\frac{1}{2^s}\right)\zeta(s)-1=\frac{1}{p_{n+1}^s}+\frac{1}{p_{n+2}^s}+\cdots.$$

Consequently,

$$\left|\left(1-\frac{1}{p_n^s}\right)\cdots\left(1-\frac{1}{3^s}\right)\left(1-\frac{1}{2^s}\right)\zeta(s)-1\right|\le\left|\frac{1}{p_{n+1}^s}\right|+\left|\frac{1}{p_{n+2}^s}\right|+\cdots\to 0$$

as $n\to\infty$ uniformly for $\mathrm{Re}(s)\ge\alpha>0$, that is,

$$\lim_{n\to\infty}\left[\left(1-\frac{1}{p_n^s}\right)\cdots\left(1-\frac{1}{3^s}\right)\left(1-\frac{1}{2^s}\right)\right]\zeta(s)=1,$$

which establishes the Euler product formula. Of course, there are various other proofs.

10.14 Exercise. Prove the Euler product formula by writing each term in the infinite product

$$\frac{1}{1-\frac{1}{p_n^s}}$$

as a geometric series that converges for $\mathrm{Re}(s)>1$, so that $|1/p_n^s|<1$. Hint: Write the first few terms of the first two geometric series and multiply them together to establish a general pattern, which will become the terms of Riemann zeta function.

As an example, for $s=2$, since $\zeta(2)=\pi^2/6$, we obtain

$$\frac{\pi^2}{6}=\left(\frac{1}{1-2^{-2}}\right)\left(\frac{1}{1-3^{-2}}\right)\left(\frac{1}{1-5^{-2}}\right)\left(\frac{1}{1-7^{-2}}\right)\cdots=\frac{4\cdot 9\cdot 25\cdot 49\cdot 121\cdots}{3\cdot 8\cdot 24\cdot 48\cdot 120\cdots},$$

linking π^2 with an infinite product of all the squares of prime numbers, which seems to be a gift from the gods.

Möbius function

Of great significance in the sequel is the Möbius function investigated by August Ferdinand Möbius in 1832,[20] although it was known to Euler and touched upon by Gauss (naturally). It is a rather peculiar function, and it is a wonder that it has found any use

20 Über eine besondere Art von Umkehrung der Reihen, *J. für die reine und angewandte Math.* 9 (1832), 105–123.

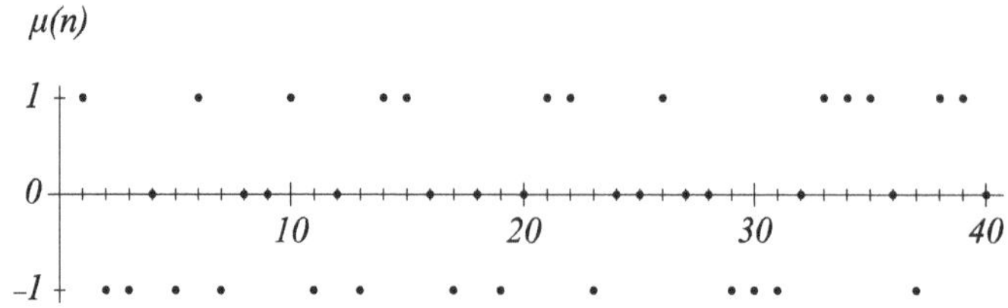

Figure 10.7: The first 40 values of the Möbius function taking the values 1, 0, −1. Courtesy Katy Metcalf.

at all, not to mention its intimate connection to the Riemann zeta function and the Riemann hypothesis. To wit,

$$\begin{aligned} &\mu(1) = 1; \\ &\mu(n) = 0 \quad \text{if } p^2 \mid n \text{ for some prime } p; \\ &\mu(n) = (-1)^j \quad \text{if } n \text{ is the product of } j \text{ distinct primes.} \end{aligned}$$

The first 40 values of the Möbius function are depicted in Figure 10.7.

There is another expression for the Möbius function that does not involve primes but rather the *primitive* roots of unity:[21]

$$\mu(n) = \sum_{\substack{1 \le k \le n \\ \gcd(k,n)=1}} e^{i\frac{2\pi k}{n}}.$$

This leads to the following fundamental property of the Möbius function which we will make extensive use of:

$$\sum_{d|n} \mu(d) = \begin{cases} 1 & n = 1 \\ 0 & n > 1, \end{cases} \tag{10.21}$$

which follows from the preceding expression for $\mu(n)$, since each nth root of unity is a primitive dth root for just one value of d that divides n, and the nth roots of unity sum to zero. Sometimes, (10.21) is written for $n > 1$ as

$$\sum_{d|n} \mu\left(\frac{n}{d}\right) = 0,$$

since if d is a divisor of n, then so is n/d and vice versa.

21 Cf. Apostol (2010, p. 48).

10.15 Exercise. Let $f(n)$ be an *arithmetical function*, that is, one that maps the positive integers to $\mathbb{C}$. Show that if g is an arithmetical function given by

$$g(n) = \sum_{d|n} f(d),$$

then

$$f(n) = \sum_{d|n} \mu(d) g\left(\frac{n}{d}\right).$$

This is the *Möbius inversion formula*. An infinite version will be presented in the sequel.

It is remarkable that the Möbius function also has a strong connection with the Riemann zeta function, namely,

$$\frac{1}{\zeta(s)} = \prod_{p \text{ prime}} \left(1 - \frac{1}{p^s}\right) = \sum_{n=1}^{\infty} \frac{\mu(n)}{n^s} \tag{10.22}$$

for $\text{Re}(s) > 1$. The first equality follows from the Euler product formula for $\zeta(s)$, and for the second, in view of the absolute convergence, we can expand the product

$$\begin{aligned}
\prod_{p \text{ prime}} \left(1 - \frac{1}{p^s}\right) &= \left(1 - \frac{1}{p_1^s}\right)\left(1 - \frac{1}{p_2^s}\right)\left(1 - \frac{1}{p_3^s}\right)\left(1 - \frac{1}{p_4^s}\right)\cdots \\
&= 1 - \left(\frac{1}{p_1^s} + \frac{1}{p_2^s} + \frac{1}{p_3^s} + \frac{1}{p_4^s} + \cdots\right) \\
&\quad + \left(\frac{1}{p_1^s p_2^s} + \frac{1}{p_1^s p_3^s} + \frac{1}{p_1^s p_4^s} + \cdots + \frac{1}{p_2^s p_3^s} + \frac{1}{p_2^s p_4^s} + \cdots\right) - \cdots \\
&= 1 - \sum_{0<i} \frac{1}{p_i^s} + \sum_{0<i<j} \frac{1}{p_i^s p_j^s} - \sum_{0<i<j<k} \frac{1}{p_i^s p_j^s p_k^s} + \cdots \\
&= \sum_{n=1}^{\infty} \frac{\mu(n)}{n^s}
\end{aligned}$$

by the definition of the Möbius function, whereby establishing the result.

Mertens conjecture

Let us write the partial sum

$$M(x) = \sum_{n \le x} \mu(n),$$

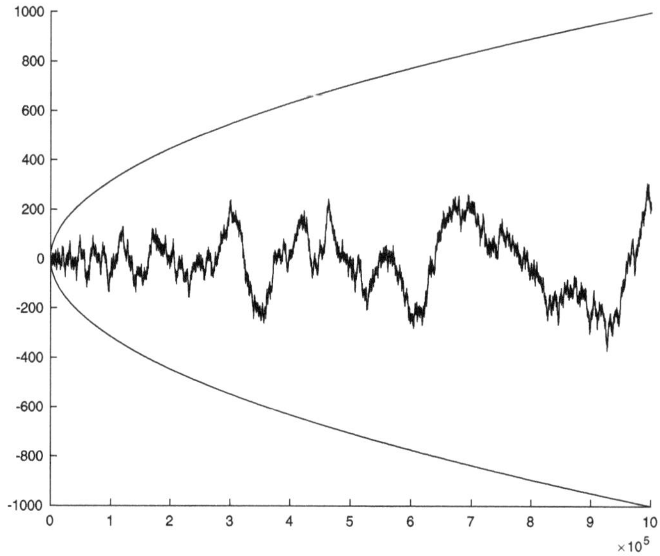

Figure 10.8: Graph of the Mertens function $M(n)$ and $\pm\sqrt{n}$ for $n \le 1{,}000{,}000$. Mertens "only" considered values of n from 1 to 10,000. Courtesy Anita Kean.

where $M(x)$ is known as the *Mertens function*. Franz Mertens (1840–1927) conjectured in 1897 that $|M(x)| \le x^{\frac{1}{2}}$, $x > 1$, which was an eminently reasonable assumption if we consider Figure 10.8.

This would have proved the Riemann hypothesis if true. To see this, we use the preceding relation (10.22) and the fact that $M(x)$ is constant on the intervals $[n, n+1)$:

$$\begin{aligned}\frac{1}{\zeta(s)} &= \sum_{n=1}^{\infty} \frac{\mu(n)}{n^s} = \sum_{n=1}^{\infty} \frac{M(n) - M(n-1)}{n^s}\\ &= \sum_{n=1}^{\infty} M(n)\left(\frac{1}{n^s} - \frac{1}{(n+1)^s}\right) = \sum_{n=1}^{\infty} M(n) \int_n^{n+1} \frac{s}{x^{s+1}}\,dx\\ &= s\sum_{n=1}^{\infty} \int_n^{n+1} \frac{M(x)}{x^{s+1}}\,dx = s\int_1^{\infty} \frac{M(x)}{x^{s+1}}\,dx.\end{aligned}$$

Assuming that $|M(x)| \le x^{\frac{1}{2}}$, $x > 1$, the last integral would converge for $\sigma = \operatorname{Re}(s) > \frac{1}{2}$ and hence would define an analytic function there. This would imply that $\frac{1}{\zeta(s)}$ has no poles, that is, $\zeta(s)$ has no zeros in $\sigma = \operatorname{Re}(s) > \frac{1}{2}$. By the Riemann functional equation there can be no other nontrivial zeros other than those on the line $\operatorname{Re}(s) = \frac{1}{2}$, so if the Mertens conjecture were true, then it would imply that RH is true as well.

However, the Mertens conjecture was disproved by A. M. Odlyzko and H. J. J. te Riele in 1985.[22] To prove the Riemann hypothesis, we actually need to establish that

$$M(x) = \sum_{n \le x} \mu(n) = O(x^{\frac{1}{2}+\varepsilon})$$

for an arbitrary $\varepsilon > 0$.

Möbius inversion

We again consider the notion of *Möbius inversion* (see Exercise 10.15), a powerful idea that is implicit in the seminal 1832 paper of Möbius and a basic tool in the arsenal of number theorists. We state it here in a manner to suit our analytical applications that subsequently follow. There are numerous other applications of Möbius inversion, especially to problems in physics, which can be found in the excellent book by Chen (2010).

10.16 Theorem (Möbius inversion). *Let $c_1, c_2, c_3, \ldots$ be a sequence of complex numbers such that*

$$\sum_{n=1}^{\infty} |c_n| < \infty,$$

and let $b_1, b_2, b_3, \ldots$ be defined by

$$b_n = \sum_{k=1}^{\infty} c_{kn}.$$

Under the assumption that

$$\sum_{k=1}^{\infty} k|c_k| < \infty,$$

we have the Möbius inversion formula

$$c_n = \sum_{k=1}^{\infty} \mu(k) b_{kn}.$$

---

22 Disproof of the Mertens conjecture, *J. für die reine und angewandte Mathematik* 357 (1985), 138–160. The authors showed that $\overline{\lim}_{x\to\infty} \frac{M(x)}{\sqrt{x}} > 1.06$ and $\underline{\lim}_{x\to\infty} \frac{M(x)}{\sqrt{x}} < -1.09$ but did not provide a specific counterexample. Further computer studies have extended the lim sup and lim inf values to 1.826054 and −1.837625, respectively, by G. Hurst.

Proof. We consider the array

$$\begin{aligned}
\mu(1)b_n &= \mu(1)c_n + \mu(1)c_{2n} + \mu(1)c_{3n} + \mu(1)c_{4n} + \cdots,\\
\mu(2)b_{2n} &= \qquad\qquad \mu(2)c_{2n} \qquad\qquad + \mu(2)c_{4n} + \cdots,\\
\mu(3)b_{3n} &= \qquad\qquad\qquad\qquad \mu(3)c_{3n} + \cdots \qquad + \cdots,\\
\mu(4)b_{4n} &= \qquad\qquad\qquad\qquad\qquad\qquad \mu(4)c_{4n} + \cdots,\\
&\vdots
\end{aligned}$$

Summing the absolute values of the jth column gives

$$C_j = \Big(\sum_{d|j} |\mu(j)|\Big)|c_{jn}| \le j|c_{jn}|,$$

and therefore summing these column values over all columns, we get

$$\sum_{j=1}^{\infty} C_j \le \sum_{j=1}^{\infty} j|c_{jn}| \le \sum_{k=1}^{\infty} k|c_k| < \infty.$$

Hence by the Weierstrass double series theorem the double series converges with the sum by columns equal to the sum by rows. Invoking the fundamental property of the Möbius function (10.21), the sum of the double series by columns is just c_n, and hence

$$c_n = \sum_{k=1}^{\infty} \mu(k) b_{kn},$$

as desired. ☐

We simply note that there are more general sufficient conditions for Möbius inversion such as

$$\sum_{n=1}^{\infty} n^{\varepsilon}|c_n| < \infty$$

for some $\varepsilon > 0$ (Wintner 1945).

10.17 Lemma. *For any positive integer j,*

$$\sum_{m=1}^{n} e^{ij(\frac{2\pi m}{n})} = \begin{cases} 0 & j \ne kn \\ n & j = kn, \end{cases}$$

for some positive integer k.

The proof is elementary and left as an exercise for the reader.

Arithmetic Fourier transform

We now consider the case of a function $f(z)$ analytic on the closed unit disk $\overline{U}$ and hence in a slightly larger open disk containing U. We will demonstrate that each of the Taylor coefficients of $f(z)$ can be determined by an infinite series of the values of the function $f(z)$ taken over a symmetrically arrayed sets of points on the boundary $\partial U : |z| = 1$. A similar formulation for Fourier cosine coefficients was found by Bruns (1903) and Wintner (1945). Without loss of generality, we will take the normalization

$$c_0 = \frac{1}{2\pi}\int_0^{2\pi} f(e^{i\theta})d\theta = 0.$$

10.18 Theorem.[23] *If $f(z)$ is analytic on the closed unit disk $\overline{U}$ with Taylor series*

$$f(z) = \sum_{j=1}^{\infty} c_j z^j,$$

then the Taylor coefficients are given by

$$c_n = \sum_{k=1}^{\infty} \frac{\mu(k)}{kn} \sum_{m=1}^{kn} f(e^{i(\frac{2\pi m}{kn})}), \quad n = 1, 2, 3, \ldots. \tag{10.23}$$

Proof. For $|z| = 1$, we can write

$$f(\theta) = f(e^{i\theta}) = \sum_{j=1}^{\infty} c_j e^{ij\theta}. \tag{10.24}$$

Define the averages $f_n(\theta)$ over a symmetrically arrayed set of points by

$$f_n(\theta) = \frac{1}{n}\sum_{m=1}^{n} f\left(\theta + \frac{2\pi m}{n}\right), \quad n = 1, 2, 3, \ldots.$$

Then by (10.24) and Lemma 10.17

$$f_n(\theta) = \frac{1}{n}\sum_{m=1}^{n}\sum_{j=1}^{\infty} c_j e^{ij(\theta + \frac{2\pi m}{n})} = \sum_{j=1}^{\infty} c_j e^{ij\theta}\frac{1}{n}\sum_{m=1}^{n} e^{ij(\frac{2\pi m}{n})} = \sum_{k=1}^{\infty} c_{kn} e^{ikn\theta}.$$

Upon setting $\theta = 0$, we now have

---

23 J. L. Schiff and W. J. Walker, A sampling theorem for analytic functions, *Proc. Amer. Math. Soc.* 99 (4) (1987), 737–740; A sampling theorem and Wintner's results on Fourier coefficients, *J. Math. Anal. Appl.* 133 (2) (1988), 466–471.

$$f_n = f_n(0) = \sum_{k=1}^{\infty} c_{kn}. \tag{10.25}$$

By the analyticity of $f(z)$,

$$\sum_{k=1}^{\infty} k|c_k| < \infty.$$

In view of (10.25) and the Möbius inversion Theorem 10.16, we obtain

$$c_n = \sum_{k=1}^{\infty} \mu(k) f_{kn} = \sum_{k=1}^{\infty} \frac{\mu(k)}{kn} \sum_{m=1}^{kn} f(e^{i(\frac{2\pi m}{kn})}),$$

proving the theorem. □

10.19 Corollary. *If $f(z)$ is analytic on the open unit disk $U : |z| < 1$ with Taylor series*

$$f(z) = \sum_{n=1}^{\infty} c_n z^n,$$

then the Taylor coefficients are given by

$$c_n = \frac{1}{r^n} \sum_{k=1}^{\infty} \frac{\mu(k)}{kn} \sum_{m=1}^{kn} f(re^{i(\frac{2\pi m}{kn})}), \quad 0 < r < 1, \ n = 1, 2, 3, \ldots .$$

The *Z-transform* given by $X(z) = \sum_{j=1}^{\infty} c_j z^{-j}$ is used to handle discrete sets of data and represents a Laurent series. By considering the function $w = 1/z$ we have the following:

10.20 Corollary. *If*

$$X(z) = \sum_{j=1}^{\infty} c_j z^{-j}$$

converges for $|z| > r$ and $r < 1$, then

$$c_n = \sum_{k=1}^{\infty} \frac{\mu(k)}{kn} \sum_{m=1}^{kn} X(e^{i(\frac{2\pi m}{kn})}), \quad n = 1, 2, 3, \ldots .$$

The series in (10.22) is known as the *arithmetic Fourier transform* (AFT) and has applications in the field of signal processing. Of course, in real-world applications, only a finite number of terms can be considered, but the resulting sum is still known by the same name. One of the AFT's virtues is that it can be readily computed by parallel processing and rivals the fast Fourier transform[24] in speed and efficiency.

The next theorem is a remarkable result of Aurel Wintner (1945), which, although similar, is not quite the AFT, but intimately connected with the prime number theorem. Let us first consider the following case which we state without proof.

10.21 Theorem. *Let $f(t)$ be a Riemann-integrable function of period 1. Then*

$$\int_0^1 f(t)dt = \sum_{n=1}^{\infty} \frac{1}{n} \sum_{d|n} \mu(d) f\left(\frac{nx}{d}\right) \tag{10.26}$$

for any irrational number x.

To appreciate the deep connection with the prime number theorem, consider the period 1 function for irrational x

$$f(t) = \begin{cases} 0 & 0 < t < 1,\ t \neq x \\ 1 & t = x. \end{cases}$$

The values at the endpoints $f(0) = f(1)$ are unimportant as they are not included in the sum on the right-hand side of (10.26). Note that the interior sum runs over all the divisors d of n, so that $f(\frac{nx}{d}) = 0$ except for $d = n$ since x is irrational. As $f(t) = 0$ except for $t = x$, we obtain

$$0 = \sum_{n=1}^{\infty} \frac{\mu(n)}{n},$$

and this convergence of $\sum_{n=1}^{\infty} \mu(n)/n$ to zero is a well-known equivalent of the PNT.[25]

Equality (10.26) leads directly to the more general form (Wintner 1945, p. 24) for Fourier series.

10.22 Theorem. *Let $f(t)$ be a Riemann-integrable function of period 1 with Fourier coefficients c_k, $k = 0, \pm 1, \pm 2, \ldots$. Then*

$$c_k = \sum_{n=1}^{\infty} \frac{1}{n} \sum_{d|n} \mu(d) f\left(\frac{nx}{d}\right) e^{-\frac{2\pi iknx}{d}},$$

where x is an arbitrary irrational number.

An indispensable transform in signal processing that uses complex exponentials is the discrete Fourier transform.

---

24 The fast Fourier transform is an algorithm for efficiently computing the discrete Fourier transform discussed in the sequel.

25 See Apostol (2010), who also showed that $M(x) = \sum_{n \le x} \mu(n) = o(x)$ as $x \to \infty$ if and only if $\sum_{n \le x} \frac{\mu(n)}{n} = o(1)$ as $x \to \infty$, where $M(x)$ is the Mertens function, thus providing another equivalent of the PNT.

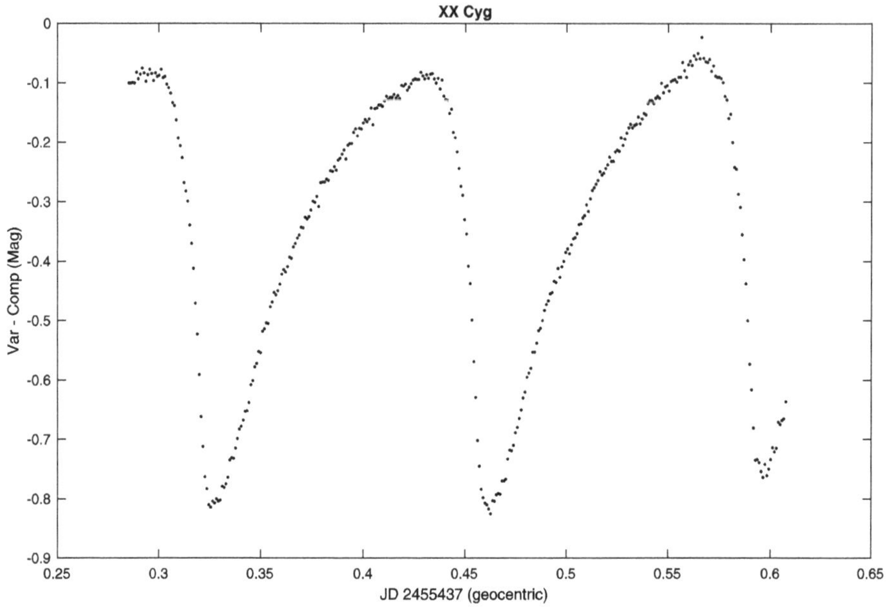

Figure 10.9: The 345 data points of the variation in light magnitude from the periodic variable star XX Cygnus. Copyright 2021 Anthony Ayiomamitis.

Discrete Fourier transform

In this scenario, we have N equally spaced sampled data points that arise from either a finite continuous or discrete signal $x(t)$ with sampled values $x(0), x(1), \ldots, x(N-1)$, where the function $x(t)$ is N-periodic as in Figure 10.9. We are interested in expressing the signal as a sum of various frequencies, similar to a Fourier series.

To this end, define the *discrete Fourier transform* (*DFT*) as the set of values[26]

$$X(k) = X(\omega_k) = \sum_{n=0}^{N-1} x(n) e^{-i\frac{2\pi kn}{N}} = \sum_{n=0}^{N-1} x(n) \cos\left(\frac{2\pi kn}{N}\right) - i \sum_{n=0}^{N-1} x(n) \sin\left(\frac{2\pi kn}{N}\right),$$

where the *frequencies* are given by $\omega_k = \frac{2\pi k}{N}$, $k = 0, 1, \ldots, N-1$. Note that

$$X(0) = \sum_{n=0}^{N-1} x(n).$$

26 Electrical engineers prefer using the letter j instead of our customary i for the imaginary number $\sqrt{-1}$. Chacun à son goût.

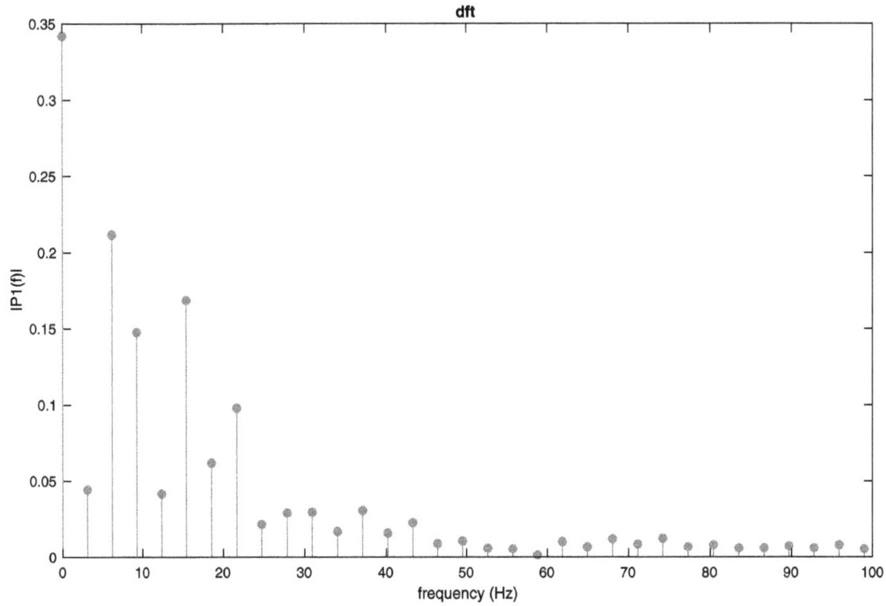

Figure 10.10: The power spectrum (vertical axis) of the DFT from the data in Figure 10.9 showing the dominant frequencies. Courtesy Nick Dudley Ward.

The *power* of each frequency component is given by $P(k) = |X(k)|^2$ and gives an indication of the relative significance of each frequency as in Figure 10.10.[27]

Moreover,

$$x(n) = \frac{1}{N}\sum_{k=0}^{N-1} X(k)e^{i\frac{2\pi kn}{N}} \tag{10.27}$$

is the *inverse DFT*(see Figure 10.11). To see this, observe that in view of Lemma 10.17,

$$\begin{aligned}\frac{1}{N}\sum_{k=0}^{N-1} X(k)e^{i\frac{2\pi kn}{N}} &= \frac{1}{N}\sum_{k=0}^{N-1}\left(\sum_{m=0}^{N-1} x(m)e^{-i\frac{2\pi km}{N}}\right)e^{i\frac{2\pi kn}{N}}\\ &= \sum_{m=0}^{N-1} x(m)\left(\frac{1}{N}\sum_{k=0}^{N-1} e^{-i\frac{2\pi k(m-n)}{N}}\right) = \sum_{m=0}^{N-1} x(m)\delta_{n,m} = x(n),\end{aligned}$$

where $\delta_{n,m}$ is the Kroncker delta function, which equals 1 for $n = m$ and zero otherwise. Note that representation (10.27) is in fact periodic of period N and thus extends the signal values to be periodic and is equal to the signal values on $[0, N-1]$. Each $X(k)$

27 For the sampling rate f_s (samples/sec), the frequency is given by $f = \frac{kf_s}{N}$, which allows the power spectrum to be plotted against frequency.

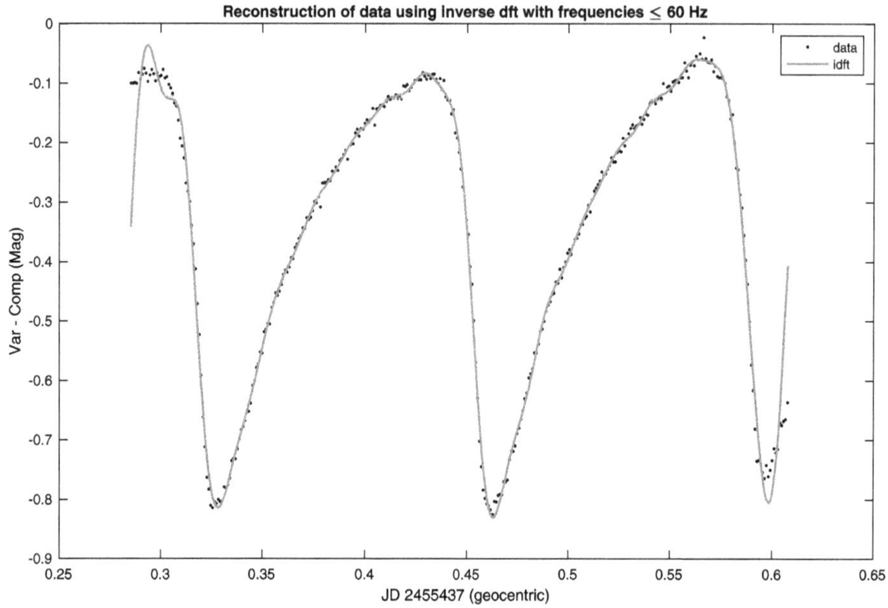

Figure 10.11: A reconstruction of the original data from the inverse DFT with a cutoff of frequencies greater than 60 Hz. Courtesy Nick Dudley Ward.

represents the coefficients of the inverse $x(n)$, encoding both the amplitude and phase of a sinusoidal wave with frequency $\frac{2\pi k}{N}$.

10.23 Exercise. Prove Parseval's theorem for the DFT:

$$\sum_{n=0}^{N-1} |x(n)|^2 = \frac{1}{N} \sum_{k=0}^{N-1} |X(k)|^2,$$

that is, the *energy* of the signal in the time domain (given by the left-hand side) equals the energy of the transformed signal in the frequency domain (given by the right-hand side).

Laplace transform

Integral transforms date back to the work of Léonard Euler (1763 and 1769), and the Laplace transform has turned out to be one of the most useful tools in analysis. Placing it in the complex domain goes back to Poincaré and Pincherle, which opens up the entire calculus of residues.

We deal with real-valued functions f that are piecewise continuous on $[0, \infty)$ and satisfy the growth condition of *exponential order* $\alpha \geq 0$, that is,

$$|f(t)| \le Me^{\alpha t}, \quad t \ge t_0,$$

for some constant $M > 0$ and some $t_0 \ge 0$. Since $e^{\alpha t} < e^{\beta t}$ for $\alpha < \beta$, generally, the smallest α for which the inequality holds is considered. Often, the specific value of α is not significant.

Then we have the following basic result, where $s = x + iy$ is a complex variable.

10.24 Theorem. *If f is piecewise continuous on $[0, \infty)$ of exponential order α, then the Laplace transform*

$$F(s) = \mathcal{L}(f) = \int_0^\infty e^{-st} f(t)dt$$

is well-defined and converges uniformly for $\mathrm{Re}(s) \ge x_0 > \alpha$ and absolutely for $\mathrm{Re}(s) > \alpha$ and represents an analytic function there.

Proof. Assuming that

$$|f(t)| \le Me^{\alpha t}, \quad t \ge t_0,$$

for $x = \mathrm{Re}(s) > \alpha$, we have

$$\left|\int_{t_0}^\infty e^{-st} f(t)dt\right| \le M \int_{t_0}^\infty e^{-(x-\alpha)t} dt = \frac{Me^{-(x-\alpha)t_0}}{(x-\alpha)}.$$

Now for $x \ge x_0 > \alpha$, we have for the last term

$$\frac{Me^{-(x-\alpha)t_0}}{(x-\alpha)} \le \frac{Me^{-(x_0-\alpha)t_0}}{(x_0-\alpha)},$$

and taking t_0 sufficiently large, the right-hand side can be made arbitrarily small. This establishes the uniform convergence of the Laplace transform integral in the region $\mathrm{Re}(s) \ge x_0 > \alpha$ and, consequently, represents an analytic function for $\mathrm{Re}(s) > \alpha$ (why?). The absolute convergence is evident. □

10.25 Corollary. *If f is piecewise continuous on $[0, \infty)$ of exponential order α, then $F(s) \to 0$ as $\mathrm{Re}(s) \to \infty$.*

10.26 Exercise. Show that
(a) $\mathcal{L}(e^{at}) = \frac{1}{s-a}$, $\mathrm{Re}(s) > a$;
(b) $\mathcal{L}(t) = \frac{1}{s^2}$, $\mathrm{Re}(s) > 0$;
(c) $\mathcal{L}(t^n) = \frac{n!}{s^{n+1}}$, $\mathrm{Re}(s) > 0$, for $n = 1, 2, 3, \ldots$;
(d) $\mathcal{L}(t^\nu) = \frac{\Gamma(\nu+1)}{s^{\nu+1}}$, $s > 0, \nu > -1$.

10.27 Example. We are presently interested in evaluating the integral

$$I = \int_{-\infty}^{\infty} \frac{1}{(1+ax^2)^\nu} dx$$

for $\nu > 1/2$.

Let us write the result in Exercise 10.26(d) as

$$\frac{1}{\Gamma(\nu)} \int_0^\infty e^{-st} t^{\nu-1} dt = \frac{1}{s^\nu}. \tag{10.28}$$

Using (10.28) and setting $s = 1 + ax^2$ (a real), we get

$$\begin{aligned} I = \int_{-\infty}^{\infty} \frac{1}{(1+ax^2)^\nu} dx &= \frac{1}{\Gamma(\nu)} \int_{-\infty}^{\infty} \left(\int_0^\infty e^{-t(1+ax^2)} t^{\nu-1} dt \right) dx \\ &= \frac{1}{\Gamma(\nu)} \int_0^\infty e^{-t} t^{\nu-1} \left(\int_{-\infty}^{\infty} e^{-tax^2} dx \right) dt \\ &= \frac{1}{\Gamma(\nu)} \sqrt{\frac{\pi}{a}} \int_0^\infty e^{-t} t^{\nu-\frac{3}{2}} dt = \sqrt{\frac{\pi}{a}} \frac{\Gamma(\nu - \frac{1}{2})}{\Gamma(\nu)}, \end{aligned}$$

where we leave justification of reversing the order of integration to the reader.

As well, we can relate the Laplace transform to the Euler beta function. To this end, let us consider a very important integral operator that has many widespread applications. For functions f and g defined on $[0, \infty)$, the integral

$$(f * g)(t) = \int_0^t f(\tau) g(t-\tau) d\tau$$

is called the *convolution* of f and g. The integral certainly exists in our present setting of f and g being piecewise continuous, and it is readily verified that the convolution satisfies elementary algebraic properties.[28]

---

28 Namely,

$$\begin{aligned} c(f * g) &= cf * g = f * cg, \quad c > 0; \\ f * (g * h) &= (f * g) * h; \\ f * (g + h) &= f * g + f * h. \end{aligned}$$

Their verification is left as an exercise.

For example, considering the Laplace transform (with $a \neq b$)

$$\mathcal{L}(e^{at} * e^{bt}) = \mathcal{L}\left(\int_0^t e^{a\tau} e^{b(t-\tau)}\,d\tau\right) = \mathcal{L}\left(\frac{e^{at} - e^{bt}}{a-b}\right),$$

the latter term happens to be the product $\mathcal{L}(e^{at}) \cdot \mathcal{L}(e^{bt})$ of the respective Laplace transforms.

For a convolution, whenever f and g are piecewise continuous on $[0, \infty)$ of exponential order α, indeed it is always the case that:

$$\mathcal{L}(f * g)(t) = \mathcal{L}(f(t))\mathcal{L}(g(t)) \quad (\mathrm{Re}(s) > \alpha).$$ [29]

Given this *convolution theorem*, let us take $f(t) = t^{a-1}$ and $g(t) = t^{b-1}$ for $a, b > 0$, so that

$$(f * g)(t) = \int_0^t \tau^{a-1}(t-\tau)^{b-1}\,d\tau.$$

With a substitution $\tau = ut$, we obtain

$$(f * g)(t) = t^{a+b-1}\int_0^1 u^{a-1}(1-u)^{b-1}\,du = t^{a+b-1}B(a, b),$$

the latter term encompassing the Euler beta function. Therefore

$$\mathcal{L}(t^{a+b-1}B(a, b)) = \mathcal{L}(f * g)(t) = \frac{\Gamma(a)\Gamma(b)}{s^{a+b}}$$

by the convolution theorem and Exercise 10.26(d). The left-hand side of the preceding expression equals

$$\frac{\Gamma(a+b)}{s^{a+b}}B(a, b),$$

whence we obtain the Euler formula for the beta function for real numbers $a, b > 0$,

$$B(a, b) = \frac{\Gamma(a)\Gamma(b)}{\Gamma(a+b)}.$$

---

29 The proof is straightforward and can be found in Schiff (1999, pp. 92–93).

Inverse Laplace transform

The algorithm employed with the Laplace transform involves to take the inverse transform to retrieve the original function:

$$\mathcal{L}^{-1}(F(s)) = f(t), \quad t \geq 0.$$

Note however that altering the function $f(t)$ at a finite number of points will yield the same Laplace transform, and therefore the inverse $\mathcal{L}^{-1}(F(s))$ is not uniquely defined. However, restricting ourselves to continuous functions, we have the following:

10.28 Lerch theorem.[30] *Distinct continuous functions on* $[0, \infty)$ *have distinct Laplace transforms.*

For example,

$$\mathcal{L}^{-1}\left(\frac{\omega}{s^2 - \omega^2}\right) = \sinh \omega t,$$

but of course, altering the function $f(t) = \sinh \omega t$ at a finite number of points making it piecewise continuous will give the same inverse. However, since most of the applications deal with solutions to differential equations, we will take the inverse to be continuous. On the other hand, one of the great virtues of the Laplace transform is that it can be applied to discontinuous functions, and we must bear in mind the caveat about the inverse.

10.29 Example. A classical function in electrical systems is the *Heaviside (step) function*

$$u_a(t) = \begin{cases} 1 & t > a \\ 0 & t < a, \end{cases}$$

for $a \geq 0$. Commonly also, $u_a(t) = 1$ for $t \geq a$ and $u_a(t) = 0$ otherwise. The Laplace transform in either case is easily determined to be

$$\mathcal{L}(u_a(t)) = \frac{e^{-as}}{s} \quad (\mathrm{Re}(s) > 0),$$

and thus we can write

$$\mathcal{L}^{-1}\left(\frac{e^{-as}}{s}\right) = u_a(t).$$

30 Sur un point de la théorie des fonctions génératrices d'Abel, *Acta Math.* 27 (1903), 339–351. M. Lerch in fact proved that the inverse Laplace transform is uniquely determined with the proviso that we identify any two functions that differ only on a set of Lebesgue measure zero.

As it is clear that the Laplace transform is linear, the utility of the transform comes from computing the transform for each term of say a linear differential equation, simplifying the algebraic terms and then determining the inverse function from a table of standard Laplace transforms if possible. Therefore we need the following:

10.30 Derivative theorem. *Suppose that f is continuous on $(0,\infty)$ of exponential order α. If f' is piecewise continuous on $[0,\infty)$, then*

$$\mathcal{L}(f'(t)) = s\mathcal{L}(f(t)) - f(0^+).$$

The proof follows is via integration by parts for $\mathrm{Re}(s) > \alpha$:

$$\int_0^\infty e^{-st} f'(t)dt = \lim_{\substack{\varepsilon\to 0\\ \tau\to\infty}} \int_\varepsilon^\tau e^{-st} f'(t)dt = \lim_{\substack{\varepsilon\to 0\\ \tau\to\infty}} \left(e^{-s\tau} f(\tau) - e^{-s\varepsilon} f(\varepsilon) + s\int_\varepsilon^\tau e^{-st} f(t)dt \right)$$

$$= s\int_0^\infty e^{-st} f(t)dt - f(0^+).$$

The fact that $f(0^+)$ exists follows from the equality

$$\int_\varepsilon^\kappa f'(t)dt = f(\kappa) - f(\varepsilon)$$

for κ sufficiently small and letting $\varepsilon \to 0$.

Similarly, we have the following:

10.31 Integral theorem. *If f is piecewise continuous on $[0,\infty)$ of exponential order α, then the function*

$$g(t) = \int_0^t f(\tau)d\tau$$

has the Laplace transform

$$\mathcal{L}(g(t)) = \frac{1}{s}\mathcal{L}(f(t)), \quad \mathrm{Re}(s) > \alpha.$$

Indeed, again integrating by parts gives

$$\int_0^\infty e^{-st} g(t)dt = \lim_{\tau\to\infty} \left(g(t)\frac{e^{-st}}{-s}\Bigg|_0^\tau + \frac{1}{s}\int_0^\tau e^{-st} f(t)dt \right).$$

Then the first term vanishes (why?) so that the result follows. Similarly, the result also holds for $\alpha = 0$.

Computing the inverse Laplace transform via residues

Here we discuss a very useful technique for finding the inverse Laplace transform based on complex variable theory. To this end, let us assume that our function $f(t)$ is continuous on $[0, \infty)$, which we extend to $(-\infty, \infty)$ by defining $f(t) = 0$ for $t < 0$. Moreover, we assume that f has exponential order α and that f' is piecewise continuous on $[0, \infty)$.

Therefore, for $s = x + iy$, $x > \alpha$,

$$\mathcal{L}(f(t)) = \int_{-\infty}^{\infty} e^{-iyt}(e^{-xt}f(t))dt = F(x, y),$$

where $F(x, y)$ is now the *Fourier transform* of the function $g(t) = e^{-xt}f(t)$. Here we merely state the critical result concerning the inversion of the Fourier transform:[31]

10.32 Fourier inversion theorem. *Suppose that f and f' are piecewise continuous on $(-\infty, \infty)$ and that f is absolutely integrable. Then for a real parameter λ, the Fourier transform of $f(t)$ is given by*

$$F(\lambda) = \int_{-\infty}^{\infty} e^{-i\lambda t}f(t)dt,$$

and its inverse by

$$f(t) = \frac{1}{2\pi}\int_{-\infty}^{\infty} e^{i\lambda t}F(\lambda)d\lambda$$

at each point where $f(t)$ is continuous. At a jump discontinuity t,

$$\frac{1}{2\pi}\int_{-\infty}^{\infty} e^{i\lambda t}F(\lambda)d\lambda = \frac{f(t^+) + f(t^-)}{2},$$

where $f(t^+)$ and $f(t^-)$ are the right- and left-hand limits of $f(t)$, respectively.

Note that in the present setting, by Theorem 10.24, $g(t) = e^{-xt}f(t)$ is absolutely integrable. It follows from the Fourier inversion formula that

$$e^{-xt}f(t) = \frac{1}{2\pi}\int_{-\infty}^{\infty} e^{iyt}F(x, y)dy$$

31 See A. Jerri, *Integral and Discrete Transforms with Applications to Error Analysis*, Marcel Dekker Inc., 1992.

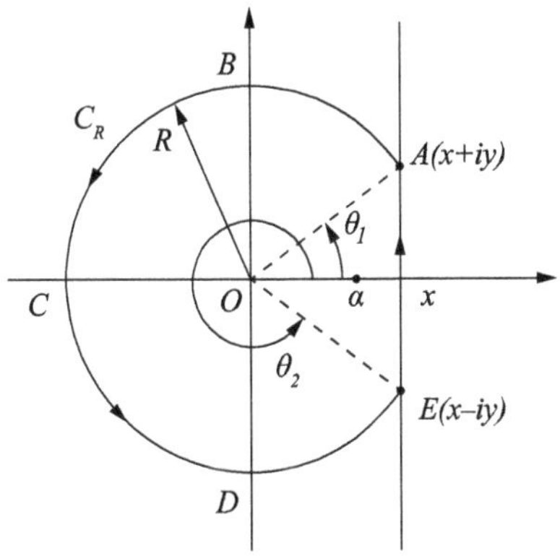

Figure 10.12: The contour Γ_R and its various components, including the Bromwich line. Courtesy Katy Metcalf.

or, in other words,

$$f(t) = \frac{1}{2\pi} \int_{-\infty}^{\infty} e^{xt} e^{iyt} F(x, y) dy, \quad t > 0. \tag{10.29}$$

Since $dy = (\frac{1}{i})ds$, for fixed $x > \alpha$, we can write (10.29) as a contour integral

$$f(t) = \frac{1}{2\pi i} \int_{x-i\infty}^{x+i\infty} e^{st} F(s) ds = \lim_{y\to\infty} \frac{1}{2\pi i} \int_{x-iy}^{x+iy} e^{st} F(s) ds \tag{10.30}$$

for $t > 0$, where the integration is to be taken along the vertical line at $x > \alpha$ known as the *Bromwich line*, and the integral in (10.30) is known as the *Fourier–Mellin inversion formula*.

We proceed to evaluate the integral in (10.30) via contour integration as in Figure 10.12. Our contour is $\Gamma_R = C_R + EA$, where the circular part is $C_R = ABCDE$, so that

$$\frac{1}{2\pi i} \int_{\Gamma_R} e^{st} F(s) ds = \frac{1}{2\pi i} \int_{C_R} e^{st} F(s) ds + \frac{1}{2\pi i} \int_{EA} e^{st} F(s) ds. \tag{10.31}$$

As the Laplace transform $F(s)$ is analytic in the region $\mathrm{Re}(s) > \alpha$, any possible singularities of $F(s)$ must lie to the left of the Bromwich line. We first assume that $F(s)$ is analytic in the region $\mathrm{Re}(s) < \alpha$ except for finitely many poles at the points $a_1, a_2, \ldots, a_n$, which we may assume lie inside the contour Γ_R. Thus by the Cauchy residue theorem

$$\frac{1}{2\pi i} \int_{\Gamma_R} e^{st} F(s) ds = \sum_{i=1}^{n} \mathrm{Res}(a_i),$$

where $\operatorname{Res}(a_i)$ is the residue of $e^{st}F(s)$ at a_i. Hence by (10.31)

$$\frac{1}{2\pi i}\int_{C_R} e^{st}F(s)ds + \frac{1}{2\pi i}\int_{x-iy}^{x+iy} e^{st}F(s)ds = \sum_{i=1}^{n}\operatorname{Res}(a_i). \tag{10.32}$$

Our next goal is to show that the first integral tends to zero as $R \to \infty$. To achieve this, we must impose a mild condition on the growth of the function $F(s)$, which turns out to be the typical growth of most Laplace transforms anyway.

10.33 Lemma. *If $F(s)$ satisfies the growth condition $|F(s)| \le \sigma_R$ and $\sigma_R \to 0$ as $R \to \infty$ uniformly for all $s \in C_R$, then*

$$\int_{C_R} e^{st}F(s)ds \to 0 \quad (t > 0)$$

as $R \to \infty$.

Proof. For s on C_R, let $s = Re^{i\theta}$ so that $|F(s)| \le \sigma_R$. As a consequence, for the arc BCD of Figure 10.12,

$$\left|\int_{BCD} e^{st}F(s)ds\right| \le R\sigma_R \int_{\frac{\pi}{2}}^{\frac{3\pi}{2}} e^{Rt\cos\theta}d\theta,$$

which after the substitution $\phi = \theta - (\pi/2)$ becomes

$$\left|\int_{BCD} e^{st}F(s)ds\right| \le R\sigma_R \int_0^{\pi} e^{-Rt\sin\phi}d\phi = 2R\sigma_R \int_0^{\frac{\pi}{2}} e^{-Rt\sin\phi}d\phi.$$

We can obtain a bound for the last integral by noting that $\sin\phi \ge 2\phi/\pi$ on the interval $[0, \frac{\pi}{2}]$. Therefore, for $|s| = R$ and $t > 0$,

$$\left|\int_{BCD} e^{st}F(s)ds\right| \le 2R\sigma_R \int_0^{\frac{\pi}{2}} e^{-2Rt\phi/\pi}d\phi = \sigma_R\frac{\pi}{t}(1 - e^{-Rt}) \to 0$$

as $R \to \infty$.

To treat the integrals over the arcs AB and DE, note that their lengths remain bounded as $R \to \infty$, so that

$$\left|\int_{AB+DE} e^{st}F(s)ds\right| \le \sigma_R \int_{AB} e^{xt}|ds| + \sigma_R \int_{DE} e^{xt}|ds| \to 0$$

as $R \to \infty$ for any fixed $t > 0$. This concludes the proof of the lemma. □

10.34 Exercise. Compute the Laplace transform of the function $f(t) = \sinh at$ and show that it satisfies the condition of the preceding lemma.

We have just demonstrated from (10.32) and Lemma 10.33 the following exceptionally useful result.

10.35 Theorem. *Suppose that f is continuous of exponential order α on $[0, \infty)$ such that f' is also piecewise continuous there. Assuming that the Laplace transform $F(s) = \mathcal{L}(f(t))$ exists for $\operatorname{Re}(s) > \alpha$ and satisfies on $C_R : |s| = R$ the growth condition $|F(s)| \le \sigma_R \to 0$ as $R \to \infty$ uniformly for all $s \in C_R$, then for $F(s)$ meromorphic in $\mathbb{C}$ except for poles at the points $a_1, a_2, \ldots, a_n$, the inverse transform is given by*

$$f(t) = \lim_{y\to\infty} \frac{1}{2\pi i} \int_{x-iy}^{x+iy} e^{st} F(s) ds = \sum_{i=1}^{n} \operatorname{Res}(a_i) \quad (t > 0),$$

where $\operatorname{Res}(a_i)$ is the residue of the function $e^{st}F(s)$ at $s = a_i$.

According to the Fourier inversion formula, we also obtain the following:

10.36 Corollary. *If f is only piecewise continuous on $[0, \infty)$ as above, then the value of $f(t)$ at any jump discontinuity $t > 0$ is given by*

$$f(t) = \frac{f(t^+) + f(t^-)}{2}.$$

10.37 Example. Let us consider the form of numerous Laplace transforms, the quotient of two polynomials

$$\begin{aligned} F(s) = \frac{P(s)}{Q(s)} &= \frac{\alpha_n s^n + \alpha_{n-1} s^{n-1} + \cdots + \alpha_0}{\beta_m s^m + \beta_{m-1} s^{m-1} + \cdots + \beta_0} \quad (\alpha_n, \beta_m \neq 0) \\ &= \frac{\alpha_n + \frac{\alpha_{n-1}}{s} + \cdots + \frac{\alpha_0}{s^n}}{s^{m-n}(\beta_m + \frac{\beta_{m-1}}{s} + \cdots + \frac{\beta_0}{s^m})} = \frac{p(s)}{q(s)}, \end{aligned}$$

where $P(s)$ and $Q(s)$ have no common roots, and $m > n$. Expressing $Q(s)$ in terms of its m linear factors, we have m simple poles of $F(s)$ at the points $a_1, a_2, \ldots, a_m$. Note that for all $|s| = R$ sufficiently large,

$$|p(s)| \le |\alpha_n| + |\alpha_{n-1}| + \cdots + |\alpha_0| = c_1,$$

and

$$|q(s)| \ge R^{m-n} \left| \frac{\beta_m}{2} \right| = R^{m-n} c_2.$$

Therefore

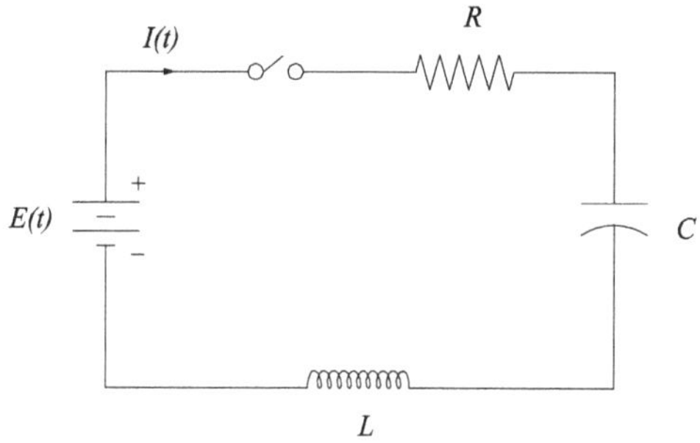

Figure 10.13: An example of an RCL circuit. Courtesy Katy Metcalf.

$$|F(s)| \leq \frac{C_1/C_2}{R^{m-n}} \to 0$$

as $R \to \infty$. Regarding the poles for $i = 1, 2, \ldots, m$, we obtain

$$\operatorname{Res}(a_i) = \frac{e^{a_i t} P(a_i)}{Q'(a_i)},$$

so that by Theorem 10.35

$$f(t) = \sum_{i=1}^{m} \frac{e^{a_i t} P(a_i)}{Q'(a_i)} \qquad (t > 0).$$

10.38 Example. An *RCL* circuit consists of a resistor (*R*), a capacitor (*C*), and an inductor (*L*), where $I(t)$ represents the current as in Figure 10.13.[32]

Let us assume then that the current $I(t)$ satisfies the equation[33]

$$L\frac{dI}{dt} + \frac{1}{C}\int_0^t I(\tau)d\tau = E_0,$$

where L, C, E_0 are constants, and $I(0) = I_0$. Taking the Laplace transform of both sides, we obtain

$$Ls\mathcal{L}(I(t)) - LI_0 + \frac{1}{Cs}\mathcal{L}(I(t)) = \frac{E_0}{s},$$

which yields

32 These quantities are related by *Kirchoff's voltage law*:

$$L\frac{dI}{dt} + RI + \frac{1}{C}\int_0^t I(\tau)d\tau = E(t),$$

where $E(t)$ is the impressed voltage.

33 Thus there is no resistor in this circuit example.

$$\mathcal{L}(I(t)) = F(s) = \frac{E_0 + LI_0 s}{L(s^2 + 1/LC)},$$

so that $O(F(s)) = 1/s \to 0$ as $|s| \to \infty$.

Since $F(s)$ has simple poles at $s = \frac{i}{\sqrt{LC}}$ and $s = -\frac{i}{\sqrt{LC}}$, direct calculation (exercise) shows that

$$I(t) = \sum \text{Res} = E_0\sqrt{\frac{C}{L}} \sin\frac{t}{\sqrt{LC}} + I_0 \cos\frac{t}{\sqrt{LC}}$$

by Theorem 10.35.

10.39 Exercise. Find the current $I(t)$, $t > 0$, if

$$L\frac{dI}{dt} + RI + \frac{1}{C}\int_0^t I(\tau)d\tau = \sin t$$

for $L = 1$, $R = 1$, $C = 1$, and $I(0) = 0$. Use $\mathcal{L}(\sin t) = \frac{1}{s^2+1}$.

In the case that $F(s)$ has infinitely many poles at $a_1, a_2, a_3, \ldots$ to the left of the Bromwich line with $|a_1| \le |a_2| \le \ldots$ such that $|a_n| \to \infty$ as $n \to \infty$, we have to take a sequence of contours $\Gamma_n = C_n \cup [x_0 - iy_n, x_0 + iy_n]$ such that each Γ_n encloses just the poles $a_1, a_2, a_3, \ldots, a_n$. Once again, we have

$$\frac{1}{2\pi i}\int_{C_n} e^{st}F(s)ds + \frac{1}{2\pi i}\int_{x_0-iy_n}^{x_0+iy_n} e^{st}F(s)ds = \sum_{i=1}^{n} \text{Res}(a_i),$$

and therefore we have to show *in each particular case* that

$$\frac{1}{2\pi i}\int_{C_n} e^{st}F(s)ds \to 0$$

as $n \to \infty$ for suitably chosen contours C_n that straddle the poles. Once this is achieved then one obtains the following elegant formulation of the inversion formula:

$$f(t) = \frac{1}{2\pi i}\int_{x-i\infty}^{x+i\infty} e^{st}F(s)ds = \sum_{i=1}^{\infty} \text{Res}(a_i),$$

where $\text{Res}(a_i)$ is the residue of the function $e^{st}F(s)$ at $s = a_i$. See Schiff (1999, p. 160), for examples.

Bibliography

Ahlfors, L., *Complex Analysis*, 3rd ed., McGraw-Hill, 1979.
Anderson, A. W., *Hyperbolic Geometry*, Springer-Verlag, London, 2005.
Apostol, T. M., *Introduction to Analytic Number Theory*, Springer, 2010.
Axler, S., Bourdon, P., and Ramey, W., *Harmonic Function Theory*, 2nd ed., Springer, 2001.
Chen, N.-X., *Möbius Inversion in Physics*, World Sci., New Jersey, London, Singapore, 2010.
Collingwood, E. F. and Lohwater, A. J., *Theory of Cluster Sets*, Cambridge University Press, 1966.
Courant, R., *Dirichlet's Principle, Conformal Mapping, and Minimal Surfaces*, Interscience Publ., 1950.
Derbyshire, J., *Prime Obsession: Bernhard Riemann and the Greatest Unsolved Problem in Mathematics*, John Henry Press, 2003.
Garabedian, P. R., *Partial Differential Equations*, John Wiley & Sons, 1964.
Garnett, J. B., *Applications of Harmonic Measure*, Wiley-Interscience/John Wiley & Sons, 1986.
Hayman, W. K., *Meromorphic Functions*, Oxford University Press, London 1964.
Hayman, W. K. and Kennedy, P. B., *Subharmonic Functions*, Vol. 1, Academic Press, London, 1976.
Helms, L. L., *Introduction to Potential Theory*, Wiley-Interscience, New York, 1969.
Hille, E., *Analytic Function Theory*, Vol. II, Ginn Publ. Co., 1962.
Miller, K. S. and Ross, B., *An Introduction to the Fractional Calculus and Fractional Differential Equations*, John Wiley & Sons, 1993.
Monna, A. F., *Dirichlet's Principle – A Mathematical Comedy of Errors and Its Influence on the Development of Analysis*, Oosthoek, Scheltema and Holkema, Utrecht, 1975.
Montel, P., *Leçons sur les familles normales de fonctions analytique et leurs applications*, Gauthier-Villars, Paris, 1927. Reprinted by Chelsea Publ. Co., New York, 1974.
Nehari, Z., *Conformal Mappings*, McGraw-Hill, 1952.
Nevanlinna, R., *Analytic Functions*, Springer-Verlag, New York, 1970. Translated from the 2nd German edition by Phillip Emig.
Peitgen, H.-O. and Richter, P. H., *The Beauty of Fractals – Images of Complex Dynamical Systems*, Springer-Verlag, 1986.
Pommerenke, C., *Univalent Functions*, Vandenhoeck and Ruprecht, Göttingen, 1975.
Radó, T., *Subharmonic Functions*, Springer-Verlag, 1937, reprint 1971.
Rubel, L. A. and Colliander, J. E., *Entire and Meromorphic Functions*, Springer, 1996.
Schiff, J. L., *Normal Families*, Springer-Verlag, 1993.
Schiff, J. L., *The Laplace Transform – Theory and Applications*, Springer-Verlag, 1999.
Tsuji, M., *Potential Theory in Modern Function Theory*, Maruzen Co. Ltd., 1959.
Wintner, A., *An Arithmetical Approach to Ordinary Fourier Series*, Waverly Press, 1945.

Further Reading

Behnke, H. and Sommer, F., *Theorie der analytischen Funktionen einer komplexen Veränderlichen*, 3rd ed., Springer-Verlag, Berlin, 1965.
Burckel, R. B., *An Introduction to Classical Complex Analysis*, Vol. 1, Birkhäuser Verlag, Basel, 1979.
Ford, L. R., *Automorphic Functions*, Chelsea Publ., 1951.
Goluzin, G. M., *Geometric Theory of Functions of a Complex Variable*, Translations of Mathematical Monographs, Vol. 26, Amer. Math. Soc., Providence, RI, 1969.
Titchmarsh, E. C., *The Theory of Functions*, 2nd ed. Oxford University Press, 1964.
Veech, W. A., *A Second Course in Complex Analysis*, W. A. Benjamin, Inc. 1967.
Whittaker, E. T. and Watson, G. N., A Course in Modern Analysis, 3rd ed., Cambridge University Press, 1920.

https://doi.org/10.1515/9783110757828-011

Index

https://doi.org/10.1515/9783110757828-012

De Gruyter Studies in Mathematics

Volume 66
Igor V. Nikolaev
Noncommutative Geometry. A Functorial Approach, 2nd Edition, 2022
ISBN 978-3-11-078860-0, e-ISBN 978-3-11-078870-9,
e-ISBN (ePUB) 978-3-11-078881-5

Volume 87
René Erlin Castillo
Fourier Meets Hilbert and Riesz. An Introduction to the Corresponding Transforms, 2022
ISBN 978-3-11-078405-3, e-ISBN 978-3-11-078409-1,
e-ISBN (ePUB) 978-3-11-078412-1

Volume 86
Flavia Smarrazzo, Alberto Tesei
Measure Theory and Nonlinear Evolution Equations, 2022
ISBN 978-3-11-055600-1, e-ISBN 978-3-11-055690-2,
e-ISBN (ePUB) 978-3-11-055604-9

Volume 85
Shai Dekel
Pointwise Variable Anisotropic Function Spaces on $\mathbb{R}^n$, 2022
ISBN 978-3-11-076176-4, e-ISBN 978-3-11-076179-5,
e-ISBN (ePUB) 978-3-11-076187-0

Volume 84
Marko Kostić
Selected Topics in Almost Periodicity, 2021
ISBN 978-3-11-076322-5, e-ISBN 978-3-11-076352-2,
e-ISBN (ePUB) 978-3-11-076360-7

Volume 83
Lydia Außenhofer, Dikran Dikranjan, Anna Giordano Bruno
Topological Groups and the Pontryagin-van Kampen Duality. An Introduction, 2021
ISBN 978-3-11-065349-6, e-ISBN 978-3-11-065493-6,
e-ISBN (ePUB) 978-3-11-065355-7

www.ingramcontent.com/pod-product-compliance
Ingram Content Group UK Ltd.
Pitfield, Milton Keynes, MK11 3LW, UK
UKHW051022091026
15624UKWH00012B/630